AF581873

Oxidative Stress and Inflammation as Targets for Novel Preventive and Therapeutic Approaches in Non-Communicable Diseases II

Oxidative Stress and Inflammation as Targets for Novel Preventive and Therapeutic Approaches in Non-Communicable Diseases II

Editors

Chiara Nediani
Monica Dinu

MDPI • Basel • Beijing • Wuhan • Barcelona • Belgrade • Manchester • Tokyo • Cluj • Tianjin

Editors
Chiara Nediani
University of Florence
Italy

Monica Dinu
University of Florence
Italy

Editorial Office
MDPI
St. Alban-Anlage 66
4052 Basel, Switzerland

This is a reprint of articles from the Special Issue published online in the open access journal *Antioxidants* (ISSN 2076-3921) (available at: https://www.mdpi.com/journal/antioxidants/special_issues/oxidative_stress_inflammation_II).

For citation purposes, cite each article independently as indicated on the article page online and as indicated below:

LastName, A.A.; LastName, B.B.; LastName, C.C. Article Title. *Journal Name* **Year**, *Volume Number*, Page Range.

ISBN 978-3-0365-4363-5 (Hbk)
ISBN 978-3-0365-4364-2 (PDF)

Contents

About the Editors

Chiara Nediani

Chiara Nediani is Associate Professor of Clinical Biochemistry and Clinical Molecular Biology at the University of Florence, Italy. She studied Biological Sciences and then completed her studies, obtaining her post-graduate diploma in Clinical Biochemistry. She currently works at the Dipartimento di Scienze Biomediche, Sperimentali e Cliniche "Mario Serio", University of Florence. Her research topics are: (a) the regulatory mechanism of calcium homeostasis in cardiac and skeletal muscle; (b) biochemical aspects and molecular mechanisms of myocardial damage induced by ischemia-reperfusion in cell models and in pigs undergoing volume and pressure overload, and (c) oxidative stress mechanisms in end-stage human failing hearts. Now, she is working on the antioxidant and anti-inflammatory properties of Oleuropein and Oleocathal (polyphenols present in extra virgin olive oil) in in vitro, on human neuroblastoma and melanoma cells, murine cardiomyocytes, and in vivo on TgCRND8 mice, animal model of amyloid deposition, and PIRC rat, colon cancer model. She has been responsible for and a component of various research project units of Relevant National Interest (PRIN) and private foundations (Telethon and Cassa di Risparmio di Firenze). Currently, she is Scientific Responsible of Unit of the Project Biosynol PS-GO 2017 Partenariato Europeo per l'Innovazione in materia di produttività e sostenibilità dell'agricoltura.

Monica Dinu

Monica Dinu is Assistant Professor in Human Nutrition at the Department of Experimental and Clinical Medicine, University of Florence, Italy. She studied Biological Sciences and then completed her studies by obtaining a Master's degree in Nutrition Sciences. Three years later, she completed a PhD in Biomedical Sciences—Physiological and Nutritional Sciences Curriculum. She is secretary of the regional section of the Italian Society of Human Nutrition (SINU) and coordinator of the SINU youth working group. Her research activities are: (a) the design and implementation of clinical trials to evaluate the role of food and dietary patterns on health-related biomarkers in primary and secondary prevention; (b) the design and implementation of observational studies to investigate the association between diet and disease; (c) the development of systematic reviews and meta-analyses on the role of diet in the prevention and treatment of overweight, obesity and chronic diseases; and (d) the development and validation of questionnaires to assess food consumption and adherence to dietary patterns. She has participated in many research projects of international and national interest. She is the author of more than 50 scientific articles, published in international peer-reviewed journals, and two book chapters.

MDPI

Editorial

Oxidative Stress and Inflammation as Targets for Novel Preventive and Therapeutic Approaches in Non-Communicable Diseases II

Chiara Nediani [1,*] and Monica Dinu [2]

1 Department of Experimental and Clinical Biomedical Sciences "Mario Serio", University of Florence, 50134 Florence, Italy

2 Department of Experimental and Clinical Medicine, University of Florence, 50134 Florence, Italy; monica.dinu@unifi.it

* Correspondence: chiara.nediani@unifi.it

Citation: Nediani, C.; Dinu, M. Oxidative Stress and Inflammation as Targets for Novel Preventive and Therapeutic Approaches in Non-Communicable Diseases II. *Antioxidants* **2022**, *11*, 824. https://doi.org/10.3390/antiox11050824

Received: 22 April 2022
Accepted: 22 April 2022
Published: 23 April 2022

Non-communicable diseases (NCDs) are non-infectious chronic pathologies—including obesity, metabolic syndrome, chronic kidney disease (CKD), cardiovascular (CV) diseases, cancer, and chronic respiratory diseases—which represent the main cause of death and disability for the general population [1]. Their growing prevalence is related to the increasing age of the population; urbanization; and lifestyle changes [1]. As previously reported, oxidative stress and inflammation induce and modulate several signaling pathways that play a crucial role in the pathophysiology and progression of these diseases [2]. Thus, they represent a good target for the development of several therapeutic strategies. This Special Issue consists of 12 articles providing different approaches to elucidate the underlying pathogenesis and treatment mechanisms of conditions related to oxidative stress and inflammation.

In the review by Fibbi et al. [3] the authors explore the role of oxidative stress in both osmolality-dependent and -independent impairment of cell and tissue functions observed in hyponatremic conditions. Hyponatremia is defined as a serum sodium concentration ([Na+]) < 136 mEq/L and has been associated with augmented morbidity and mortality. Following the description of neurological and systemic manifestations even in mild and chronic hyponatremia, the authors show how reduced extracellular [Na+] is associated with detrimental effects on cellular homeostasis independently of hypoosmolality, and how most of these alterations are elicited by oxidative stress. They also review a range of basic and clinical research showing that oxidative stress is a common denominator of degenerative processes linked to aging, neurocognitive deficits, osteoporosis, and cancer progression. Given all the evidence indicating that hyponatremia plays a part in exacerbating multiple manifestations of senescence and decreasing survival in cancer patients, they conclude by stressing the need for further studies to fully elucidate the specific molecular pathways triggered by reduced extracellular [Na+] and responsible for oxidative damage. Alemany-Cosme et al. [4] on the other hand, summarize the main findings regarding the oxidant and antioxidant mechanisms involved in Crohn's disease (CD), their role in the immunological response, the environment's effects on oxidative stress status, and its involvement in epigenetic changes/modifications. To date, it is known that in CD, oxidative stress is present not only locally in the most affected tissues, but also at a systemic level, and is associated with an unbalanced immune response and dysbiosis. This review highlights the need for further studies to determine environmental- and oxidative-stress-induced epigenetic changes that may contribute to the onset and development of CD.

Conditions causing predisposition to pro-inflammatory state and oxidative stress include obesity, which stimulates adipose tissue to release inflammatory mediators such as tumor necrosis factor (TNF)-α and interleukin (IL)-6 [5]. Obesity is also an important risk factor for breast cancer [6]. In the study by Martinez-Bernabe et al. [7] authors examined the effects of obesity-related inflammation on mitochondrial functionality in breast cancer cell lines and breast tumors, focusing on estrogen receptors (ER) ratio. The analysis of mitochondrial activity after 17β-estradiol, leptin, IL-6, and TNF-α exposure, which aimed to stimulate the hormonal conditions of a postmenopausal obese woman, showed that ER-β maintained mitochondrial functionality and avoided invasiveness in breast-cancer cell lines. Moreover, the authors found a strong correlation between IL-6 receptor gene expression and inflammation, mitochondrial functionality, and oxidative stress markers, as well as with ER-β. Overall, these findings confirm that under an obesity-related inflammation condition, the presence of ER-β allows the maintenance of mitochondrial functionality, reduced production of ROS, and high expression of antioxidant enzymes which result in a less aggressive phenotype.

The presence of high levels of ROS is also a hallmark of idiopathic pulmonary fibrosis (IPF), mainly due to the H_2O_2-generating enzyme NADPH oxidase 4 (NOX4) [8]. In turn, H_2O_2 is a substrate of the di-tyrosine peroxidase (DT) cross-linking, which appears to be involved in many diseases, including IPF. Recent studies have documented a significant increase in DT in IPF, but whether DT is formed in the lungs of IPF patients, and how its levels and localization contribute to the IPF pathogenesis [9], is unknown. In the study by Blaskovic et al. [10] authors wanted to deepen the role of DT and NOX4 in IPF with the perspective to find new therapies for IPF, since those in current use (Nintedanib and Pirfenidone) have adverse effects. They performed immunohistochemical staining for DT and NOX4 in pulmonary tissue from patients with IPF and controls. In IPF, both DT and NOX4 were present, whereas in the healthy lung DT showed little or no staining, and NOX4 was present mainly in normal vascular endothelium. The link between NOX4 and DT was addressed in MRC5 lung fibroblasts deficient in NOX4 activity (mutation in the CYBA gene). Induction of NOX4 by transforming growth factor beta 1 (TGFβ1) in fibroblasts led to moderate DT staining after the addition of a heme-containing peroxidase in control cells, but not in the fibroblasts deficient for NOX4 activity. These results indicate that DT is a histological marker of IPF and that NOX4 can generate enough H_2O_2 for DT formation in vitro. On the other hand, the absence of NOX4 and DT in all lung regions suggests that NOX4-dependent DT formation could be limited to the fibrotic foci.

To discover new biomarkers for the progression of CKD, a systemic disease to which development chronic oxidative stress and inflammation contribute, Vida et al. [11] analyzed several redox state markers in plasma of advanced CKD patients and isolated peripheral polymorphonuclear (PMNs) and mononuclear (MNs) leukocytes. Study patients were divided into healthy controls, non-dialysis-dependent-CKD (NDD-CKD) patients, hemodialysis (HD) and peritoneal dialysis (PD) patients and were characterized for the presence of some co-morbidities, for the etiology of CKD and for the treatment received. The analysis revealed increased oxidative stress and damage in plasma, PMNs and MNs from NDD-CKD, HD and PD patients compared to controls. Interestingly, PD patients showed greater oxidative stress than HD patients, especially in MNs. Based on these results, the authors encourage the evaluation of PMNs and MNs in CKD patients to follow both CKD progression and dialysis procedures.

In animal model studies, several approaches have been used to assess the damage induced by oxidative stress. Chang et al. [12] used a rodent model of stroke to demonstrate that glycerol is capable of alleviating post-stroke brain injury and associated acute kidney injury (AKI). The authors evaluated blood–brain barrier (BBB) integrity parameters and revealed that glycerol was useful in alleviating BBB disruption and reducing hemorrhagic stroke brain damage. In addition, they showed an improvement in kidney markers and a decrease in stress hormones levels after glycerol injection. Finally, they analyzed morphological alteration induced by hemorrhagic stroke in kidney structure and discovered a glycerol-induced modulation on cytokine-induced neutrophil chemoattractant 1 (CINC-1) and malondialdehyde (MDA), two urinary markers of oxidative stress overexpressed in AKI. Lazar et al. [13] also focused on kidney function, but in type I and II diabetes. In their study, they investigated the possibility of treating the manifestations of diabetic kidney disease (DKD) by inhibiting histone acetyltransferase p300/CBP, which regulates many genes and may be responsible for Nox expression, ROS production, inflammation and fibrosis, all manifestations of DKD. They used STZ-induced diabetic mice treated with C646 and showed a significant decrease in H3K27ac—an epigenetic mark of active gene expression, reduced ROS production and reduced inflammation-driving molecules and mesangial extracellular matrix (ECM) components responsible for fibrosis. Furthermore, using human embryonic kidney cells, they found a down-regulation of the luciferase level and a decrease in glomerular hypertrophy. All these data confirm the implication of p300/CBP in DKD and reveal the possibility of using a p300/CBP inhibitor to alleviate DKD.

Two other articles in this Special Issue used mouse models, in this case to assess cardiac function and the role of inflammation in the development of hypertension. Liu et al. [14] used wild-type (WT) and $p47^{phox}$ knockout (KO) mice to investigate $p47^{phox}$-dependent oxidant signaling in Angiotensin II (AngII) infusion-induced cardiac hypertrophy and cardiomyocyte apoptosis. In fact, the signaling pathways of $p47^{phox}$ in the heart are still unclear. In their experiment, the authors showed how AngII infusion resulted in high blood pressure and cardiac hypertrophy in WT mice, whereas these pathological changes were significantly reduced in $p47^{phox}$ KO mice. These findings confirm that $p47^{phox}$ is a key player in mediating the AngII-induced oxidative stress signaling cascade from the phosphorylation of ASK1, MKK3/6 and MAPKs to the activation of H2AX and p53 and suggest that targeting $p47^{phox}$ could have great therapeutic potential for preventing or treating AngII-induced cardiac dysfunction and damages. On the other hand, Sun et al. [15] demonstrated that an increase in endogenous μ-opioids in the nucleus tractus solitarius (NTS) induces a neurotoxicity cascade with enhanced Ang II binding to the AT1R receptor and activates the microglia which induces superoxide production. Furthermore, they showed how the increase in endogenous μ-opioids induces the formation of μOR/AT1R heterodimers and the TLR4-dependent inflammatory response, which attenuate the nitric oxide (NO)-dependent depressor effect. These results imply an important link between neurotoxicity and superoxide and deepen our understanding of μOR as a novel candidate for intervention in hypertensive conditions.

The search for antioxidants to be used in therapy and prevention of NCDs also relies on plant compounds. In this respect, oleuropein—a phenolic compound found in *Olea europaea* L. fruits and leaves—is one of the most studied bioactive compounds in the context of the Mediterranean diet. In previous studies, oleuropein exhibited a wide range of antioxidant, anti-inflammatory, antidiabetic, neuro- and cardioprotective, antimicrobial and immunomodulatory activities [16,17]. Furthermore, it was able to reduce crypt dysplasia in a short-term colon carcinogenesis experiment in rats and showed protective effects in colitis-associated colorectal cancer (CRC) in mice, suggesting that this molecule may reduce colon tumorigenesis [18]. However, whether these protective effects can be extended to already developed colon tumors it is not yet known. Ruzzolini et al. [19] evaluated the effect of oleuropein-rich leaf extracts (ORLE) for the first time in already-developed colon tumors arising in Apc (adenomatous polyposis coli) -mutated PIRC rats. The authors assessed whether one-week low-dose treatment with an ORLE-enriched diet could exert a beneficial effect against established colon cancer lesions and local and systemic inflammation. Although in vivo experiments were performed with a limited number of PIRC rats fed with ORLE, the overall results disclose a significant increase in tumor apoptosis together with a downregulation of proliferation associated with the inhibition of NO and relative pro-inflammatory mediators expressed by tumor cells and inflammatory cells of the tumor microenvironment. These findings suggest the possibility of testing ORLE as a complementary therapy in combination with standard anticancer drugs.

Columbianadin (CBN), another plant compound, was also tested. It is a natural coumarin isolated from *Angelica decursiva*, which has showed anticancer and platelet-aggregation-inhibiting properties. In their study, Jayakumar et al. [20] demonstrated that CBN exhibits compelling anti-inflammatory and hepatoprotective effects by preventing free-radical formation and decreasing the expression of mitogen-activated protein kinase

(MAPK), followed by the suppression of the nuclear factor kappa B (NF-κB) pathways. This, in turn, led to inhibition of NO, inducible nitric oxide synthase (iNOS), TNF-α, and IL-1β in LPS-activated RAW cells and mouse liver. Although further study of the principal mechanisms is required, these data suggested that CBN represents a valid anti-inflammatory and hepatoprotective agent, and thus is a valid candidate for treating inflammation-mediated diseases. Finally, Li et al. [21] explored the effects of rebaudioside A (Reb A), a natural non-nutritive sweetener obtained from the extracts of *Stevia rebaudiana*. In previous studies, this compound has demonstrated many biological properties, such as anti-inflammatory, antioxidant, antifibrotic and anticancer properties [22]. In their study, Li et al. [21] assessed the effect of Reb A on the health span and lifespan of nematodes and investigated the potential mechanisms underlying Reb A-induced metabolic changes using a combination of transcriptomics and lipidomic approaches. In the model organism *C. elegans*, Reb A prolonged lifespan, enhanced oxidative stress resistance, and improved lipid metabolism. These findings serve as a ground for future studies exploring the potential medicinal and beneficial effects of Reb A as a replacement for caloric sugars in human foods and beverages.

The Guest Editors would like to thank all the authors, the reviewers who contributed to the success of this Special Issue, and the *Antioxidants* team for their valuable and constant support.

Funding: This research received no external funding.

Conflicts of Interest: The authors declare no conflict of interest.

References

1. Wang, Y.; Wang, J. Modelling and prediction of global non-communicable diseases. *BMC Public Health* **2020**, *20*, 822. [CrossRef]
2. Seyedsadjadi, N.; Grant, R. The Potential Benefit of Monitoring Oxidative Stress and Inflammation in the Prevention of Non-Communicable Diseases (NCDs). *Antioxidants* **2020**, *10*, 15. [CrossRef]
3. Fibbi, B.; Marroncini, G.; Anceschi, C.; Naldi, L.; Peri, A. Hyponatremia and Oxidative Stress. *Antioxidants* **2021**, *10*, 1768. [CrossRef]
4. Alemany-Cosme, E.; Sáez-González, E.; Moret, I.; Mateos, B.; Iborra, M.; Nos, P.; Sandoval, J.; Beltrán, B. Oxidative Stress in the Pathogenesis of Crohn's Disease and the Interconnection with Immunological Response, Microbiota, External Environmental Factors, and Epigenetics. *Antioxidants* **2021**, *10*, 64. [CrossRef] [PubMed]
5. Ellulu, M.S.; Patimah, I.; Khazaai, H.; Rahmat, A.; Abed, Y. Obesity & inflammation: The linking mechanism & the complications. *Arch. Med. Sci.* **2017**, *13*, 851–863. [PubMed]
6. Macciò, A.; Madeddu, C. Obesity, inflammation, and postmenopausal breast cancer: Therapeutic implications. *Sci. World J.* **2011**, *11*, 2020–2036. [CrossRef] [PubMed]
7. Martinez-Bernabe, T.; Sastre-Serra, J.; Ciobu, N.; Oliver, J.; Pons, D.; Roca, P. Estrogen Receptor Beta (ERβ) Maintains Mitochondrial Network Regulating Invasiveness in an Obesity-Related Inflammation Condition in Breast Cancer. *Antioxidants* **2021**, *10*, 1371. [CrossRef]
8. Veith, C.; Boots, A.W.; Idris, M.; Van Schooten, F.-J.; Van Der Vliet, A. Redox Imbalance in Idiopathic Pulmonary Fibrosis: A Role for Oxidant Cross-Talk Between NADPH Oxidase Enzymes and Mitochondria. *Antioxid. Redox Signal.* **2019**, *31*, 1092–1115. [CrossRef]
9. Pennathur, S.; Vivekanandan-Giri, A.; Locy, M.L.; Kulkarni, T.; Zhi, D.; Zeng, L.; Byun, J.; de Andrade, J.A.; Thannickal, V.J. Oxidative Modifications of Protein Tyrosyl Residues Are Increased in Plasma of Human Subjects with In-terstitial Lung Disease. *Am. J. Respir. Crit. Care Med.* **2016**, *193*, 861–868. [CrossRef]
10. Blaskovic, S.; Donati, Y.; Ruchonnet-Metrailler, I.; Seredenina, T.; Krause, K.; Pache, J.; Adler, D.; Barazzone-Argiroffo, C.; Jaquet, V. Di-Tyrosine Crosslinking and NOX4 Expression as Oxidative Pathological Markers in the Lungs of Patients with Idiopathic Pulmonary Fibrosis. *Antioxidants* **2021**, *10*, 1833. [CrossRef]
11. Vida, C.; Oliva, C.; Yuste, C.; Ceprián, N.; Caro, P.; Valera, G.; González de Pablos, I.; Morales, E.; Carracedo, J. Oxidative Stress in Patients with Advanced CKD and Renal Replacement Therapy: The Key Role of Peripheral Blood Leukocytes. *Antioxidants* **2021**, *10*, 1155. [CrossRef]
12. Chang, C.; Pan, P.; Li, J.; Ou, Y.; Liao, S.; Chen, W.; Kuan, Y.; Chen, C. Glycerol Improves Intracerebral Hemorrhagic Brain Injury and Associated Kidney Dysfunction in Rats. *Antioxidants* **2021**, *10*, 623. [CrossRef] [PubMed]
13. Lazar, A.; Vlad, M.; Manea, A.; Simionescu, M.; Manea, S. Activated Histone Acetyltransferase p300/CBP-Related Signalling Pathways Mediate Up-Regulation of NADPH Oxidase, Inflammation, and Fibrosis in Diabetic Kidney. *Antioxidants* **2021**, *10*, 1356. [CrossRef] [PubMed]
14. Liu, F.; Fan, L.; Geng, L.; Li, J. p47phox-Dependent Oxidant Signalling through ASK1, MKK3/6 and MAPKs in Angiotensin II-Induced Cardiac Hypertrophy and Apoptosis. *Antioxidants* **2021**, *10*, 1363. [CrossRef] [PubMed]

15. Sun, G.; Tse, J.; Hsu, Y.; Ho, C.; Tseng, C.; Cheng, P. μ-Opioid Receptor-Mediated AT1R–TLR4 Crosstalk Promotes Microglial Activation to Modulate Blood Pressure Control in the Central Nervous System. *Antioxidants* **2021**, *10*, 1784. [CrossRef] [PubMed]
16. Nediani, C.; Ruzzolini, J.; Romani, A.; Calorini, L. Oleuropein, a Bioactive Compound from *Olea europaea* L., as a Potential Preventive and Therapeutic Agent in Non-Communicable Diseases. *Antioxidants* **2019**, *8*, 578. [CrossRef]
17. Romani, A.; Ieri, F.; Urciuoli, S.; Noce, A.; Marrone, G.; Nediani, C.; Bernini, R. Health effects of phenolic compounds found in extra-virgin olive oil, by-products and leaf of *Olea europaea* L. *Nutrients* **2019**, *11*, 1776. [CrossRef]
18. Sepporta, M.V.; Fuccelli, R.; Rosignoli, P.; Ricci, G.; Servili, M.; Fabiani, R. Oleuropein Prevents Azoxymethane-Induced Colon Crypt Dysplasia and Leukocytes DNA Damage in A/J Mice. *J. Med. Food* **2016**, *19*, 983–989. [CrossRef]
19. Ruzzolini, J.; Chioccioli, S.; Monaco, N.; Peppicelli, S.; Andreucci, E.; Urciuoli, S.; Romani, A.; Luceri, C.; Tortora, K.; Calorini, L.; et al. Oleuropein-Rich Leaf Extract as a Broad Inhibitor of Tumour and Macrophage iNOS in an Apc Mutant Rat Model. *Antioxidants* **2021**, *10*, 1577. [CrossRef]
20. Jayakumar, T.; Hou, S.; Chang, C.; Fong, T.; Hsia, C.; Chen, Y.; Huang, W.; Saravanabhavan, P.; Manubolu, M.; Sheu, J.; et al. Columbianadin Dampens In Vitro Inflammatory Actions and Inhibits Liver Injury via Inhibition of NF-κB/MAPKs: Impacts on ·OH Radicals and HO-1 Expression. *Antioxidants* **2021**, *10*, 553. [CrossRef]
21. Li, P.; Wang, Z.; Lam, S.; Shui, G. Rebaudioside A Enhances Resistance to Oxidative Stress and Extends Lifespan and Healthspan in Caenorhabditis elegans. *Antioxidants* **2021**, *10*, 262. [CrossRef] [PubMed]
22. Casas-Grajales, S.; Ramos-Tovar, E.; Chávez-Estrada, E.; Alvarez-Suarez, D.; Hernández-Aquino, E.; Reyes-Gordillo, K.; Cerda-García-Rojas, C.M.; Camacho, J.; Tsutsumi, V.; Lakshman, M.R.; et al. Antioxidant and immunomodulatory activity induced by stevioside in liver damage: In Vivo, In Vitro and in silico assays. *Life Sci.* **2019**, *224*, 187–196. [CrossRef] [PubMed]

 antioxidants

Review

Hyponatremia and Oxidative Stress

Benedetta Fibbi [1,2], Giada Marroncini [1,2], Cecilia Anceschi [2], Laura Naldi [2] and Alessandro Peri [1,2,*]

[1] Pituitary Diseases and Sodium Alterations Unit, AOU Careggi, 50139 Florence, Italy; benedetta.fibbi@unifi.it (B.F.); giada.marroncini@unifi.it (G.M.)
[2] Endocrinology, Department of Experimental and Clinical Biomedical Sciences "Mario Serio", University of Florence, AOU Careggi, 50139 Florence, Italy; cecilia.anceschi@unifi.it (C.A.); laura.naldi1@stud.unifi.it (L.N.)
* Correspondence: alessandro.peri@unifi.it

Abstract: Hyponatremia, i.e., the presence of a serum sodium concentration ([Na^+]) < 136 mEq/L, is the most frequent electrolyte imbalance in the elderly and in hospitalized patients. Symptoms of acute hyponatremia, whose main target is the central nervous system, are explained by the "osmotic theory" and the neuronal swelling secondary to decreased extracellular osmolality, which determines cerebral oedema. Following the description of neurological and systemic manifestations even in mild and chronic hyponatremia, in the last decade reduced extracellular [Na^+] was associated with detrimental effects on cellular homeostasis independently of hypoosmolality. Most of these alterations appeared to be elicited by oxidative stress. In this review, we focus on the role of oxidative stress on both osmolality-dependent and -independent impairment of cell and tissue functions observed in hyponatremic conditions. Furthermore, basic and clinical research suggested that oxidative stress appears to be a common denominator of the degenerative processes related to aging, cancer progression, and hyponatremia. Of note, low [Na^+] is able to exacerbate multiple manifestations of senescence and to decrease progression-free and overall survival in oncologic patients.

Keywords: hyponatremia; osmotic stress; oxidative stress; ROS; senescence; cancer; osteoporosis; neuronal cells homeostasis

Citation: Fibbi, B.; Marroncini, G.; Anceschi, C.; Naldi, L.; Peri, A. Hyponatremia and Oxidative Stress. *Antioxidants* **2021**, *10*, 1768. https://doi.org/10.3390/antiox10111768

Academic Editors: Chiara Nediani, Monica Dinu and Edward E. Schmidt

Received: 30 September 2021
Accepted: 3 November 2021
Published: 4 November 2021

1. Introduction

Hyponatremia, defined as a serum sodium concentration ([Na^+]) < 136 mEq/L, is the most frequent electrolyte disorder encountered in clinical practice, especially in hospitalized patients and in the elderly [1]. Since hyponatremia is associated with increased morbidity and mortality even in mildly affected patients, it represents an economic burden in terms of hospitalization and health care costs [2–8]. From this view, a prompt and appropriate correction of this electrolytic imbalance is critical to prevent short- and long-term complications. However, treating hyponatremia is not always perceived as crucial by clinicians, especially when [Na^+] is only slightly reduced, but it is potentially associated with negative impacts on body functions [9]. For this reason, the comprehension of molecular mechanisms involved in the pathogenesis of symptoms related to hyponatremia might help to raise the awareness about the importance of correcting even chronic and mild reductions of [Na^+].

2. Hyponatremia and Health

Although it has long been thought that persistently but slightly reduced [Na^+] was completely inconsequential on health, and therefore did not require any correction [1], nowadays chronic hyponatremia is known to have adverse outcomes on several organs and systems [10].

If prolonged over time, the perturbation of internal homeostasis can lead to permanent injuries of biological functions and potentially life-threatening events, as demonstrated by the association of [Na^+] even mildly below the normal range with increased mortality [4,11–14].

This correlation was confirmed in cross-sectional studies performed on both inpatient and outpatient cohorts [11,12], and in a meta-analysis including 81 studies and 147,948 participants, which estimated an overall mortality risk ratio of 2.60 (95% confidence interval [CI], 2.31–2.93) in hyponatremic compared to normonatremic patients [3]. A worse prognosis was observed in patients affected by mild hyponatremia and heart, liver, brain, kidney and lung diseases [8,15–17], but this electrolyte imbalance was initially considered as a mere marker of disease severity rather than accelerating patient deterioration [18]. In some clinical settings, such as heart failure [19] and cirrhosis [20], inadequate circulation determines a non-osmotic trigger to vasopressin secretion, aimed to preserve blood pressure and circulating volume; moreover, the release of antidiuretic hormone can be stimulated in response to stress and hypothalamic-pituitary-adrenal axis activation [21]. The simultaneous measurement of plasma sodium and copeptin (a molecule co-released with vasopressin) in 6962 patients, revealed a significant association of all cause 30-day mortality with hyponatremia even independently of copeptin (and consequently of vasopressin) levels [22].

Besides the presence of a correlation between hyponatremia and mortality, clinical manifestations of chronic hyponatremia also include neurocognitive deficits and bone fractures/osteoporosis. In a cohort of 122 patients admitted to an emergency department and affected by apparently asymptomatic hyponatremia, the frequency of falls was significantly higher than age-matched normonatremic controls [9]; low [Na^+] was demonstrated to induce gait disturbances similar to ethanol, which were more severe in patients older than 65 years than in younger subjects [23]. These alterations normalized after hyponatremia correction [9]. More recently, the analysis of data from 5435 patients included in the Osteoporotic Fractures in Men Study revealed a significant association between mild hyponatremia and cognitive impairment and decline at the baseline, evaluated by Mini-Mental Status Examination score or Trail Making Test Part B time. The correlation was much stronger for the second test, which measures executive functions (attention and inhibition control, cognitive flexibility, working memory) [24]. Even if dementia may alter vasopressin secretion [25], the maintenance of the association between cognitive impairment and hyponatremia in patients undergoing hemodialysis and peritoneal dialysis supports a mechanism independent of antidiuretic hormone release [26,27]. Bone fractures are more frequent in mild hyponatremic patients than in normonatremic ones [28–30], with a higher risk of hospitalization [12,20,31], increased lengths of stay in the hospital and mortality [32]. The increased rate of fractures associated with chronic hyponatremia is not only due to more frequent falls, but also to a higher prevalence of osteoporosis. Epidemiological data highlighted that the increased risk of bone demineralization depends on both severity and duration of the electrolyte imbalance: the lower the serum [Na^+] is and the more it is maintained over time, the higher the patients risk of developing osteoporosis is [33]. Several studies have now confirmed a strong correlation between a decrease in serum [Na^+] by as little as 2 to 4 mEq/L below the normal range and osteoporosis and fragility fractures, exceeding the risk related to the use of corticosteroid drugs or smoking [30,34–36]. The direct effect of low [Na^+] on human health and body homeostasis is confirmed by the reversibility of clinical abnormalities secondary to mild and moderate chronic hyponatremia after appropriate correction. A statistically significant association between an increase in serum [Na^+] in hyponatremic patients and a reduction in mortality, with a calculated odds ratio of 0.57 (95% CI, 0.40–0.81) was demonstrated in a meta-analysis of 15 studies [37]. Treatment of marked hyponatremia with the vasopressin antagonist tolvaptan improved mental health (SALT-1 and SALT-2) [38], psychomotor speed domain (INSIGHT) [39], Timed Up and Go test, nerve conduction velocities and F-wave latencies [40], and gait disturbances [9]. In the INSIGHT trial, the improvement of bone frailty after 22 days of tolvaptan treatment was also demonstrated [39]. Finally, the improvement of bone density was reported in a young man after removal of a vasopressin-secreting esthesioneuroblastoma of the maxillary sinus and normalization of [Na^+] [41].

Pulmonary diseases are a frequent cause of hyponatremia, occurring in about 30% of patients affected by pneumonia [42]. Retrospective analysis of hyponatremia occurrence in

COVID-19 patients during the first pandemic period demonstrated a prevalence of 22.9% at hospital admission, a worse respiratory performance (evaluated as P/F, i.e., the ratio of the partial pressure of oxygen in arterial blood PaO_2 to the inspired oxygen fraction FiO_2), and higher IL-6 levels in hyponatremic rather than in normonatremic hospitalized patients [43]. Since IL-6 is able to induce vasopressin secretion by a direct hypothalamic stimulation and by inducing alveolar basement membrane injury and pulmonary hypoxia and vasoconstriction [44–47], the pro-inflammatory cytokine may represent the common denominator of both acute respiratory insufficiency and syndrome of inappropriate antidiuresis (SIAD)-related hyponatremia. A very recent metanalysis of 8 studies and 11,493 patients showed a correlation of hyponatremia with COVID-19 poor outcomes (a composite of mortality, prolonged hospitalization and severe COVID-19, defined as severe pneumonia and/or needing intensive care unit support/invasive mechanical ventilation; OR 2.65 [1.89, 3.72], $p < 0.001$; I2: 67.2%, $p = 0.003$), with a 37% sensitivity and 82% specificity; while normal serum [Na^+] is associated with a 16% post-test probability of a worse prognosis, the presence of hyponatremia increases this probability up to 33% [48].

An intriguing and unexpected association was also observed between chronic hyponatremia and overall and progression-free survival in cancer patients. An underlying tumor is responsible for about 14% of hyponatremias [49], whose prevalence in the oncologic setting varies with tumor type and treatment protocols, as well as serum [Na^+] threshold employed. The most frequent cause of chronic hyponatremia in cancer patients is SIAD, mainly due to ectopic vasopressin secretion by cancer cells [50–53]. Several clinical evidences reported hyponatremia as an independent, negative prognostic factor in different types of blood and solid tumors (e.g., lymphoma [54], gastrointestinal cancers [55,56], hepatocellular carcinoma [57,58], mesothelioma [59], renal cell carcinoma [60,61], and small cell lung cancer [62,63]), and a concordant improvement of overall and progression-free survival after the appropriate correction of reduced serum [Na^+] [24], even in patients with extensive and terminal disease [64]. Accordingly, hyponatremia has been proposed as a biomarker able to identify high-risk subjects affected by lung cancer [65].

3. Osmotically-Induced Oxidative Stress

The "osmotic theory" was the first one formulated to explain neurologic symptoms associated with low extracellular [Na^+]. When hyponatremia occurs, the resulting decrease in plasma osmolality (except for the rare cases of non-hypoosmotic hyponatremia) causes water movement into the brain by osmotic gradient, thus causing cerebral oedema [1,66]. The cellular elements most involved in swelling are astrocytes, namely glial cells which are a constituent of the blood-brain barrier and have a fundamental role in maintaining the fluid and electrolyte concentration of the extracellular space in the central nervous system [67]. In the brain, the intracellular/extracellular ionic homeostasis is particularly important, since excitatory and inhibitory synaptic events are driven by ionic gradients, which regulate the resting potential and the discharge pattern of neurons [68]. Sparing neurons from hypoosmolar stress is functional to preserve brain excitability, which is increased both directly (swelling-induced release of excitatory neurotransmitters) and indirectly (restrained diffusion of neurotransmitters and depolarizing agents due to the reduction of extracellular space volume) by swelling [68]. Therefore, an acute decrease in external osmolality determines an initial astrocyte swelling as a result of water movement from the extracellular to intracellular compartment, thus preventing the same phenomenon from occurring in neuronal cells [69] and limiting brain swelling. This first-line defense mechanism is quickly followed by a process known as volume regulatory decrease. This ancient adaptive mechanism, which is able to counteract cell volume alterations and consequently perturbations of cell functions (cell-cycle progression, proliferation, apoptosis, excitability and metabolism) [70], consists in extruding intracellular solutes (electrolytes and organic osmolytes) together with osmotically obligated water [71]. This phenomenon is crucial in the brain, in which the physical restriction of the skull limits the expansion and may determine a life-threatening increase in intracranial pressure. In the first hours, cells mainly lose inorganic ions (first Na^+ and Cl^-, then K^+), as a result of an energy-dependent mechanism based on the activation of the Na^+-K^+ ATPase pump (the first signaling pathway of osmotransduction activated by cell swelling), Ca^{2+}-dependent and -independent K^+ channels, K^+/Cl^- co-transporters and volume-sensitive Cl^- channels [66,71–73]. In cells exposed to sustained hyponatremia, a delayed loss of small organic osmolytes also starts: myoinisotol, betaine, creatine and amino acids (taurine, glycine, aspartate, glutamine and glutamate) are progressively extruded [74], and their efflux is maintained as long as low $[Na^+]$ persists as an essential adaptive mechanism in chronic hyponatremia. The completion of inorganic solute extrusion within 48 h defines the empirical threshold for acute (<48 h) and chronic (>48 h) hyponatremia [75].

While chronic hyponatremia has been traditionally defined as asymptomatic because of cell volume adaptation, in the last decade several studies demonstrated that even a mild chronic reduction of $[Na^+]$ may be associated with neurological signs and symptoms, i.e. gait impairment, attention and memory deficit, and increased risk of falls [3,9,28,76,77]. Accordingly, the correction of low $[Na^+]$ may effectively counteract the reduced cognitive performances observed in hyponatremic patients compared to normonatremic subjects [37,38,76]. The mechanisms that potentially cause these alterations are not completely understood, but the impairment of excitatory neurotransmitters may be involved. As previously mentioned, glutamate is one of the most important organic osmolytes involved in cellular adaptation to hyponatremia [71]. In physiologic conditions, the extracellular glutamate concentration is kept low to avoid an excessive activation of its receptors and glutamate neurotoxicity (GNT), a condition characterized by time-dependent damage of many cell components leading to cell death and prevented through the astrocytic re-uptake mediated by the Na^+-dependent glial glutamate transporters GLT-1 and GLAST [78]. While the cerebral extracellular concentration of glutamate is increased under acute hypoosmotic conditions [79], its brain content decreases by about 40% after 14 days of sustained hyponatremia in rats [75], thus suggesting an impairment of synaptic excitatory neurotransmission due to chronic hyponatremia. Moreover, it was demonstrated that the sustained reduction of serum $[Na^+]$ induces gait disturbances and memory impairment in murine models by decreasing astrocytic glutamate re-uptake (through inhibition of GLT-1 and GLAST activities), and consequently long-term potentiation (LTP) at hippocampal synapses [80]. Nowadays, it is well established that GNT is a result of neuronal Ca^{2+} overloading, which is triggered by acute neuronal swelling (the cellular uptake of extracellular Na^+ and Cl^- causes plasma membrane depolarization, and subsequently Ca^{2+} channel opening) and initiates a cascade-like effect leading to cell death [81]. Beyond mitochondria accumulation of Ca^{2+}, the generation of Ca^{2+}-dependent reactive oxygen species (ROS) (e.g., hydrogen peroxides and superoxides, hydroxyl radicals and oxygen radicals) undoubtedly takes place in GNT [82–86], which is usually associated with marked oxidative stress [87,88]. ROS trigger peroxidative degradation of lipid membranes and modify the redox state of proteins involved in osmotransduction, specifically osmotically-activated tyrosine kinases (ERK1/2, p38, FAK, members of the Src family), which further increase their activity and alter cellular homeostasis [89,90]. Increased ROS formation after exposure to glutamate is divided in two phases: an early ROS production coupled to xanthyne-oxidase activation [91,92], and a later one mostly due to mitochondria as a by-product of glucose metabolism and ATP generation [85,93,94]. Therefore, some authors concluded that in the early GNT, non-mitochondrial ROS generation triggers a cell defense mechanism, while the delayed superoxide production, as well as in apoptosis, occurs secondary to a defect in mitochondrial electron transport and is a result of mitochondrial damage, which acts as a self-propagating process leading to cell dysfunction and death. In particular, the initial oxidative stress could impair mitochondrial energy production and promote depletion of energy stores, thus affecting intracellular homeostatic and protective mechanisms [84].

Beyond neuroactive solutes depletion by neurons, additional mechanisms are hypothesized to explain neurological alterations observed in chronically hyponatremic patients. It is noteworthy that the increased ROS production is also expected to deplete cellular antioxidant defenses, which in turn amplify oxidative stress and radical-mediated injury [95]. As Schultz et al. demonstrated for the first time in vivo, a disturbance of the antioxidant glutathione homeostasis is linked to both excitotoxic neuronal injury and neurodegeneration [96]. Among organic osmolytes, the antioxidants taurine and glutathione are also extruded by neuronal cells in response to extracellular hypoosmolality [97], and the adaptive decrease in their cell content was supposed to make neurons more susceptible to oxidative injury. In fact, glutathione depletion induced by treatment with buthionine sulfoximine or diethylmaleate exacerbates brain injury due respectively to middle cerebral artery ligation in rats [98], and hyperbaric hyperoxia in humans [99]. Using in vivo and in vitro murine and human models, Clark and colleagues demonstrated that brain tissues and cell cultures reduced their content of taurine and glutathione in response to hypoosmolality, and that the depletion was reverted by a slow normalization of serum [Na^+] [100]. Regarding intracellular functions involved in the reduced availability of antioxidants, the authors observed an osmotically-induced decrease in the synthetic rate of glutathione (whose direct transport across the blood–brain barrier is supposed to be preserved), and an increased release of taurine from cells into the extracellular medium [100]. It has also been suggested that glutathione produced by astrocytes might be a disposal pathway for glutamate, and that decreased synthesis of the antioxidant due to hypoosmolality could exacerbate the injury induced by neurotransmitter accumulation [100,101]. In agreement with a role of the osmotic depletion of antioxidants in the pathogenesis of hyponatremia-related brain injury, the incidence of cerebral infarction in patients with subarachnoid hemorrhage who developed this electrolyte imbalance was significantly higher than in eunatremic subjects [102]. The same mechanism may also play a role in the pathogenesis of the osmotic demyelination syndrome [100].

In addition to inorganic and organic solutes extrusion, activation of phospholipases (particularly the isoforms A2 and D) is an intracellular pathway involved in osmotransduction signaling, as demonstrated by mobilization of arachidonic acid and lysophosphatidylcholine (LPC) in association with hypoosmotic swelling [37,38]. Arachidonic acid contributes to the regulation of K^+ and Cl^- channel activity and organic osmolyte efflux, and similarly to LPC, promotes the generation of ROS [91]. Interestingly, arachidonic acid and ROS were found to inhibit glutamate uptake in astrocytes [103].

The main mechanisms triggered by hyponatremia and involved in osmotically-induced production of ROS are summarized in Figure 1.

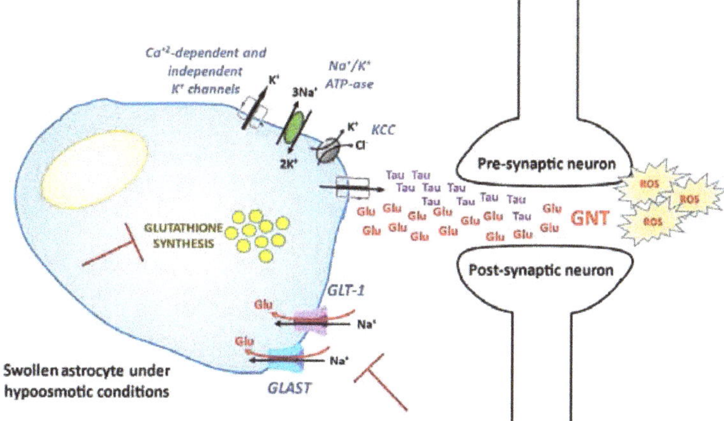

Figure 1. Non osmotically-induced effects of hyponatremia and oxidative stress. GLT-1 and GLAST: Na^+-dependent glial glutamate transporters; ROS: reactive oxygen species; Glu: glutamate; Tau: taurine; GNT: glutamate neurotoxicity; KCC: K^+/Cl^- co-transporters.

4. Non Osmotically-Induced Oxidative Stress

Nowadays, it is well accepted that the central nervous system is not the only target of low [Na^+]. Indeed, mild chronic hyponatremia has also been associated with detrimental effects on bone, specifically increased risk of osteoporosis and fractures independently of bone demineralization [12,30,31,36,104]. Bone matrix is a large reservoir of the body's Na^+, storing approximately one-third of this electrolyte [105]; in dogs, it is an osmotically inactive compartment from which Na^+ is released during prolonged dietary deprivation [106]. As demonstrated in a rat model of SIAD, hyponatremia-related osteoporosis is due to increased osteoclastic activity, in the absence of other metabolic or hormonal alterations able to explain the accelerated bone resorption (i.e., sex steroid deficiency, metastasis-induced osteolysis, calcium-mediated signals, etc.) [36].

The described detrimental systemic effects secondary to chronic hyponatremia, traditionally defined as an asymptomatic or mildly symptomatic disorder, open a new scenario in understanding the pathophysiology of this condition and its clinical sequelae. In fact, neurological and extra-neurological alterations observed in chronic hyponatremia are explained in principle neither by the "osmotic theory" nor by the homeostatic mechanisms that counteract cell swelling in the presence of extracellular hypotonicity. Therefore, the intriguing hypothesis that hyponatremia could directly impair cellular homeostasis—and hence health status—was postulated. With regard to this point, Barsony et al. first showed that sustained low extracellular [Na^+] activates osteoclastogenesis and osteoclastic bone matrix resorption in rats both in vivo and in vitro, independently of reduced osmolality [104]. In their view, this response is likely necessary to mobilize bone-stored Na^+ in the attempt to restore normal extracellular [Na^+]. Moreover, low [Na^+] is able to stimulate the differentiation of early-stage osteoclast progenitors compared to normonatremic conditions, by increasing their sensitivity to growth factors (in particular M-CSF) through oxidative stress. These findings suggest the existence of a Na^+-sensing mechanism or receptor on osteoclasts, as hypothesized also in the central nervous system and in the kidney. Interestingly, the authors found that the activity of the Na^+-dependent vitamin C transporter is inhibited by low extracellular [Na^+] in a dose-dependent manner, thus resulting in a reduced uptake of ascorbic acid. As well as playing a central role in setting the equilibrium between osteoclastogenenesis and osteoblastic functions, ascorbic acid is also a key scavenger of oxidative stress [107,108]. As expected, reduced ascorbic acid uptake observed in the above-mentioned model of chronic hyponatremia is associated with increased accumulation of ROS in osteoclastic cells and oxidative DNA damage product 8-OHdg in the sera of hyponatremic rats compared to controls, in agreement with the excessive production of free radicals and osteoclastic bone reabsorption observed in other forms of osteoporosis (e.g., estrogen/androgen deficiency and chronic inflammation) [104]. By developing an in vitro model able to mimic chronic hyponatremia, we further assessed the correlation between hyponatremia and bone health and analyzed the second process involved in bone remodeling alongside resorption, namely neoformation of bone matrix. We showed that reduced extracellular [Na^+] disrupts gene expression, proliferation, migration, and cytokine production in human mesenchymal stromal cells (hMSC) [109], which are precursors of mesodermal cell types (including adipocytes and osteoblasts of bone matrix) exhibiting different degrees of stemness [110]. In post-menopausal osteoporosis and other conditions characterized by bone loss, the bone marrow shows an imbalance between adipogenesis and osteogenesis, with an accumulation of adipose tissue at the expense of the osteoblastic compartment [111]. In our in vitro model, low [Na^+] impairs osteoblast activity and differentiation of hMSC, which are shifted toward the adipogenic phenotype at the expense of the osteogenic one [109].

Oxidative stress is also a well-recognized mediator of degenerative processes related to senescence, other than osteoporosis [112], especially in the brain [113]. It is then not inconceivable to speculate that chronic hyponatremia might play a direct role in the pathogenesis of degenerative diseases, in particular aging-related multi-organ pathologies, and that its combination with comorbidities in old people might critically weaken the defense against oxidative stress. As a consequence, sustained low [Na^+] might accelerate the aging process and represent an independent risk factor for the development and progression of age-related infirmities. In fact, the prevalence of hyponatremia increases progressively with aging, and its major impact (in terms of morbidity and mortality) is exerted in the elderly [114]. The link between chronic hyponatremia and senescence is supported by evidence that chronic hyponatremia (also in this case regardless of hypoosmolality) accelerates and exacerbates multiple manifestations of senescence, including osteoporosis, hypogonadism with testicular fibrosis and arrest of spermatogenesis, reduced adiposity, cardiomyopathy with left ventricular hypertrophy and fibrosis, and sarcopenia, in male rats [115]. Consistently with these data, primary cultures of neonatal rat cardiomyocytes exposed to low extracellular [Na^+] (but compensated hypoosmolality) and hearts isolated from hyponatremic animals showed increased ROS production and intracellular Ca^{2+} concentrations compared to control cells and tissues [116]. This results in a greater vulnerability of cells against oxidative stress and an exacerbation of myocardial injury due to ischemia/reperfusion, as evidenced by significantly larger infarct size and lower left ventricular developed pressure after exposure to global hypoxia in rats with hyponatremia compared to normonatremic ones [116]. Reoxygenation of cells triggers a burst of ROS, and their increment in low Na^+ conditions may amplify mitochondrial permeability transition pore opening and induce cell death [117]. Swelling and enlargement of mitochondria and destruction of cristae in cardiomyocytes exposed to low [Na^+] might be the result of increased ROS content, which in turn could be secondary to intracellular Ca^{2+} overload and activation of Ca^{2+}-dependent ROS-generating enzymes [118].

Understanding the potential direct effects of low extracellular [Na^+] is of particular interest also in the brain, which is one of the main targets of both chronic hyponatremia and senescence. In the last decade, our laboratory demonstrated that low extracellular [Na^+] directly impairs cellular homeostasis in an in vitro neuronal model of chronic hyponatremia [119]. Sustained low extracellular [Na^+] was demonstrated to induce cell distress by affecting cell viability and adhesion, expression of anti-apoptotic genes (Bcl-2, DHCR24) and ability to differentiate into a mature neuronal phenotype, even in the presence of compensated osmolality. As a result of a comprehensive microarray analysis, we showed that cell functions involved in "cell death and survival" are the most altered in the presence of reduced [Na^+] compared to controls, and that the expression of the heme oxygenase-1 (HMOX-1) gene is the most increased [119] (Figure 2).

SK-N-AS 127 vs 153 mEq/L		
Symbol	Fold Reg	Gene
HMOX1	199,8565	Heme oxygenase (decycling) 1
CTGF	90,4959	Connective tissue growth factor
MAP2K2	65,8208	Mitogen-activated protein kinase kinase 2
IL8	59,3783	Interleukin 8
VEGFA	27,7917	Vascular endothelial growth factor A
EGR1	17,0175	Early growth response 1
GADD45G	14,1729	Growth arrest and DNA-damage-inducible, gamma
TGFA	13,8432	Transforming growth factor, alpha
SLC2A1	13,5732	Solute carrier family 2 (facilitated glucose transporter), member 1
NPR1	13,0995	Natriuretic peptide receptor A
JUN	12,763	Jun proto-oncogene
DUSP1	11,3025	Dual specificity phosphatase 1
AKT1	9,9431	V-akt murine thymoma viral oncogene homolog 1
SLC9A3	8,8017	Solute carrier family 9 (sodium/hydrogen exchanger), member 3
GADD45A	8,4745	Growth arrest and DNA-damage-inducible, alpha
EGR3	7,8234	Early growth response 3
AKR1B1	7,6538	Aldo-keto reductase family 1, member B1 (aldose reductase)
AQP3	5,9858	Aquaporin 3 (Gill blood group)
AQP1	5,4952	Aquaporin 1 (Colton blood group)
ATP1A1	5,1788	ATPase, Na+/K+ transporting, alpha 1 polypeptide
PCK2	4,9527	Phosphoenolpyruvate carboxykinase 2 (mitochondrial)
MAPK1	3,9791	Mitogen-activated protein kinase 1
TRPV4	3,8599	Transient receptor potential cation channel, subfamily V, member 4
TPM4	3,8012	Tropomyosin 4
PAK2	3,7283	P21 protein (Cdc42/Rac)-activated kinase 2
ATF4	3,6314	Activating transcription factor 4 (tax-responsive enhancer element B67)
AGT	3,4103	Angiotensinogen (serpin peptidase inhibitor, clade A, member 8)
NFAT5	3,1266	Nuclear factor of activated T-cells 5, tonicity-responsive
SRC	2,9139	V-src sarcoma (Schmidt-Ruppin A-2) viral oncogene homolog (avian)

SH-SY5Y 115 vs 153 mEq/L		
Symbol	Fold Reg	Gene
HMOX1	26,7003	Heme oxygenase (decycling) 1
CD9	19,0635	CD9 molecule
AQP3	7,2638	Aquaporin 3 (Gill blood group)
GADD45G	6,8802	Growth arrest and DNA-damage-inducible, gamma
TP53	5,6538	Tumor protein p53
HSPA1A	5,5579	Heat shock 70kDa protein 1A
FOS	5,1322	FBJ murine osteosarcoma viral oncogene homolog
EGR1	4,95	Early growth response 1
VIM	4,7351	Vimentin
AKT1	4,6797	V-akt murine thymoma viral oncogene homolog 1
EGR3	4,3594	Early growth response 3
MAP2K2	4,0722	Mitogen-activated protein kinase kinase 2
AKR1B1	3,9334	Aldo-keto reductase family 1, member B1 (aldose reductase)
SLC2A1	3,6008	Solute carrier family 2 (facilitated glucose transporter), member 1
VEGFA	3,3438	Vascular endothelial growth factor A
ODC1	3,235	Ornithine decarboxylase 1
AQP1	3,1375	Aquaporin 1 (Colton blood group)
PCK2	2,7703	Phosphoenolpyruvate carboxykinase 2 (mitochondrial)
CALR	2,4686	Calreticulin
HSPA4L	2,3821	Heat shock 70kDa protein 4-like
PDIA4	2,3641	Protein disulfide isomerase family A, member 4
NOS3	2,3462	Nitric oxide synthase 3 (endothelial cell)
TGFA	2,2321	Transforming growth factor, alpha
HPRT1	2,1569	Hypoxanthine phosphoribosyltransferase 1
ATP1A1	2,1346	ATPase, Na+/K+ transporting, alpha 1 polypeptide
HSP90AA1	2,1074	Heat shock protein 90kDa alpha (cytosolic), class A member 1
HSPA4	2,0918	Heat shock 70kDa protein 4
B2M	2,0881	Beta-2-microglobulin
DUSP1	2,0259	Dual specificity phosphatase 1

Figure 2. List of differentially expressed genes in two in vitro neuronal models (SH-SY5Y and SKN-AS cell lines), maintained at reduced (115 mmol/L and 127 mmol/L, respectively) or normal (153 mmol/L) [Na^+], as assessed by microarray analysis. Positive fold-regulations are reported in red, negative fold-regulations are in blue. Yellow marked genes are commonly regulated in both cell lines.

HMOX-1 is an inducible stress protein with a metabolic function in heme turnover [120] and potent anti-apoptotic and antioxidant activities in different cells, including neurons [121]. In the brain, induction of HMOX-1 by intracellular factors that directly or indirectly generate ROS, preserves neurons from oxidative injury secondary to cerebral ischemia [122] or ethanol intoxication [123]. Elicitation of oxidative stress in the presence of low [Na^+] was confirmed by cytofluorimetric analysis of total intracellular ROS and ROS-induced lipid peroxidation [124]. These findings reinforce the hypothesis that chronic hyponatremia, through increased oxidative stress and ROS generation, may have a role in brain distress and aging by reducing neuronal differentiation ability, a well-known co-factor in the etiopathogenesis of neurodegenerative diseases such as Alzheimer's disease [125]. Finally, we also demonstrated that the correction of sustained low extracellular [Na^+] may not be able to revert all the cell alterations associated with reduced [Na^+], specifically the expression level of the anti-apoptotic genes Bcl-2 and DHCR24 or of the HMOX-1 gene, even when [Na^+] was gradually increased [124]. Admittedly, these data appear to reinforce the recommendation to carefully diagnose and treat patients with hyponatremia because a prompt intervention aimed to correct serum [Na^+] might prevent possible residual abnormalities.

It is now widely accepted that hyponatremia represents a negative independent prognostic factor in oncologic patients, and is associated with poor progression-free and overall survival in several cancers [54–63]. The direct contribution of this electrolyte imbalance (which cannot be considered a mere surrogate marker of the severity of clinical conditions) is supported by the observation that the correction of serum [Na^+] may reduce the overall mortality rate in hyponatremic patients [37]. We recently demonstrated, for the first time, that the reduction of extracellular [Na^+] is able to alter the homeostasis of different human cancer cell lines, thus affecting cell functions (i.e., proliferation, adhesion and invasion) distinctive of a more malignant behavior able to increase cell tumorigenicity [126]. The three steps of carcinogenesis (initiation, promotion, and progression) and the resistance to treatment are strongly impaired by an imbalance between ROS and antioxidant production [127,128]. In fact, oxidative stress regulates cell growth, cytoskeleton remodeling and migration, excitability, exocytosis and endocytosis, autophagy, hormone signaling, necrosis, and apoptosis, namely cell properties deregulated in cancer [127,129]. Furthermore, ROS involvement in carcinogenesis, local invasiveness and metastatization is displayed by their ability to induce genomic instability and/or transcriptional errors [130], and to activate pro-survival and pro-metastatic pathways [129]. Our demonstration of an increased expression of HMOX-1 in cancer cell lines cultured in low extracellular [Na^+], compared to normal Na^+ conditions, validates the role of oxidative stress as the molecular basis of hyponatremia-associated poorer outcomes in oncologic patients [126]. Cancer cells have great abilities to adapt to perturbation of cellular homeostasis, including the imbalanced redox status secondary to their high metabolism and local hypoxia. Through a fine regulation of both ROS production and ROS scavenging pathways (the theory of ROS rheostat), they show a high antioxidant capacity, allowing oxidative stress levels compatible with cellular functions even if higher than in normal cells [131]. Recent studies reported an increased expression of ROS scavengers and low ROS levels in liver and breast cancer stem cells [132,133], whose maintenance is crucial for the survival of pre-neoplastic foci. In this view, chemotherapy and radiotherapy, which strongly induce ROS synthesis, are often able to eliminate the bulk of cancer cells but not to definitely cure cancer, because of the up-regulated levels of antioxidants in stem cells, which are thus spared and selected for in the presence of high ROS. An additional mechanism responsible for therapeutic failure is ROS-dependent accumulation of DNA mutations, leading to drug resistance [131]. In this very complex scenario, antioxidant inhibitors are considered a promising therapeutic tool in cancer treatment, especially regarding glutathione metabolism. Since glutathione is a key regulator of the redox balance and protects cancer cells from stress due to hypoxia and nutrient deficiency in solid tumors, the combination of glutathione inhibitors with radiotherapy or chemotherapy could improve the effects of radiation or drugs. However, other enzymes with a scavenging effect on oxidative stress (HSP90, thioredoxin, enzyme poly-ADP-ribose polymerase or *PARP*) may be targeted for anticancer treatments, and are currently under study [131]. Since cancer-related hyponatremia adversely affects the response to chemotherapy and everolimus in metastatic renal cell carcinoma [61,134], correction of low serum [Na^+] may exert its role in improving cancer survival [135,136] by regulating cancer cell ROS rheostat.

The direct effects of reduced extracellular [Na^+] on cells and tissues are summarized in Figure 3.

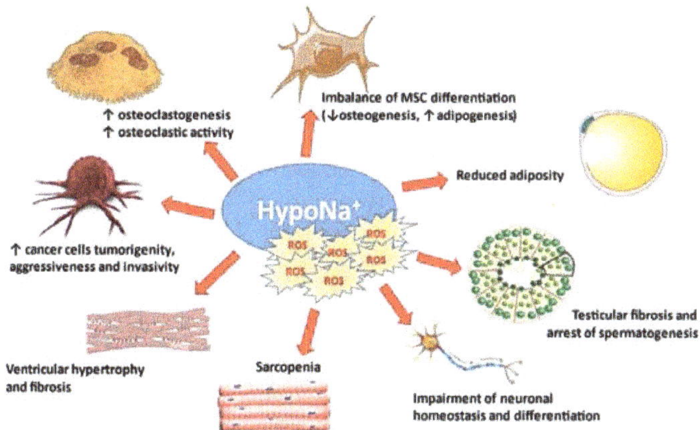

Figure 3. Osmotically-independent effects of hyponatremia and oxidative stress. MSC: mesenchymal stromal cells.

5. Conclusions

In the last decade, several in vitro and in vivo studies suggested that neurological and extra-neurological symptoms observed in hyponatremic patients might be dependent on a perturbation of cellular homeostasis. Specifically, low extracellular [Na^+] impairs cellular functions (e.g., gene protein expression, proliferation, migration, and invasion) involved in senescence and carcinogenesis, thus amplifying tissue injuries related to aging and promoting cancer progression. In this scenario, oxidative stress seems to be the common denominator of intracellular events common to both processes (Figure 4). The studies published in recent years opened a new, unpredicted scenario. Further data will be necessary to fully elucidate the specific molecular pathways triggered by reduced extracellular [Na^+] and responsible for oxidative damage, and to comprehensively understand all potential implications of long-term exposure to hyponatremic conditions.

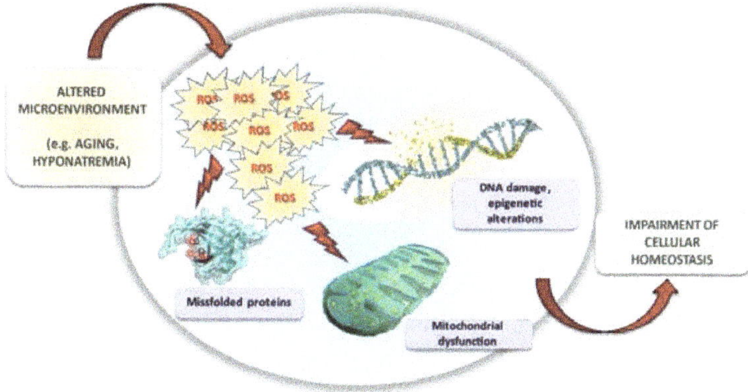

Figure 4. Effects of hyponatremia on cell and tissue homeostasis.

Author Contributions: Writing—review and editing, B.F., G.M., C.A., L.N., A.P. All authors have read and agreed to the published version of the manuscript.

Funding: This research received no external funding. The APC was funded by PRIN 2017R5ZE2C.

Acknowledgments: The authors wish to thank other collaborators, who have been involved in past years in the studies presented in this review, and in particular Susanna Benvenuti, Paola Luciani, Cristiana Deledda, Alice Errico, Federica Baldanzi.

Conflicts of Interest: The authors declare no conflict of interest.

References

1. Adrogué, H.J.; Madias, N.E. Hyponatremia. *N. Engl. J. Med.* **2000**, *342*, 1581–1589. [CrossRef]
2. Wald, R.; Jaber, B.L.; Price, L.L.; Upadhyay, A.; Madias, N.E. Impact of hospital-associated hyponatremia on selected outcomes. *Arch. Intern. Med.* **2010**, *170*, 294–302. [CrossRef]
3. Corona, G.; Giuliani, C.; Parenti, G.; Norello, D.; Verbalis, J.G.; Forti, G.; Maggi, M.; Peri, A. Moderate hyponatremia is associated with increased risk of mortality: Evidence from a meta-analysis. *PLoS ONE* **2013**, *8*, e80451. [CrossRef] [PubMed]
4. Mohan, S.; Gu, S.; Parikh, A.; Radhakrishnan, J. Prevalence of hyponatremia and association with mortality: Results from NHANES. *Am. J. Med.* **2013**, *126*, 1127–1137. [CrossRef]
5. Luca, A.; Angermayr, B.; Bertolini, G.; Koenig, F.; Vizzini, G.; Ploner, M.; Peck-Radosavljevic, M.; Gridelli, B.; Bosch, J. An integrated MELD model including serum sodium and age improves the prediction of early mortality in patients with cirrhosis. *Liver Transpl.* **2007**, *13*, 1174–1180. [CrossRef] [PubMed]
6. Grodin, J.L. Pharmacologic approaches to electrolyte abnormalities in heart failure. *Curr. Heart Fail. Rep.* **2016**, *13*, 181–189. [CrossRef] [PubMed]
7. Rossi, J.; Bayram, M.; Udelson, J.E.; Lloyd-Jones, D.; Adams, K.F.; Oconnor, C.M.; Gattis Stough, W.; Ouyang, J.; Shin, D.D.; Orlandi, C.; et al. Improvement in hyponatremia during hospitalization for worsening heart failure is associated with improved outcomes: Insights from the Acute and Chronic Therapeutic Impact of a Vasopressin Antagonist in Chronic Heart Failure (ACTIV in CHF) trial. *Acute Card. Care* **2007**, *9*, 82–86. [CrossRef] [PubMed]
8. Nair, V.; Niederman, M.S.; Masani, N.; Fishbane, S. Hyponatremia in community-acquired pneumonia. *Am. J. Nephrol.* **2007**, *27*, 184–190. [CrossRef]
9. Renneboog, B.; Musch, W.; Vandemergel, X.; Manto, M.U.; Decaux, G. Mild Chronic Hyponatremia is Associated with Falls, Unsteadiness, and Attention Deficits. *Am. J. Med.* **2006**, *119*, 71.e1–71.e8. [CrossRef] [PubMed]
10. Decaux, G. Is asymptomatic hyponatremia really asymptomatic? *Am. J. Med.* **2006**, *119*, S79–S82. [CrossRef]
11. Rondon-Berrios, H.; Berl, T. Mild chronic hyponatremia in the ambulatory setting: Significance and management. *Clin. J. Am. Soc. Nephrol.* **2015**, *10*, 2268–2278. [CrossRef] [PubMed]
12. Hoorn, E.J.; Rivadeneira, F.; van Meurs, J.B.; Ziere, G.; Stricker, B.H.; Hofman, A.; Pols, H.A.; Zietse, R.; Uitterlinden, A.G.; Zillikens, M.C. Mild hyponatremia as a risk factor for fractures: The Rotterdam Study. *J. Bone Miner. Res.* **2011**, *26*, 1822–1828. [CrossRef]
13. Sajadieh, A.; Binici, Z.; Mouridsen, M.R.; Nielsen, O.W.; Hansen, J.F.; Haugaard, S.B. Mild hyponatremia carries a poor prognosis in community subjects. *Am. J. Med.* **2009**, *122*, 679–686. [CrossRef]
14. Holland-Bill, L.; Christiansen, C.F.; Heide-Jørgensen, U.; Ulrichsen, S.P.; Ring, T.; Jørgensen, J.O.L.; Sørensen, H.T. Hyponatremia and mortality risk: A Danish cohort study of 279 508 acutely hospitalized patients. *Eur. J. Endocrinol.* **2015**, *173*, 71–81. [CrossRef] [PubMed]
15. Ma, Q.Q.; Fan, X.D.; Li, T.; Hao, Y.Y.; Ma, F. Short- and long-term prognostic value of hyponatremia in patients with acute coronary syndrome: A systematic review and meta-analysis. *PLoS ONE* **2018**, *13*, e0193857. [CrossRef] [PubMed]
16. Rodrigues, B.; Staff, I.; Fortunato, G.; McCullough, L.D. Hyponatremia in the prognosis of acute ischemic stroke. *J. Stroke Cerebrovasc. Dis.* **2014**, *23*, 850–854. [CrossRef]
17. Curhan, G.C.; Brunelli, S.M. Mortality associated with low serum sodium concentration in maintenance hemodialysis. *Am. J. Med.* **2011**, *124*, 77–84.
18. Chawla, A.; Sterns, R.H.; Nigwekar, S.U.; Cappuccio, J.D. Mortality and serum sodium: Do patients die from or with hyponatremia? *Clin. J. Am. Soc. Nephrol.* **2011**, *6*, 960–965. [CrossRef]
19. Velavan, P.; Khan, N.K.; Goode, K.; Rigby, A.S.; Loh, P.H.; Komajda, M.; Follath, F.; Swedberg, K.; Madeira, H.; Cleland, J.G.F. Predictors of short term mortality in hert failure—insights from the Euro Heart Failure Survey. *Int. J. Cardiol.* **2010**, *138*, 63–69. [CrossRef]
20. Arroyo, V.; Rodes, J.; Gutierrez-Lizarraga, M.A.; Revert, L. Prognostic value of spontaneous hyponatremia in cirrhosis with ascites. *Am. J. Dig. Dis.* **1976**, *21*, 249–256. [CrossRef]
21. Schrier, R.W. Water and sodium retention in edematous disorders: Role of vasopressin and aldosterone. *Am. J. Med.* **2006**, *119*, S47–S53. [CrossRef]

22. Eckart, A.; Hausfater, P.; Amin, D.; Amin, A.; Haubitz, S.; Bernard, M.; Baumgartner, A.; Struja, T.; Kutz, A.; Christ-Crain, M.; et al. Hyponatremia and activation of vasopressin secretion are both independently associated with 30-day mortality: Results of a multicenter, observational study. *J. Intern. Med.* **2018**, *284*, 270–281. [CrossRef]
23. Renneboog, B.; Sattar, L.; Decaux, G. Attention and postural balance are much more affected in older than in younger adults with mild or moderate chronic hyponatremia. *Eur. J. Intern. Med.* **2017**, *41*, e25–e26. [CrossRef]
24. Nowak, K.L.; Yaffe, K.; Orwoll, E.S.; Ix, J.H.; You, Z.; Barrett-Connor, E.; Hoffman, A.R.; Chonchol, M. Serum sodium and cognition in older community-dwelling men. *Clin. J. Am. Soc. Nephrol.* **2018**, *13*, 366–374. [CrossRef] [PubMed]
25. Nilsson, E.D.; Melander, O.; Elmståhl, S.; Lethagen, E.; Minthon, L.; Pihlsgård, M.; Nägga, K. Copeptin, a marker of vasopressin, predicts vascular dementia but not Alzheimer's disease. *J. Alzheimers Dis.* **2016**, *52*, 1047–1053. [CrossRef] [PubMed]
26. Shavit, L.; Mikeladze, I.; Torem, C.; Slotki, I. Mild hyponatremia is associated with functional and cognitive decline in chronic hemodialysis patients. *Clin. Nephrol.* **2014**, *82*, 313–319. [CrossRef] [PubMed]
27. Duan, L.P.; Dong, J. Hyponatremia and cognitive impairment in patients treated with peritoneal dialysis. *Clin. J. Am. Soc. Nephrol.* **2015**, *10*, 1806–1813.
28. Corona, G.; Norello, D.; Parenti, G.; Sforza, A.; Maggi, M.; Peri, A. Hyponatremia, falls and bone fractures: A systematic review and meta-analysis. *Clin. Endocrinol.* **2018**, *89*, 505–513. [CrossRef]
29. Negri, A.L.; Ayus, J.C. Hyponatremia and bone disease. *Rev. Endocr. Metab. Disord.* **2017**, *18*, 67–78. [CrossRef]
30. Gankam, K.F.; Andres, C.; Sattar, L.; Melot, C.; Decaux, G. Mild hyponatremia and risk of fracture in the ambulatory elderly. *QJM Int. J. Med.* **2008**, *101*, 583–588. [CrossRef]
31. Kinsella, S.; Moran, S.; Sullivan, M.O.; Molloy, M.G.; Eustace, J.A. Hyponatremia independent of osteoporosis is associated with fracture occurrence. *Clin. J. Am. Soc. Nephrol.* **2010**, *5*, 275–280. [CrossRef]
32. Kuo, S.C.; Kuo, P.; Rau, C.; Wu, S.; Hsu, S.; Hsieh, C. Hyponatremia is associated with worse outcomes from fall injuries in the elderly. *Int. J. Environ. Res. Public Health* **2017**, *14*, 460. [CrossRef] [PubMed]
33. Usala, R.L.; Fernandez, S.J.; Mete, M.; Cowen, L.; Shara, N.M.; Barsony, J.; Verbalis, J.G. Hyponatremia is associated with increased osteoporosis and bone fractures in a large US health system population. *J. Clin. Endocrinol. Metab.* **2015**, *100*, 3021–3031. [CrossRef]
34. Van Staa, T.; Leufkens, H.; Cooper, C. The epidemiology of corticosteroid-induced osteoporosis: A meta-analysis. *Osteoporos. Int.* **2002**, *13*, 777–787. [CrossRef] [PubMed]
35. Kruse, C.; Eiken, P.; Verbalis, J.; Vestergaard, P. The effect of chronic mild hyponatremia on bone mineral loss evaluated by retrospective national Danish patient data. *Bone* **2016**, *84*, 9–14. [CrossRef]
36. Verbalis, J.G.; Barsony, J.; Sugimura, Y.; Tian, Y.; Adams, D.J.; Carter, E.A.; Resnick, H.E. Hyponatremia-induced osteoporosis. *J. Bone Miner. Res.* **2010**, *25*, 554–563. [CrossRef] [PubMed]
37. Corona, G.; Giuliani, C.; Verbalis, J.G.; Forti, G.; Maggi, M.; Peri, A. Hyponatremia improvement is associated with a reduced risk of mortality: Evidence from a meta-analysis. *PLoS ONE* **2015**, *10*, e0124105. [CrossRef]
38. Schrier, R.W.; Gross, P.; Gheorghiade, M.; Berl, T.; Verbalis, J.G.; Czerwiec, F.S.; Orlandi, C.; SALT Investigators. Tolvaptan, a selective oral vasopressin V2-receptor antagonist, for hyponatremia. *N. Engl. J. Med.* **2006**, *355*, 2099–2112. [CrossRef]
39. Verbalis, J.G.; Ellison, H.; Hobart, M.; Krasa, H.; Ouyang, J.; Czerwiec, F.S.; Investigation of the Neurocognitive Impact of Sodium Improvement in Geriatric Hyponatremia: Efficacy and Safety of Tolvaptan (INSIGHT) Investigators. Tolvaptan and neurocognitive function in mild to moderate chronic hyponatremia: A randomized trial (INSIGHT). *Am. J. Kidney Dis.* **2016**, *67*, 893–901. [CrossRef]
40. Gosch, M.; Joosten-Gstrein, B.; Heppner, H.J.; Lechleitner, M. Hyponatremia in geriatric inhospital patients: Effects on results of a comprehensive geriatric assessment. *Gerontology* **2012**, *58*, 430–440. [CrossRef]
41. Sejling, A.; Thorsteinsson, A.; Pedersen-Bjergaard, U.; Eiken, P. Recovery from SIADH-associated osteoporosis: A case report. *J. Clin. Endocrinol. Metab.* **2014**, *99*, 3527–3530. [CrossRef]
42. Cuesta, M.; Slattery, D.; Goulden, E.L.; Gupta, S.; Tatro, E.; Sherlock, M.; Tormey, W.; O'Neill, S.; Thompson, C.J. Hyponatraemia in patients with community-acquired pneumonia; prevalence and aetiology, and natural history of SIAD. *Clin. Endocrinol.* **2019**, *90*, 744–752. [CrossRef]
43. Berni, A.; Malandrino, D.; Corona, G.; Maggi, M.; Parenti, G.; Fibbi, B.; Poggesi, L.; Bartoloni, A.; Lavorini, F.; Fanelli, A.; et al. Serum sodium alterations in SARS CoV-2 (COVID-19) infection: Impact on patient outcome. *Eur. J. Endocrinol.* **2021**, *185*, 137–144. [CrossRef] [PubMed]
44. Mehta, P.; McAuley, D.F.; Brown, M.; Sanchez, E.; Tattersall, R.S.; Manson, J.J.; HLH Across Speciality Collaboration, UK. COVID19: Consider cytokine storm syndromes and immunosuppression. *Lancet* **2020**, *395*, 1033–1034. [CrossRef]
45. Park, S.J.; Shin, J.I. Inflammation and hyponatremia: An underrecognized condition? *Korean, J. Pediatr.* **2013**, *56*, 519–522. [CrossRef] [PubMed]
46. Qin, C.; Zhou, L.; Hu, Z.; Zhang, S.; Yang, S.; Tao, Y.; Xie, C.; Ma, K.; Shang, K.; Wang, W.; et al. Dysregulation of immune response in patients with COVID-19 in Wuhan, China. *Clin. Infect. Dis.* **2020**, *71*, 762–768. [CrossRef]
47. Mastorakos, G.; Weber, J.S.; Magiakou, M.A.; Gunn, H.; Chrousos, G.P. Hypothalamic-pituitary-adrenal axis activation and stimulation of systemic vasopressin secretion by recombinant interleukin-6 in humans: Potential implications for the syndrome of inappropriate vasopressin secretion. *J. Clin. Endocrinol. Metab.* **1994**, *79*, 934–939.

48. Akbar, M.R.; Pranata, R.; Wibowo, A.; Irvan; Sihite, T.A.; Martha, J.W. The Prognostic Value of Hyponatremia for Predicting Poor Outcome in Patients With COVID-19: A Systematic Review and Meta-Analysis. *Front. Med.* **2021**, *14*, 666949. [CrossRef]
49. Castillo, J.J.; Vincent, M.; Justice, E. Diagnosis and management of hyponatremia in cancer patients. *Oncologist* **2012**, *17*, 756–765. [CrossRef] [PubMed]
50. Zogheri, A.; Di Mambro, A.; Mannelli, M.; Serio, M.; Forti, G.; Peri, A. Hyponatremia and pituitary adenoma: Think twice about the etiopathogenesis. *J. Endocrinol. Investig.* **2006**, *29*, 750–753. [CrossRef]
51. Padfield, P.L.; Morton, J.J.; Brown, J.J.; Lever, A.F.; Robertson, J.I.; Wood, M.; Fox, R. Plasma arginine vasopressin in the syndrome of antidiuretic hormone excess associated with bronchogenic carcinoma. *Am. J. Med.* **1976**, *61*, 825–831. [CrossRef]
52. Shapiro, J.; Richardson, G.E. Hyponatremia of malignancy. *Crit. Rev. Oncol. Hematol.* **1995**, *18*, 129–135. [CrossRef]
53. Sorensen, J.B.; Andersen, M.K.; Hansen, H.H. Syndrome of inappropriate secretion of antidiuretic hormone (SIADH) in malignant disease. *J. Intern. Med.* **1995**, *238*, 97–110. [CrossRef]
54. Dhaliwal, H.S.; Rohatiner, A.Z.; Gregory, W.; Richards, M.A.; Johnson, P.W.; Whelan, J.S.; Gallagher, C.J.; Matthews, J.; Ganesan, T.S.; Barnett, M.J. Combination chemotherapy for intermediate and high grade non-Hodgkin's lymphoma. *Br. J. Cancer* **1993**, *68*, 767–774. [CrossRef]
55. Zhou, M.H.; Wang, Z.H.; Zhou, H.W.; Liu, M.; Gu, Y.J.; Sun, J.Z. Clinical outcome of 30 patients with bone marrow metastases. *J. Cancer Res. Ther.* **2018**, *14*, S512–S515.
56. Choi, J.S.; Bae, E.H.; Ma, S.K.; Kweon, S.S.; Kim, S.W. Prognostic impact of hyponatraemia in patients with colorectal cancer. *Colorectal Dis.* **2015**, *17*, 409–416. [CrossRef] [PubMed]
57. Gines, P.; Guevara, M. Hyponatremia in cirrhosis: Pathogenesis, clinical significance, and management. *Hepatology* **2008**, *48*, 1002–1010. [CrossRef]
58. Cescon, M.; Cucchetti, A.; Grazi, G.L.; Ferrero, A.; Viganò, L.; Ercolani, G.; Zanello, M.; Ravaioli, M.; Capussotti, L.; Pinna, A.D. Indication of the extent of hepatectomy for hepatocellular carcinoma on cirrhosis by a simple algorithm based on preoperative variables. *Arch. Surg.* **2009**, *144*, 57–63. [CrossRef] [PubMed]
59. Berardi, R.; Caramanti, M.; Fiordoliva, I.; Morgese, F.; Savini, A.; Rinaldi, S.; Torniai, M.; Tiberi, M.; Ferrini, C.; Castagnani, M.; et al. Hyponatraemia is a predictor of clinical outcome for malignant pleural mesothelioma. *Support Care Cancer* **2015**, *23*, 621–626. [CrossRef]
60. Vasudev, N.S.; Brown, J.E.; Brown, S.R.; Rafiq, R.; Morgan, R.; Patel, P.M.; O'Donnell, D.; Harnden, P.; Rogers, M.; Cocks, K.; et al. Prognostic factors in renal cell carcinoma: Association of preoperative sodium concentration with survival. *Clin. Cancer Res.* **2008**, *14*, 1775–1781. [CrossRef]
61. Schutz, F.A.; Xie, W.; Donskov, F.; Sircar, M.; McDermott, D.F.; Rini, B.I.; Agarwal, N.; Pal, S.K.; Srinivas, S.; Kollmannsberger, C.; et al. The impact of low serum sodium on treatment outcome of targeted therapy in metastatic renal cell carcinoma: Results from the International Metastatic Renal Cell Cancer Database Consortium. *Eur. Urol.* **2014**, *65*, 723–730. [CrossRef]
62. Gandhi, L.; Johnson, B.E. Paraneoplastic syndromes associated with small cell lung cancer. *J. Natl. Compr. Canc. Netw.* **2006**, *4*, 631–638. [CrossRef]
63. Rawson, N.S.; Peto, J. An overview of prognostic factors in small cell lung cancer. A report from the Subcommittee for the Management of Lung Cancer of the United Kingdom Coordinating Committee on Cancer Research. *Br. J. Cancer* **1990**, *61*, 597–604. [CrossRef] [PubMed]
64. Balachandran, K.; Okines, A.; Gunapala, R.; Morganstein, D.; Popat, S. Resolution of severe hyponatraemia is associated with improved survival in patients with cancer. *BMC Cancer* **2015**, *15*, 163. [CrossRef]
65. Kasi, P.M. Proposing the use of hyponatremia as a marker to help identify high risk individuals for lung cancer. *Med. Hypotheses* **2012**, *79*, 327–328. [CrossRef]
66. Ayus, C.; Achinger, S.G.; Arieff, A. Brain cell volume regulation in hyponatremia: Role of sex, age, vasopressin, and hypoxia. *Am. J. Physiol. Renal Physiol.* **2008**, *295*, F619–F624. [CrossRef] [PubMed]
67. Kimelberg, H.K. Water homeostasis in the brain: Basic concepts. *Neuroscience* **2004**, *129*, 851–860. [CrossRef] [PubMed]
68. Pasantes-Morales, H.; Tuz, K. Volume changes in neurons: Hyperexcitability and neuronal death. *Contrib. Nephrol.* **2006**, *152*, 221–240.
69. Simard, M.; Nedergaard, M. The neurobiology of glia in the context of water and ion homeostasis. *Neuroscience* **2004**, *129*, 877–896. [CrossRef] [PubMed]
70. Okada, Y.; Sato, K.; Numata, T. Pathophysiology and puzzles of the volume-sensitive outwardly rectifying anion channel. *J. Physiol.* **2009**, *15*, 2141–2149.
71. Verbalis, J.G. Brain volume regulation in response to changes in osmolality. *Neuroscience* **2010**, *168*, 862–870. [CrossRef] [PubMed]
72. Pasantes-Morales, H.; Franco, R.; Ordaz, B.; Ochoa, L.D. Mechanisms Counteracting Swelling in Brain Cells During Hyponatremia. *Arch. Med. Res.* **2002**, *33*, 237–244. [CrossRef]
73. Okada, Y. Volume expansion-sensing outward-rectifier Cl-channel: Fresh start to the molecular identity and volume sensor. *Am. J. Physiol.* **1997**, *273*, C755–C789. [CrossRef]
74. Fisher, S.K.; Heacock, A.M.; Keep, R.F.; Foster, D.J. Receptor regulation of osmolyte homeostasis in neural cells. *J. Physiol.* **2010**, *18*, 3355–3364. [CrossRef]
75. Verbalis, J.G.; Gullans, S.R. Hyponatremia causes large sustained reductions in brain content of multiple organic osmolytes in rats. *Brain Res.* **1991**, *567*, 274–282. [CrossRef]

76. Miyazaki, T.; Ohmoto, K.; Hirose, T.; Fujiki, H. Chronic hyponatremia impairs memory in rats: Effects of vasopressin antagonist tolvaptan. *J. Endocrinol.* **2010**, *206*, 105–111. [CrossRef] [PubMed]
77. Suárez, V.; Norello, D.; Sen, E.; Todorova, P.; Hackl, M.J.; Hüser, C.; Grundmann, F.; Kubacki, T.; Becker, I.; Peri, A.; et al. Impairment of Neurocognitive Functioning, Motor Performance, and Mood Stability in Hospitalized Patients with Euvolemic Moderate and Profound Hyponatremia. *Am. J. Med.* **2020**, *133*, 986–993.e5. [CrossRef]
78. Danbolt, N.C. Glutamate uptake. *Prog. Neurobiol.* **2001**, *65*, 1–105. [CrossRef]
79. Haskew-Layton, R.E.; Rudkouskaya, A.; Jin, Y.; Feustel, P.J.; Kimelberg, H.K.; Mongin, A.A. Two distinct modes of hypoosmotic medium-induced release of excitatory amino acids and taurine in the rat brain in vivo. *PLoS ONE* **2008**, *3*, e3543. [CrossRef]
80. Fujisawa, H.; Sugimura, Y.; Takagi, H.; Mizoguchi, H.; Takeuchi, H.; Izumida, H.; Nakashima, K.; Ochiai, H.; Takeuchi, S.; Kiyota, A.; et al. Chronic Hyponatremia Causes Neurologic and Psychologic Impairments. *J. Am. Soc. Nephrol.* **2016**, *27*, 766–780. [CrossRef]
81. Nicholls, D.G.; Budd, S.L. Mitochondria and neuronal survival. *Physiol. Rev.* **2000**, *80*, 315–360. [CrossRef]
82. Lafon-Cazal, M.; Pietri, S.; Culcasi, M.; Boeckaert, J. NMDA-dependent superoxide production and neurotoxicity. *Nature* **1993**, *364*, 535–537. [CrossRef]
83. Budd, S.L.; Nicholls, D.G. Mitochondria, calcium regulation, and acute glutamate excitotoxicity in cultured cerebellar granule cells. *J. Neurochem.* **1996**, *67*, 2282–2291. [CrossRef] [PubMed]
84. Atlante, A.; Gagliardi, S.; Minervini, G.M.; Ciotti, M.T.; Marra, E.; Calissano, P. Glutamate neurotoxicity in rat cerebellar granule cells: A major role for xanthine oxidase in oxygen radical formation. *J. Neurochem.* **1997**, *68*, 2038–2045. [CrossRef]
85. Castilho, R.F.; Nicholls, D.G. Oxidative stress, mitochondrial function, and acute glutamate excitotoxicity in cultured cerebellar granule cells. *J. Neurochem.* **1999**, *72*, 1394–1401. [CrossRef]
86. Pereira, C.F.; Oliveira, C.R. Oxidative glutamate toxicity involves mitochondrial dysfunction and perturbation of intracellular Ca2+ homeostasis. *Neurosci. Res.* **2000**, *37*, 227–236. [CrossRef]
87. Cadenas, E.; Davies, K.J. Mitochondrial free radical generation, oxidative stress, and aging. *Free Radic. Biol. Med.* **2000**, *29*, 222–230. [CrossRef]
88. Chan, P.H. Reactive oxygen radicals in signaling and damage in the ischemic brain. *J. Cereb. Blood Flow Metab.* **2001**, *21*, 2–14. [CrossRef]
89. Pasantes-Morales, H.; Lezama, R.; Ramos-Mandujano, G.; Tuz, K.L. Mechanisms of cell volume regulation in hypo-osmolality. *Am. J. Med.* **2006**, *119*, S4–S11. [CrossRef] [PubMed]
90. Lambert, I.H. Regulation of the cellular content of the organic osmolyte taurine in mammalian cells. *Neurochem. Res.* **2004**, *29*, 27–63. [CrossRef]
91. Dykens, J.A.; Stern, A.; Trenkner, E. Mechanism of kainate toxicity to cerebellar neurons in vitro is analogous to reperfusion tissue injury. *J. Neurochem.* **1987**, *49*, 1222–1228. [CrossRef] [PubMed]
92. Dykens, J.A. Isolated cerebral and cerebellar mitochondria produce free radicals when exposed to elevated CA2+ and Na+: Implications for neurodegeneration. *J. Neurochem.* **1994**, *63*, 584–591. [CrossRef]
93. Ankarcrona, M.; Dypbukt, J.M.; Bonfoco, E.; Zhivotovsky, B.; Orrenius, S.; Lipton, S.A.; Nicotera, P. Glutamate-induced neuronal death: A succession of necrosis or apoptosis depending on mitochondrial function. *Neuron* **1995**, *15*, 961–973. [CrossRef]
94. Carriedo, S.G.; Yin, H.Z.; Sensi, S.L.; Weiss, J.H. Rapid Ca^{2+} entry through Ca^{2+}-permeable AMPA/Kainate channels triggers marked intracellular Ca^{2+} rises and consequent oxygen radical production. *J. Neurosci.* **1998**, *18*, 7727–7738. [CrossRef]
95. Ciani, E.; Groneng, L.; Voltattorni, M.; Rolseth, V.; Contestabile, A.; Paulsen, R.E. Inhibition of free radical production or free radical scavenging protects from the excitotoxic cell death mediated by glutamate in cultures of cerebellar granule neurons. *Brain Res.* **1996**, *728*, 1–6. [CrossRef]
96. Schulz, J.B.; Lindenau, J.; Seyfried, J.; Dichgans, J. Glutathione, oxidative stress and neurodegeneration. *Eur. J. Biochem.* **2000**, *267*, 4904–4911. [CrossRef] [PubMed]
97. Lien, Y.H.; Shapiro, J.I.; Chan, L. Study of brain electrolytes and organic osmolytes during correction of chronic hyponatremia. *J. Clin. Invest.* **1991**, *88*, 303–309. [CrossRef]
98. Mizui, T.; Kinouchi, H.; Chan, P.H. Depletion of brain glutathione by buthionine sulfoximine enhances cerebral ischemic injury in rats. *Am. J. Physiol.* **1992**, *262*, H313–H317. [CrossRef]
99. Weber, C.A.; Duncan, C.A.; Lyons, M.J.; Jenkinson, S.G. Depletion of tissue glutathione with diethyl maleate enhances hyperbaric oxygen toxicity. *Am. J. Physiol.* **1990**, *258*, L308–L312. [CrossRef]
100. Clark, E.C.; Thomas, D.; Baer, J.; Sterns, R.H. Depletion of glutathione from brain cells in hyponatremia. *Kidney Int.* **1996**, *49*, 470–476. [CrossRef]
101. Yudkoff, M.; Pleasure, D.; Cregar, L.; Lin, Z.P.; Nissim, I.; Stern, J.; Nissim, I. Glutathione turnover in cultured astrocytes: Studies with [15N]glutamate. *J. Neurochem.* **1990**, *55*, 137–145. [CrossRef]
102. Hasan, D.; Wudicks, E.F.M.; Vermeulen, M. Hyponatremia is associated with cerebral ischemia in patients with aneurysmal subarachnoid hemorrhage. *Ann. Neurol.* **1990**, *27*, 106–108. [CrossRef]
103. Volterra, A.; Trotti, D.; Racagni, G. Glutamate uptake is inhibited by arachidonic acid and oxygen radicals via two distinct and additive mechanisms. *Mol. Pharmacol.* **1994**, *46*, 986–992.
104. Barsony, J.; Sugimura, Y.; Verbalis, J.G. Osteoclast response to low extracellular sodium and the mechanism of hyponatremia-induced bone loss. *J. Biol. Chem.* **2011**, *286*, 10864–10875. [CrossRef] [PubMed]

105. Bergstrom, W.H.; Wallace, W.M. Bone as a sodium and potassium reservoir. *J. Clin. Investig.* **1954**, *33*, 867–873. [CrossRef] [PubMed]
106. Seeliger, E.; Ladwig, M.; Reinhardt, H.W. Are large amounts of sodium stored in an osmotically inactive form during sodium retention? Balance studies in freely moving dogs. *Am. J. Physiol. Regul. Integr. Comp. Physiol.* **2006**, *290*, R1429–R1435. [CrossRef] [PubMed]
107. Wu, X.; Itoh, N.; Taniguchi, T.; Nakanishi, T.; Tatsu, Y.; Yumoto, N.; Tanaka, K. Zinc-induced sodium-dependent vitamin C transporter 2 expression: Potent roles in osteoblast differentiation. *Arch. Biochem. Biophys.* **2003**, *420*, 114–120. [CrossRef] [PubMed]
108. Xiao, X.H.; Liao, E.Y.; Zhou, H.D.; Dai, R.C.; Yuan, L.Q.; Wu, X.P. Ascorbic acid inhibits osteoclastogenesis of RAW264.7 cells induced by receptor activated nuclear factor kappaB ligand (RANKL) in vitro. *J. Endocrinol. Investig.* **2005**, *28*, 253–260. [CrossRef]
109. Fibbi, B.; Benvenuti, S.; Giuliani, C.; Deledda, C.; Luciani, P.; Monici, M.; Mazzanti, B.; Ballerini, C.; Peri, A. Low extracellular sodium promotes adipogenic commitment of human mesenchymal stromal cells: A novel mechanism for chronic hyponatremia-induced bone loss. *Endocrine* **2016**, *52*, 73–85. [CrossRef]
110. Baksh, D.; Song, L.; Tuan, R.S. Adult mesenchymal stem cells: Characterization, differentiation, and application in cell and gene therapy. *J. Cell. Mol. Med.* **2004**, *8*, 301–316. [CrossRef]
111. Rosen, C.J.; Ackert-Bicknell, C.; Rodriguez, J.P.; Pino, A.M. Marrow fat and the bone microenvironment: Developmental, functional and pathological implications. *Crit. Rev. Eukariot. Gene Expr.* **2009**, *19*, 109–124. [CrossRef]
112. Sohal, R.S.; Weindruch, R. Oxidative stress, caloric restriction, and aging. *Science* **1996**, *273*, 59–63. [CrossRef] [PubMed]
113. Yakner, B.A.; Lu, T.; Loerch, P. The Aging Brain. *Annu. Rev. Pathol.* **2008**, *3*, 41–66. [CrossRef] [PubMed]
114. Upadhyay, A.; Jaber, B.L.; Madias, N.E. Incidence and prevalence of hyponatremia. *Am. J. Med.* **2006**, *119*, S30–S35. [CrossRef]
115. Barsony, J.; Manigrasso, M.B.; Xu, Q.; Tam, H.; Verbalis, J.G. Chronic hyponatremia exacerbates multiple manifestations of senescence in male rats. *Age* **2013**, *35*, 271–288. [CrossRef] [PubMed]
116. Oniki, T.; Teshima, Y.; Nishio, S.; Ishii, Y.; Kira, S.; Abe, I.; Yufu, K.; Takahashi, N. Hyponatremia aggravates cardiac susceptibility to ischemia/reperfusion injury. *Int. J. Exp. Path.* **2019**, *100*, 350–358. [CrossRef]
117. Crompton, M. The mitochondrial permeability transition pore and its role in cell death. *Biochem. J.* **1999**, *341*, 233–249. [CrossRef]
118. Goordeva, A.V.; Zvyagilskaya, R.A.; Labas, Y.A. Cross-talk between reactive oxygen species and calcium in living cells. *Biochemistry* 2003, 68:1077-1080; Brookes PS. Mitochondrial nitric oxide synthase. *Mitochondrion* **2004**, *3*, 187–204.
119. Benvenuti, S.; Deledda, C.; Luciani, P.; Modi, G.; Bossio, A.; Giuliani, C.; Fibbi, B.; Peri, A. Low extracellular sodium causes neuronal distress independently of reduced osmolality in an experimental model of chronic hyponatremia. *Neuromol. Med.* **2013**, *15*, 493–503. [CrossRef]
120. Mancuso, C. Heme oxygenase and its products in the nervous system. *Antioxid. Redox Signal* **2004**, *6*, 878–887.
121. Chen, K.; Gunter, K.; Maines, M.D. Neurons overexpressing heme oxygenase-1 resist oxidative stress-mediated cell death. *J. Neurochem.* **2000**, *75*, 304–313. [CrossRef]
122. Takizawa, S.; Hirabayashi, H.; Matsushima, K.; Tokuoka, K.; Shinohara, Y. Induction of heme oxygenase protein protects neurons in cortex and striatum, but not in hippocampus, against transient forebrain ischemia. *J. Cereb. Blood Flow Metab.* **1998**, *18*, 559–569. [CrossRef] [PubMed]
123. Ku, B.M.; Joo, Y.; Mun, J.; Roh, G.S.; Kang, S.S.; Cho, G.J. Heme oxygenase protects hippocampal neurons from ethanol-induced neurotoxicity. *Neurosci. Lett.* **2006**, *405*, 168–171. [CrossRef] [PubMed]
124. Benvenuti, S.; Deledda, C.; Luciani, P.; Giuliani, C.; Fibbi, B.; Muratori, M.; Peri, A. Neuronal distress induced by low extracellular sodium in vitro is partially reverted by the return to normal sodium. *J. Endocrinol. Investig.* **2016**, *39*, 177–184. [CrossRef] [PubMed]
125. Shruster, A.; Melamed, E.; Offen, D. Neurogenesis in the aged and neurodegenerative brain. *Apoptosis* **2010**, *15*, 1415–1421. [CrossRef]
126. Marroncini, G.; Fibbi, B.; Errico, A.; Grappone, C.; Maggi, M.; Peri, A. Effects of low extracellular sodium on proliferation and invasive activity of cancer cells in vitro. *Endocrine* **2020**, *67*, 473–484. [CrossRef]
127. Hanahan, D.; Weinberg, R.A. Hallmarks of cancer: The next generation. *Cell* **2011**, *144*, 646–674. [CrossRef]
128. Panieri, E.; Santoro, M.M. ROS homeostasis and metabolism: A dangerous liason in cancer cells. *Cell Death Dis.* **2016**, *7*, e2253. [CrossRef]
129. Zelenka, J.; Koncosova, M.; Ruml, T. Targeting of stress response pathways in the prevention and treatment of cancer. *Biotechnol. Adv.* **2018**, *36*, 583–602. [CrossRef]
130. Marnett, L.J. Oxyradicals and DNA damage. *Carcinogenesis* **2000**, *21*, 361–370. [CrossRef]
131. Gorrini, C.; Harris, I.S.; Mak, T.W. Modulation of oxidative stress as an anticancer strategy. *Nat. Rev. Drug Discov.* **2013**, *12*, 931–947. [CrossRef]
132. Diehn, M.; Cho, R.W.; Lobo, N.A.; Kalisky, T.; Dorie, M.J.; Kulp, A.N.; Qian, D.; Lam, J.S.; Ailles, L.E.; Wong, M.; et al. Association of reactive oxygen species levels and radioresistance in cancer stem cells. *Nature* **2009**, *458*, 780–783. [CrossRef]
133. Kim, H.M.; Haraguchi, N.; Ishii, H.; Ohkuma, M.; Okano, M.; Mimori, K.; Eguchi, H.; Yamamoto, H.; Nagano, H.; Sekimoto, M.; et al. Increased CD13 expression reduces reactive oxygen species, promoting survival of liver cancer stem cells via an epithelial-mesenchymal transition-like phenomenon. *Ann. Surg. Oncol.* **2012**, *19*, 539–548. [CrossRef]

134. Jeppesen, A.N.; Jensen, H.K.; Donskov, F.; Marcussen, N.; von der Maase, H. Hyponatremia as a prognostic and predictive factor in metastatic renal cell carcinoma. *Br. J. Cancer* **2010**, *102*, 867–872. [CrossRef] [PubMed]
135. Penttila, P.; Bono, P.; Peltola, K.; Donskov, F. Hyponatremia associates with poor outcome in metastatic renal cell carcinoma patients treated with everolimus: Prognostic impact. *Acta Oncol.* **2018**, *57*, 1580–1585. [CrossRef] [PubMed]
136. Berardi, R.; Santoni, M.; Newsom-Davis, T.; Caramanti, M.; Rinaldi, S.; Tiberi, M.; Morgese, F.; Torniai, M.; Pistelli, M.; Onofri, A.; et al. Hyponatremia normalization as an independent prognostic factor in patients with advanced non-small cell lung cancer treated with first-line therapy. *Oncotarget* **2017**, *8*, 23871–23879. [CrossRef] [PubMed]

MDPI

Review

Oxidative Stress in the Pathogenesis of Crohn's Disease and the Interconnection with Immunological Response, Microbiota, External Environmental Factors, and Epigenetics

Ester Alemany-Cosme [1,†], Esteban Sáez-González [2,†], Inés Moret [2,3], Beatriz Mateos [2], Marisa Iborra [2,3], Pilar Nos [2,3], Juan Sandoval [1,4,*,‡] and Belén Beltrán [2,3,*,‡]

1 Biomarkers and Precision Medicine Unit, Medical Research Institute Hospital La Fe (IIS La Fe), 46026 Valencia, Spain; ester_alemany@iislafe.es
2 Inflammatory Bowel Disease Research Group, Medical Research Institute Hospital La Fe (IIS La Fe), 46026 Valencia, Spain; saez_estgon@gva.es (E.S.-G.); ines.moret@uv.es (I.M.); beatriz_mateos@externos.iislafe.es (B.M.); iborra_mis@gva.es (M.I.); nos_pil@gva.es (P.N.)
3 Center for Biomedical Research and Network in the Thematic Area of Liver and Digestive Diseases (CIBEREHD), 28029 Madrid, Spain
4 Epigenomics Core Facility, Medical Research Institute Hospital La Fe (IIS La Fe), 46026 Valencia, Spain
* Correspondence: epigenomica@iislafe.es (J.S); belenbeltranniclos@gmail.com or beltran_bel@gva.es (B.B.)
† Both authors should be considered as first authors.
‡ Both authors should be considered as last authors.

Abstract: Inflammatory bowel disease (IBD) is a complex multifactorial disorder in which external and environmental factors have a large influence on its onset and development, especially in genetically susceptible individuals. Crohn's disease (CD), one of the two types of IBD, is characterized by transmural inflammation, which is most frequently located in the region of the terminal ileum. Oxidative stress, caused by an overabundance of reactive oxygen species, is present locally and systemically in patients with CD and appears to be associated with the well-described imbalanced immune response and dysbiosis in the disease. Oxidative stress could also underlie some of the environmental risk factors proposed for CD. Although the exact etiopathology of CD remains unknown, the key role of oxidative stress in the pathogenesis of CD is extensively recognized. Epigenetics can provide a link between environmental factors and genetics, and numerous epigenetic changes associated with certain environmental risk factors, microbiota, and inflammation are reported in CD. Further attention needs to be focused on whether these epigenetic changes also have a primary role in the pathogenesis of CD, along with oxidative stress.

Keywords: Crohn's disease; oxidative stress; antioxidants; pathogenesis; inflammation; microbiota; dysbiosis; environmental factors; epigenetics

Citation: Alemany-Cosme, E.; Sáez-González, E.; Moret, I.; Mateos, B.; Iborra, M.; Nos, P.; Sandoval, J.; Beltrán, B. Oxidative Stress in the Pathogenesis of Crohn's Disease and the Interconnection with Immunological Response, Microbiota, External Environmental Factors, and Epigenetics. *Antioxidants* **2021**, *10*, 64. https://doi.org/10.3390/antiox10010064

Received: 15 December 2020
Accepted: 4 January 2021
Published: 7 January 2021

Publisher's Note: MDPI stays neutral with regard to jurisdictional claims in published maps and institutional affiliations.

1. Introduction

Inflammatory bowel disease (IBD) is a complicated and multifactorial disorder characterized by relapsing and remitting inflammation that can involve the entire gastrointestinal tract. Crohn's disease (CD) and ulcerative colitis (UC), the two types of IBD, are recognized worldwide as major contributors to gastrointestinal disease. The location of the inflammation and the nature of the histological disorders in the gastrointestinal tract differentiate the two diseases. IBD results from a complex interplay between genetic variation, intestinal microbiota, the host's immune system, and environmental factors such as drugs, diet, breastfeeding, and smoking, although the exact cause of the disease remains unknown. Environmental/microbiota factors can affect gene expression through epigenetic mechanisms in triggering the disease [1].

The intestinal tract is under continual attack from luminal microbes and from oxidized compounds in the diet, exposing it to recurrent oxidative changes [2]. An imbalance in redox intestinal homeostasis impairs the intestinal epithelial cells and the permeable barrier, activating dysfunctional immune responses [3,4]. Intestinal cells are the key elements in regulating the traffic of antigens toward gut-associated lymphoid tissues; discriminating between commensal and pathogenic antigens; and acting as a crossroad between immunological tolerance and the immune response. Immunological tolerance describes a diverse range of host processes that prevent potentially harmful immune responses within that host. The loss of immune tolerance allows for an exaggerated and harmful immune response [3–5]. Cell inflammation and oxidative reactions with the overproduction of reactive oxygen species (ROS) through activated leukocytes can overwhelm the tissue's antioxidant defenses and can contribute to the functional impairment of the enteric mucosa. This leads to an aberrant response to the luminal agents and the development of chronic abnormal inflammatory and dysfunctional immune responses [6]. An antioxidant intestinal environment reflects the intestinal mucosa's response, aimed at preventing oxidative damage, and is maintained by a complex dynamic recycling system in which different molecules undergo well-established oxidation–reduction reactions. A proper dietary intake of antioxidants is therefore essential for maintaining low intracellular levels of oxidative species, thereby maintaining a proper gastrointestinal redox balance [2]. Dietary compounds are therefore an important aspect of intestinal health.

Oxidative stress is reported as a pivotal factor in the pathogenesis of IBD and might be a key effector mechanism leading to cellular/molecular damage and tissue injury. There is evidence that ROS are involved in intracellular signaling and in the regulation of growth, differentiation, and cell death, as well as in inflammation [3,4]. Cells' antioxidant defenses (mainly molecules and antioxidant enzymes) avoid accumulation and the consequent cell damage is promoted by ROS. Oxidative damage was detected not only in the intestinal mucosa of patients with CD but also in peripheral blood leukocytes [5]. The immune cells that reach the mucosa in CD release a number of ROS that are potentially detrimental. The main pathological feature of CD is an infiltration of polymorphonuclear neutrophils and mononuclear cells into the affected intestinal tract. Neutrophils and other leukocytes produce noxious substances, including ROS and proinflammatory cytokines, such as interleukin (IL)-1, IL-8, and tumor necrosis factor alpha (TNF-α). An imbalance in proinflammatory and anti-inflammatory cytokine levels was shown to occur in CD [2,6]. Similarly, plasma antioxidant defenses are diminished in CD [7].

The role of oxidative stress as a potential etiological or triggering factor for IBD is the subject of increasing interest in recent years. Our group previously characterized the ROS implicated in the oxidative damage occurring in the peripheral blood of patients with CD at the beginning of the disease, prior to any treatment, as well as their antioxidative stress status and possible implications in regulating the processes in CD [6]. Mitochondria are the main organelles responsible for ROS production during physiological and pathological states. Mitochondrial dysfunction could therefore involve a combination of excess ROS production and diminished antioxidant capacity. Oxidative stress leads to mucosal layer damage and bacterial invasion, which in turn further stimulate the immune response and contribute to disease progression [7]. Environmental factors and oxidative stress can affect the disease through epigenetics. Increasing evidence suggests that oxidative stress globally affects the chromatin structure and the enzymatic and nonenzymatic post-translational modification of histones and DNA-binding proteins. A better understanding of diet–host–microbiota–environmental interactions is essential for unraveling the complex molecular basis of epigenetic and genetic interactions underlying the pathogenesis of IBD, as well as the role of oxidative stress in this complex disease.

The aim of this review is to summarize the main findings regarding the oxidant and antioxidant mechanisms involved in CD, their role in the immunological response, the environment's effects on oxidative stress status, and its involvement in epigenetic changes/modifications.

2. Oxidative Stress in Crohn's Disease

2.1. Oxidative Stress and the Impaired Immunological Response

Oxidative stress, defined as the state in which the oxidant–antioxidant homeostasis within the cell is disturbed, results from an imbalance between ROS production and the defensive system responsible for its detoxification in cells. ROS are natural by-products that include both radical and non-radical oxygen-containing molecules and are mainly produced by mitochondria during oxygen metabolism and water generation. ROS have several physiological roles (e.g., cell signaling regulating growth, differentiation, apoptosis, and inflammatory processes) [8]. However, an increase in the number of these highly reactive molecules in the state of oxidative stress can cause damage to cell components, especially membrane lipids, DNA, and proteins. The main pro-oxidant species are the ROS formed by unstable forms of oxygen—superoxide anion, hydrogen peroxide (H_2O_2), and hydroxyl radicals. In contrast, antioxidant agents include both enzymatic and nonenzymatic elements. Antioxidant enzymes are present in all cells, have a primary role in detoxification, and include the enzymes catalase, superoxide dismutase (SOD), and glutathione peroxidase (GPx) (Figure 1). Nonenzymatic antioxidants are usually located in extracellular compartments and include various molecules, such as glutathione, ascorbic acid, and vitamin E [3,9].

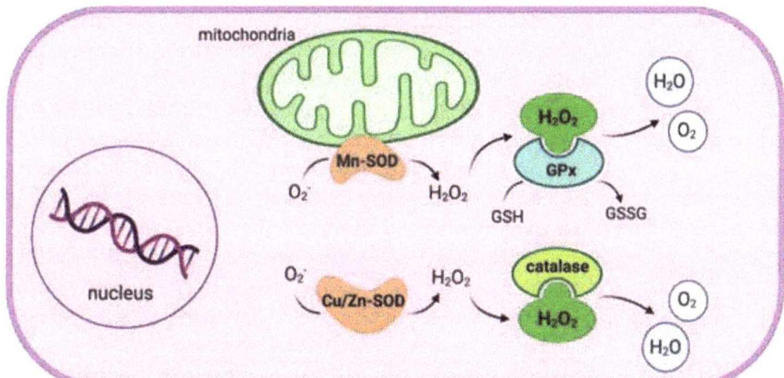

Figure 1. Intracellular antioxidant enzymes responsible for the detoxification of mitochondrial-generated reactive oxygen species. Note that there are two forms of intracellular superoxide dismutase in humans—mitochondrial (Mn-SOD) and cytosolic (Cu/Zn-SOD). These enzymes catalyze the dismutation of the highly reactive superoxide anion (O_2^-) to oxygen and hydrogen peroxide (H_2O_2). In turn, H_2O_2 serves as a substrate for both glutathione peroxidase (GPx) and catalase, which catalyze its reduction to water (figure modified from Moret I [5]).

Inflammation, the main pathological characteristic of IBD, is a process strongly linked to the generation of reactive metabolites, such as reactive nitrogen species (RNS) and ROS [10]. In CD, a massive infiltration of inflammatory cells (polymorphonuclear neutrophils and mononuclear cells) into the affected gut mucosa is reported. The activated neutrophils and macrophages that reach the mucosa stimulate the production of reactive species, including ROS, which are potentially detrimental because they can lead to oxidative stress, causing further inflammation and tissue injury [11–13]. In terms of adaptive immunity in CD, the intestinal mucosa accumulates CD4+ T cells in the lamina propria, with an immunological response of the Th1/Th17 type. These cells are resistant to apoptosis, which perpetuates the inflammatory response in the intestinal epithelium. In CD, there is an increased mucosal concentration of the proinflammatory cytokine TNF-α (even during disease remission) [14]. A study reported that numerous apoptotic stimulators, similar to TNF-α, can induce ROS generation by interacting with the respiratory chain in the mitochondria and that these ROS could be acting as mediators in apoptotic pathways [4]. Antioxidant production is the first-line defense against oxidative agents in cells; however, persistent oxidative stress can delete antioxidant cell resources and the ability to produce more antioxidants [15]. In fact, studies as far back as 2003 proposed an imbalanced and inefficient endogenous antioxidant response in the intestinal mucosa of patients with IBD [16,17]. Patients with CD show reduced activity in the main cellular antioxidant enzymes SOD and GPx, as well as reduced levels of the plasma antioxidants vitamin A, C, E, and beta-carotene in the blood and mucosa [15]. Although it is important to note that there is conflicting evidence regarding the change in antioxidant levels, the key point is that there is an imbalance in antioxidant concentrations. It is generally believed that there is excessive oxidant activity and a lower response by antioxidative compounds in CD, which sustain the oxidative stress in the disease [11,18,19].

Considerable evidence strongly suggests that the oxidative stress is coupled with an impaired inflammatory response and chronic inflammation in CD. At the molecular level, oxidative stress and redox signaling are closely involved in the upregulation of inflammatory cytokines and the increased infiltration of inflammatory cells, via the stimulation of signaling pathways (especially the redox-sensitive transcription factor, nuclear factor kappa B). Moreover, inflammation increases oxidative stress by stimulating the ROS/RNS generating systems, along with the release of myeloperoxidase from inflammatory cells [10].

At the clinical level, a recent study [20] found a positive correlation between the oxidative stress index (a general indicator of oxidative stress) and the C-reactive protein levels (a marker of inflammation) in patients with CD, indicating a putative association between higher oxidative stress levels and increased inflammation. As the authors stated, this association could be supported by a previous study [21] in which the redox status of glutathione was heavily reduced (due to increased oxidized glutathione levels in areas of inflammation, indicating greater oxidative stress) in the inflamed ileum mucosa, compared to the non-inflamed tissue of patients with CD. Similarly, in a study by Iantomasi et al. [22], higher levels of oxidized glutathione were detected in the diseased ileum than in the healthy ileum of patients with CD. However, this study and the one by Kruidenier et al. [23] reported an increase in the GPx activity (indicating antioxidant capacity) in the inflamed intestinal CD mucosa compared to the controls. A more recent study [24], however, showed reduced GPx activity in the inflamed mucosa compared to either the noninflamed CD mucosa or the healthy controls. The study also mentioned the possible methodological limitations of the previously mentioned studies. Our group established another clinical link between oxidative stress and inflammation. We found an increase in H_2O_2 in peripheral lymphocytes and monocytes that correlates significantly with certain inflammation markers (such as C-reactive protein and fibrinogen) during active CD, indicating that the inflammation is more pronounced as the H_2O_2 concentration increases in these cells. We also showed that the mitochondrial membrane potential is significantly inhibited in the immune cells (which suggests a mitochondrial source of ROS) and correlates negatively with inflammation markers [6]. The latest clinical evidence of the connection between oxidative stress and inflammation comes from Bourgonje et al. [25], who reported that plasma-free thiols (which reflect systemic oxidative stress, given that they are prime substrates for ROS) showed a negative correlation with inflammation biomarkers and were associated with favorable outcomes in CD.

Subclinical intestinal inflammation is present in a large proportion of patients with CD, even in clinical remission [26], and a recent report stated that CD in clinical remission is marked by systemic oxidative stress [25]. Although the specific mechanism through which oxidative stress is related to the characteristic inflammation in CD is not completely understood, evidence indicates that oxidative stress could have a significant role in the pathogenesis of CD [19].

2.2. Oxidative Stress as a Key Effector Mechanism in CD Pathogenesis

In the state of oxidative stress, ROS can be harmful to cell components, with especially negative effects on membrane lipids, proteins, and mitochondrial and nuclear DNA. ROS therefore have the potential of contributing to the pathogenesis in CD [3]. Lipid peroxidation caused by ROS alters the normal activity of transmembrane enzymes, membrane transporters, and receptors (by disturbing the hydrophobic lipid–lipid and lipid–protein interaction), consequently disrupting the homeostasis and cell metabolism. The end products of lipid peroxidation can cause protein damage, rendering the proteins useless [27,28]. Pelli et al. showed that excess lipid peroxidation is likely an important pathogenic factor in IBD [29], which was later proposed by Sampietro et al. for CD, in particular [30]. A study also reported that treatment with 4-hydroxynonenal (a lipid peroxidation product) exacerbates colonic inflammation through the activation of toll-like receptor 4 signaling [31]. An upward trend in serum and saliva levels of malondialdehyde (a product of lipid peroxidation) was recently reported (an increase that depends on CD severity), as well as a correlation between malondialdehyde levels and the visible symptoms of inflammation [32]. Oxidative DNA damage can cause various lesions, including single and double-strand breaks, apurinic/apyrimidinic sites, and modified pyrimidines and purines. Although DNA damage can be repaired by cellular mechanisms, chronic exposure to oxidative stress leads to the accumulation of DNA lesions, which can therefore promote mutagenesis, human pathogenesis, and loss of homeostasis [9,33]. Oxidative DNA damage was also proposed as a key player in the pathogenesis of IBD and in the associated carcinogenesis [34]. With regard to protein oxidation, ROS can cause hydroxylation or carbonylation of proteins, which can change their function considerably and even provoke their degradation [35]. Krzystek-Korpacka et al. [36] found that IBD is associated with an enhanced formation of advanced oxidation protein products, which have proinflammatory properties.

The oxidative damage of these macromolecules and the effects of pro-oxidants and antioxidants were studied over the past two decades as potential diagnostic, progression, and prognostic markers in CD, including in a recent systematic review on IBD and CD biomarkers by Krzystek-Korpacka [37]. Although a number of these markers show promise, they are mostly at the early research phase of discovery. However, the large number of studies that related CD to oxidative stress markers is evidence of the predominant role of oxidative stress in the disease.

Another finding that reiterates the primary role of oxidative stress in the pathogenesis of CD is its close connection with the main pathologic features of CD. As stated earlier, oxidative stress is related to inflammation and the immune response in CD. It was suggested that ROS overproduction by peripheral immune cells occurs before the cells reach the intestinal mucosa [6], which would link these ROS with the development of the disease. Moreover, the increased vascular density and pathological tissue hypoxia that also characterize CD might lead to increased ROS production through activation of targets of the hypoxia-inducible factor transcription factor family [38,39]. The inflamed mucosa is therefore continually exposed to the detrimental effects of oxidative substances, eventually leading to extensive cell and tissue damage, which accounts for the disease [25].

The development of new therapies targeting oxidative stress in CD also puts into perspective the pathogenic essence of this mechanism underlying the disease. A number of unconventional therapeutic methods with antioxidant effects, such as inhibitors against ROS generation, functional dietary interventions, and substances that activate antioxidant enzymes are under investigation as complementary and alternative treatments for IBD, showing promising results [40], although antioxidant therapy remains controversial [15]. Mainstream IBD treatments focus on reducing inflammation, and mainly consist of immunosuppressants, corticosteroids, and anti-TNF-α antibodies. However, it is noteworthy that the therapeutic effect of these drugs is also due to their antioxidative properties. In fact, immunosuppressants and corticosteroids possess direct free radical-scavenging abilities, and anti-TNF-α antibodies carry an indirect antioxidative effect by reducing TNF-α concentrations [18].

Oxidative stress is associated with diarrhea, a frequent symptom in IBD, given that excessive ROS production might be responsible for the excess electrolyte and water secretion that causes diarrhea [41]. The severe clinical activity in CD is reflected by systemic oxidative stress, which likely contributes to the development of the extraintestinal manifestations commonly observed in CD, such as perianal fistulas and arthritis [42]. Oxidative DNA damage might have a primordial role in the inflammation-associated tumorigenesis observed in certain patients with CD, who are at greater risk of colorectal cancer [34,43], which once again highlights the pathogenic potential of oxidative stress, which could go beyond CD.

3. The Role of the Environment in Crohn's Disease

3.1. The "In-Vironment": The Microbiota

The human gut harbors trillions of microorganisms (including bacteria, viruses, fungi, and protozoa) that constitute the gut microbiota. Intestinal bacteria are the predominant microorganisms in the microbial flora, and more than 99% belong to the Firmicutes, Bacteroidetes, Proteobacteria, or Actinobacteria phylum [44]. The microbiota symbiotically interacts with the host, exerting a variety of beneficial effects that include substrate digestion, nutrient production, metabolism, pathogen protection, and remarkably, the normal structural and functional development of the mucosal immune system [45]. The microbiota influences both the local and the systemic immune responses [46] and has a dynamic composition that changes with age and varies according to environmental factors, which is most evident in diet and food intake patterns [47]. Environmental changes can therefore be reflected through changes in microbiota, which in turn can affect the host's health, which is why the microbiota can be considered an "in-vironmental" factor—the proximate environmental influence contributes to both health and disease states [48].

The important role played by the microbiota in immunological responses is reflected in IBD. Specifically in CD, the microbiota triggers the Th1 response, with the consequent generation of interferon gamma and TNF-α, leading to inflammation and mucosal barrier damage [49]. During mucosal inflammation, intestinal epithelial cells, along with immune cells (mainly macrophages and neutrophils), produce proinflammatory cytokines that induce the production of superoxide anion, nitric oxide, and oxidant peroxynitrite, via the activation of nicotinamide adenine dinucleotide phosphate oxidase and inducible nitric oxide synthase. These reactive species are involved in the initiation and progression of CD [40].

Immune reactivity against microbial-derived antigens is reported in patients with CD. In fact, more than 10 types of antimicrobial serologic antibodies were identified as relevant to CD (such as antibodies against the outer membrane porin C of *Escherichia coli*). These serologic antibodies are associated with a more severe CD phenotype and with a higher risk for surgery [50].

Therefore, an abnormal relationship between the host and microbiota can result in an intestinal immune imbalance in CD. However, it is still unclear whether mucosal tissue damage is the result of an abnormal immune response to a normal microbiota or is the result of a normal immune response against abnormal microbiota (dysbiosis) [51].

Dysbiosis, defined as an unfavorable abnormality in the composition and function of the gut microbiota, disturbs the interaction between the host and microbiota and the host's immune system. Dysbiosis is associated with several human diseases, including CD, in which it appears to play a pivotal role in the pathogenesis [44]. The intestinal microbiota in CD is therefore characterized by decreased diversity, reduced proportions of Firmicutes, and increased proportions of Proteobacteria and Actinobacteria. Moreover, the microbiota of patients with CD is reported to be overpopulated with bacteria with proinflammatory properties (e.g., *Escherichia* and *Fusobacterium*) and reduced populations of anti-inflammatory bacteria (e.g., *Faecalibacterium*) [52,53]. Dysbiosis causes an alteration in the intercellular tight junctions that maintain the integrity of the intestinal mucosa and its permeability. Consequently, opportunistic pathogens can invade the mucosa, resulting in an activation of mucosal-associated lymphatic tissue and the inflammatory cascade (leukocytes and proinflammatory cytokines), which can cause massive tissue damage [54]. These opportunistic pathogens can therefore provoke ROS overproduction in human mucosal epithelial cells, inducing the overexpression of dual oxidase 2 [55].

The functional composition of the gut microbiome, which can be defined as the set of genomes of the microbiota, can provide a more consistent definition of dysbiosis [56], due to the increased stability that the microbiome exhibits over time and the differences in gut microbiome composition between individuals, in contrast to the similarities in phylogenetic profiling [52]. Metagenomic approaches characterizing the microbiome can provide greater insight into the function of the gut microbiota in disease. Metagenomic studies highlight that microbial metabolic pathways are more consistently perturbed in IBD than organismal abundances [57]. In the cited study, Morgan et al. showed that patients with CD show an increase in glutathione transport gene abundance. Glutathione, produced by Proteobacteria and *Enterococcus*, is involved in the maintenance of bacterial homeostasis during oxidative stress. As the authors noted, an increase in the sulfate transport, cysteine metabolism, and glutathione metabolism observed in the patients with IBD might reflect a mechanism by which the gut microbiome addresses the oxidative stress caused by inflammation.

Microbiota-induced inflammation and oxidative stress caused by ROS overproduction are strongly intertwined in CD, given that they reinforce each other and that both lead to mucosal barrier damage. This, in turn, can lead to increased mucosal permeability and loss of protection, allowing for the invasion of pathogens, which can further stimulate inflammation and ROS production, resulting in a vicious circle. Although it is still unclear whether dysbiosis is a primary or secondary phenomenon in CD, it is believed to have a key role in its pathogenesis [58,59].

Genetic findings in IBD also put the microbiota into the spotlight of disease pathogenesis [56,60]. Genome-wide association studies identified more than 160 genetic loci susceptible to conferring protection from IBD or an increased risk of developing IBD [60]. Most of these genes play an important role in the mucosal barrier function, antimicrobial recognition and function, and immune regulation [61]. Consequently, defects or certain variants of these genes can trigger an abnormal immune response to gut microbiota [56]. In CD, these include nucleotide oligomerization domain 2 (*NOD2*), autophagy-related 16-like 1 (*ATG1GL1*), and immunity-related GTPase M (*IRGM*). *ATG16L1* and *IRGM* are autophagy genes involved in the intracellular processing of bacteria [62,63]. *NOD2* was the first susceptibility gene identified for CD more than a decade ago and is known to stimulate the immune system by acting as an intracellular sensor of bacterial peptidoglycans [64,65]. *NOD2* mutations in patients with CD are associated with diminished mucosal α-defensin expression levels, which are antimicrobial peptides that play an important role in the mucosal antibacterial barrier [66]. In other clinical studies, the presence of *NOD2* risk alleles in patients with IBD was associated with changes in microbial composition, such as an increased number of Actinobacteria and Proteobacteria, leading to the idea that these *NOD2* variants could be contributing to bacterial dysbiosis [67,68]. Many of the genetic loci that confer risk in CD interact with each other, which is the case for *NOD2* and *ATG16L1*. Interestingly, NOD2 activation by bacteria and bacterial ligands provokes the ATG16L1-mediated formation of autophagic vacuoles in both epithelial and dendritic cells [69]. NOD2 thereby controls bacterial infection via the induction of autophagy [70]. It was also recently reported that NOD2 and ATG16L1 might cooperate as part of a common pathway to promote anti-inflammatory immune responses to the microbiota [71]. It is believed that *NOD2* and *ATG16L1* variants associated with CD result in the impaired induction of microbial-stimulated autophagy [51].

Human twin studies have not, however, provided much support for a host genetic influence on the gut microbiota. In fact, healthy siblings of patients with CD show an altered microbial and immune profile associated with CD that differs from their genotype-related risk [72]. Studies on twins also revealed that gastrointestinal microbial populations vary with CD phenotypes [73], which highlights the relevance of the external environment in shaping the microbiota, likely outweighing that of genes. Moreover, external factors strongly linked to changes in the gut microbiota, such as antibiotic therapy, appear to be associated with the development of IBD [74], which supports the idea that changes in the microbiota can act as "translators" of environmental factors in the development and progression of CD.

3.2. External and Environmental Factors

It is well established that external and environmental factors have an important influence on the onset and course of IBD. The fact that approximately two-thirds of patients have no identifiable genetic defects, along with the rapid increase in the incidence and prevalence of the disease (which cannot be due to genomic changes), strengthens this idea [75]. The increasing incidence of IBD in newly industrialized countries and its increasing prevalence in Western countries can be attributed to the influences of a Western lifestyle, urbanization, and industrialization, which were reported as primary risk factors for CD and UC [76]. Studies indicate that the incidence of CD increases in immigrants who migrate from regions with a lower prevalence to regions with a higher prevalence of CD within one or two generations, which further supports the massive influence of the environment on CD pathogenesis [77].

3.2.1. Western Diet Versus the Mediterranean Diet and Their Impact on Crohn's Disease

Environmental elements, such as diet, can directly affect the epithelial mucosa barrier and immune function and can act indirectly through the modulation of intestinal microbiota [78]. "Westernized diets," which are rich in saturated fatty acids and n-6 polyunsaturated fatty acids (PUFAs), animal proteins, simple sugars, and refined carbohydrates but have a low fiber content (low vegetable and fruit intake), might be a trigger for CD [79,80]. The quality and quantity of food were shown to affect gut microbiota [48], which could theoretically lead to inflammation in genetically susceptible individuals [51].

In a recent study conducted in mice, Agus et al. proved that the Western diet causes an inflammatory environment in the digestive tract associated with microbiome perturbations [81]. Previous studies (also conducted in mice) already indicated that a diet high in fat and sugars induces dysbiosis in the mucosa microbiota and is associated with a less protective mucosal layer and increased permeability, which can result in low-grade inflammation and metabolic disorders [82,83]. As Agus et al. reported, the Western diet could deregulate inflammation in the gut mucosa by affecting short-chain fatty acid (SCFA) production. SCFAs (such as acetate, propionate, and butyrate) are the main end-products of the microbial fermentation of dietary fiber. Butyrate typically constitutes 15%–20% of SCFAs in the human colon, is the predominant energy source for colonocytes, and is thought to promote intestinal barrier protection [84]. Butyrate likely augments the intestinal epithelial barrier function via the stabilization of hypoxia-inducible-factor-1 [85], which regulates the integrity of epithelial tight junctions [86]. Butyrate was also reported to act via activation of AMP-activated protein kinase [87]. Other in vitro studies proposed that low concentrations of butyrate could have a protective effect by increasing the synthesis of mucin 2 (MUC2), the main component of intestinal mucus [88,89]. It was proposed that butyrate could affect *MUC2* transcription via AP-1 and acetylation/methylation of histones at the *MUC2* promoter (a concept that is further discussed in Section 3.3 Epigenetics as a transductor of environmental factors in Crohn's disease). Patients with metabolic syndrome show increased colonic MUC2 expression, following a diet-induced increase in SCFA and butyrate production [90]. However, in vivo studies on pigs and rodents produced ambiguous results regarding the relationship between luminal butyrate (SCFA) levels and MUC2 abundance [90]. Further research is therefore needed to uncover the butyrate-mediated mechanisms in healthy individuals and in patients with IBD.

Butyrate is also attributed with anti-inflammatory properties, which could be mediated through the inhibition of nuclear factor-kappa B activation, inhibition of interferon gamma signaling, or the upregulation of peroxisome proliferator-activated receptor gamma [91]. Microbial-derived butyrate also induces functional colonic regulatory T cells [92]. Through metagenomic and proteomic studies, Erickson et al. confirmed the presence of lower overall levels of butyrate and other SCFAs in ileal CD [93]. Geirnaert et al. demonstrated that increased butyrate production by bacteria supplemented in vitro to the microbiota of a patient with CD enhanced intestinal epithelial barrier integrity [94]. A lack of butyrate and other SCFAs could also have indirect negative effects on the intestinal mucosa. By assembling a synthetic gut microbiota from fully sequenced human gut bacteria in gnotobiotic mice, Desai et al. [95] demonstrated that, in the absence of dietary fiber, mucolytic bacteria can use host mucus glycans as a source of energy and become the predominant species within the gut microbiota. Consequently, an abundance of these bacteria causes degradation of the colonic mucus layer and promotes pathogen susceptibility. Thus, an insufficiency in microbial-derived SCFA (especially butyrate) caused by a lack of dietary fiber or by dysbiosis [52] (which can also be due to diet) might be involved in the pathogenesis of CD, given it can result in impaired intestinal barrier function and inflammation.

The Western diet is characterized by a typically high consumption of n-6 PUFAs and a low consumption of chain n-3 PUFAs, leading to an imbalanced n-6/n-3 ratio, with detrimental health consequences [96]. N-6 PUFAs are considered proinflammatory compounds, given that linoleic acid (the major dietary vegetable PUFA) is a precursor for arachidonic acid, which is a precursor of inflammatory mediators such as prostaglandins and leukotrienes. In contrast, n-3 PUFAs appear to be inflammation regulators [97]. The increased consumption of n-6 PUFAs (along with the consumption of animal protein) is related to the increased incidence of CD in Japan [98]. Experimental studies indicated that a nutritional intervention with n-3 PUFAs exerts beneficial effects with regards to intestinal inflammation [97]. However, clinical trials to evaluate the effects of n-3 PUFAs for maintaining remission in CD e found no benefit from free n-3 PUFAs over placebo, on clinical relapse [99,100].

Food additives are another component typically in overabundance in the Western diet that are proposed to be proinflammatory agents. Chassaing et al. [101] conducted in vivo studies with mice and proposed that commonly used emulsifiers can disturb the host–microbiota relationship, resulting in microbiota with increased mucolytic and proinflammatory activity that promote chronic intestinal inflammation, which can manifest as colitis. Another recent study by Mu et al., also conducted in mice [102], concluded that titanium dioxide nanoparticles (another widely used food additive) could interfere with the balance of gut flora and the immune system, cause prolonged low-grade intestinal inflammation, and exacerbate the immunological response.

Unlike the Western diet, the Mediterranean diet is believed to have a positive and anti-inflammatory effect on IBD. The Mediterranean diet is characterized by a high consumption of fruit and vegetables, whole grains, oily fish, olive oil, seeds, and dried fruits [103]. In remarkable contrast to the Western diet, the Mediterranean diet provides fermentable dietary fiber, healthy monounsaturated and polyunsaturated fatty acids, with a balanced n-6/n-3 PUFA ratio, antioxidants, and vitamins originating from minimally or unprocessed food. In a case-control clinical study conducted by Souza et al. [104], a diet pattern based on vegetables, fish, olive oil, fruit, grain, and nuts (i.e., the Mediterranean diet) was inversely associated with CD. Marlow et al. reported that a Mediterranean-inspired diet appeared to benefit the health of patients with CD, showing a trend for reduced inflammation markers and for normalizing the microbiota [105]. A diet rich in vegetables and fibers has a positive impact on the microbiota, given it reduces intestinal pH and prevents the growth of potentially pathogenic bacteria (such as strains of *Escherichia coli* and other *Enterobacteriaceae)* [103]. Mediterranean-style diets also favor the proliferation of beneficial bacteria, such as lactic acid bacteria, through the high consumption of fermented foods and n-3 PUFAs [106].

The benefits of this dietary pattern could also be largely due to its antioxidant effects. Extra virgin olive oil, considered the Mediterranean "liquid gold", is rich in antioxidants (e.g., polyphenols) that cooperate to increase plasma antioxidant capacity. The consumption of this oil increases the antioxidant activity of enzymes such as catalase, SOD, and GPx, which have a primary role in preventing oxidative stress. Studies demonstrated that extra virgin olive oil (in healthy people) modulates the response against oxidative stress through antioxidant enzymes [107]. With even a greater link to IBD, recent studies indicated the strong anti-inflammatory effect of this oil in gut mucosa, due to its synergic action with other antioxidant molecules (such as hydroxytyrosol and squalene) [103]. Fruits and vegetables, also abundant in the Mediterranean diet, are not only a source of fiber (and can therefore favor SCFA production) but are also a source of vitamins, polyphenols, and other antioxidants [108] that can be useful in combating oxidative stress.

These dietary patterns are therefore an example of how diet can influence the onset and development of CD. Food and nutrients have a huge impact on microbiota, immune response-related pathways, and redox mechanisms. The Western and Mediterranean diets are probably the most studied diets with regard to CD, but despite this, there is a lack of scientific literature and clinical trials addressing their impact on CD. In addition to the Mediterranean diet, studies were conducted on the enteral exclusive nutrition diet, partial enteral nutrition diet, and supplementation with probiotics and antioxidant micronutrients as possible therapeutic strategies against IBD [40,84]. Although the study of the effects of diet is marked by conflicting results, difficulty in establishing solid conclusions, and research that is still to be undertaken, it is clear that diet could be a determinant in the pathogenesis of CD.

3.2.2. Other Lifestyle Factors and Health Conditions Relevant to the Pathogenesis

A recently published umbrella review of meta-analyses on the environmental risk factors for IBD [109] identified smoking, urban living, and having undergone appendectomy or tonsillectomy as the primary risk factors for CD, whereas physical activity, bed sharing, and high levels of vitamin D reduced the risk. Among these factors, smoking stands out because of its sizeable impact on CD. According to epidemiological data, cigarette smoking is one of the well-established risk factors for CD and probably the most widely investigated environmental factor that influences the course of CD. Smoking is believed to increase susceptibility to CD and aggravate its clinical course [77]. A recent systematic review and meta-analysis [110] revealed that smokers with CD have a more complicated disease course with greater flare-ups of disease activity and higher needs for first and second surgeries. Smokers with active CD were reported to have a clinically relevant dysbiosis of the gut microbiota [111]. Several studies suggested that smoking could suppress the innate immune response to bacteria through the direct inhibition of bacterial sensing patterns such as the recognition of lipopolysaccharide by the TLR4/MD-2 receptor [112]. As Bergeron et al. proposed, the striking anti-inflammatory and immunosuppressive effects observed in patients with CD who smoke, which are associated with compromised regulatory adaptive responses, might render cells more susceptible to persistent inflammatory and oxidant injury. Cells from patients with CD who smoke presented a defective sensitivity to anti-inflammatory or antioxidant protection. Above all, smoking most likely has a major role in promoting oxidative stress in patients with CD. Cigarette smoke affects ROS-generation pathways and has high levels of ROS, peroxynitrite, free radicals, and reactive organic compounds that ultimately produce oxidative stress [113]. The metal ions in tobacco smoke also facilitate the transformation of H_2O_2 into highly reactive hydroxyl radicals [40]. Long-term smoke exposure can therefore result in a systemic oxidant–antioxidant imbalance and ultimately systemic oxidative stress [114], which can negatively affect the gastrointestinal tract and promote the development of CD.

To focus on the impact of cigarette smoking on CD is not only relevant because of its evident negative influence but also because it is a (relatively) easy factor to control/avoid (in comparison to other environmental risk factors such as pollution or stress). All in all, cigarette smoking altogether with the previously discussed dietary patterns are lifestyle habits that could be reverted and which could modulate the predisposition or course of CD (Table 1).

Table 1. Remarkable research conducted in the last decade addressing the impact of lifestyle habits on CD—dietary patterns and cigarette smoking.

Study	Type of Study	Methodology	Main Findings
Agus et al. (2016) [81]	In vivo study	WT and CEABAC10 mice [1] were fed with a high-fat/high-sugar diet (HF/HS) ($N = 6$) vs. a conventional diet ($N = 5$) over a period of 18 weeks. Germ-Free (GF) mice were transplanted with fecal pellets of HF/HS donor mice ($N = 5$) or conventional donor mice ($N = 5$), followed by an infection with an adherent-invasive *E. coli* (AEIC) LF82 strain isolated from a CD patient	Western diet causes an inflammatory environment in the digestive tract associated with microbiome perturbations; favors the emergence of *E. coli* associated with the ileal, cecal, and colonic mucosa; and decreases the level of SCFA produced by intestinal microbiota modulating immune response. Transplantation of feces from HF/HS treated mice to GF mice increases susceptibility to AIEC infection
Martinez-Medina, M. et al. (2014) [82]	In vivo study	WT and CEABAC10 mice [1] were fed with a HF/HS diet vs. conventional diet for 12 weeks [2], and orally infected with AIEC strain LF82	Western diet induces changes in gut microbiota composition with an increase in the mucin-degrading bacterium *Ruminococcus torques* and the group *Bacteroides/Prevotella;* alters intestinal permeability, decreases barrier function, and affects the host homeostasis promoting AEIC gut colonization in genetically susceptible mice
Geirnaert et al. (2017) [94]	Clinical research—In vitro study	Butyrate-producing bacteria supplemented to the fecal microbial communities of CD patients ($N = 10$) in an in vitro system simulating the mucus- and lumen-associated microbiota, and an in vitro study of the resulting microbiota influence on epithelial barrier integrity with a Caco-2 model	In vitro supplementation of microbiota of CD patients with butyrate-producing bacteria results in a higher butyrate production, with and enhanced epithelial barrier integrity in a Caco-2 model. Supports the preclinical development of a probiotic product containing butyrate-producing species
Desai et al. (2016) [95]	In vivo study	Assembled synthetic gut microbiota from fully sequenced human gut bacteria in gnotobiotic mice were fed with fiber-rich vs. fiber-free diets [2]	In the chronic or intermittent absence of dietary fiber mucolytic bacteria become the predominant species within the gut microbiota with the consequent degradation of the colonic mucus layer and increased pathogen susceptibility
Chassaing et al. (2015) [101]	In vivo study	WT mice and two engineered strains of mice, namely $IL10^{-/-}$ and $TLR5^{-/-}$ (prone to develop shifts in microbiota composition and inflammation) exposed to emulsifiers in the drinking water or to water alone (control group) for 12 weeks [2]	Relatively low concentrations of commonly used dietary emulsifiers (carboxymethylcellulose and polysorbate-80) can disturb the host-microbiota relationship, induce low-grade inflammation and obesity/metabolic syndrome in WT hosts and promote robust colitis in mice predisposed to this disorder
Mu et al. (2019) [102]	In vivo study	WT mice and DSS-induced colitis mice treated with titanium dioxide nanoparticles vs. standard (control) diet for 3 months from weaning [2]	First demonstration that long-term dietary intake of titanium dioxide nanoparticles (which are used as food additives) results in lower body weigh along with colorectal inflammation in mice; and it aggravates DSS-induced chronic colitis and immune response *in vivo*, reduces the population of $CD4^+$ T cells, regulatory T cells, and macrophages in mesenteric lymph node

Table 1. *Cont.*

Study	Type of Study	Methodology	Main Findings
Marlow, G. et al. (2013) [105]	Clinical research	6-week intervention with a Mediterranean-inspired diet in CD patients (N = 8). Obtention of blood and fecal samples at the beginning and the end of the diet	A Mediterranean-inspired diet appears to benefit the health of CD patients: shows a trend for reducing inflammation and normalizing the microbiota
To, N. et al. (2016) [110]	Systematic review with metanalysis of the effects of smoking on disease course in CD	Search of MEDLINE, EMBASE and EMBASE classic carried out up to July 2015 (with the resulting 33 eligible studies)	Smokers, compared with non-smokers, have 55–85% higher rates of flares of disease activity, clinical recurrence rates after surgery that are two-fold higher, between 54% and 68% higher rates of need for first surgery, and are twice as likely to need a second operation. Quitting smoking appears to have a beneficial effect on CD course, especially for flare of disease activity or need for a second operation
Benjamin, J.L. et al. (2012) [111]	Clinical research	Fecal samples from patients with active CD (N = 101; 29 of whom current smokers) and healthy controls (N = 66; 8 of whom current smokers) were analyzed by fluorescent in situ hybridization (using probes targeting 16S rRNA of bacteria previously shown to be altered in active CD)	Smokers with active CD have a clinically relevant dysbiosis of the gastrointestinal microbiota; with strong and significant associations between smoking and higher bacteroides (this novel finding is also present in healthy controls)
Bergeron, V. et al. (2012) [112]	Clinical research—In vitro study	Study of mononuclear cells extracted from blood samples of CD patients (smokers N = 19, and non-smokers N = 26), UC patients (smokers N = 7, and non-smokers N = 18), and healthy controls (smokers N = 13, and non-smokers N = 18); following either in vivo or in vitro exposure to cigarette smoke	Mononuclear cells from CD patients who smoke are functionally impaired, present a defective sensitivity to anti-inflammatory or antioxidant protection, and particularly synthesize lower levels of cytoprotective Hsp70. Findings suggest that the effects of cigarette smoke are largely dependent on the oxidative stress generated rather than on the nicotine component

[1] CEABAC10 mice serve as a model of host susceptibility to adherent-invasive *Escherichia coli* (AIEC) colonization (since they express CEACAM6, which is abnormally expressed in CD patients and predisposes to AIEC colonization). [2] In these studies, the N number of mice groups varies among the conducted experiments.

We extensively discussed the interconnection of the most relevant environmental factors, from the microbiota "in-vironment" to cigarette smoking, with the impaired immunological response, genetic susceptibility, or oxidative stress in CD. However, the limitations that arise from combining such broad topics should be kept in mind, e.g., conflicting results and multiple confounding factors that prevent us from establishing robust conclusions or causative relationships. Moreover, these confounding factors also hinder the design of clinical studies. For example, cigarette smoking or the Western diet might be associated with unhealthier lifestyles; and therefore, the findings obtained when studying their influence on CD could be partly due to other factors (e.g., the amount of exercise practiced or the exposition to sunlight) that can be difficult to take entirely into account in clinical studies. We believe that ongoing research assessing these multivariable factors on CD should try to consider as many confounding aspects as possible, while remaining cautious when establishing conclusions.

3.3. Epigenetics as a Transducer of Environmental Factors in Crohn's Disease

Epigenetics is an emerging field in biomedicine and refers to the heritable alterations in gene expression that are independent of the DNA sequence. The major epigenetic mechanisms that control gene expression are DNA methylation, histone modifications (such as acetylation and methylation), and small, non-coding RNAs. Epigenetic mechanisms are dynamic, reversible, and influenced by exposure to environmental factors [115]. Given that these mechanisms are involved in proper cell development, differentiation, function, and homeostasis, their dysregulation is proposed to play a key role in the onset and development of several diseases, especially cancer [116]. In terms of IBD, epigenetic mechanisms are shown to play a potentially primary role in its pathogenesis [117,118]. Epigenetics can provide a link between genetics and the environment, including the "in-vironment" (microbiota), acting as a transducer of environmental risk factors or even extending the inflammation and oxidative stress that characterize CD.

Epigenetic mechanisms, especially DNA methylation and microRNA expression, are identified as dysregulated in CD and are proposed as candidate biomarkers of the disease [115]. DNA methylation is the most studied epigenetic modification and consists of the covalent addition of a methyl group to the 5′ carbon of the cytosine ring, in the context of CpG dinucleotides. DNA methylation regulates gene transcription in such a way that the methylation restrains gene expression [119]. Our group recently identified an epigenetic methylation signature that allows for the characterization of patients with CD and supports the involvement of the environment and immune system in the pathogenesis of CD [120]. A previous study defined a global methylation profile characteristic of ileal CD, in which the targets of epigenetic modification appeared to be involved in immunity-related pathways [121]. Blood-derived DNA methylation signatures of CD were described that correlate with the severity of the intestinal inflammation [122]. In the latter study, the DNA methylation signatures were a result of the inflammatory features of the disease (given that, with treatment, the DNA methylation patterns resembled the patterns observed in patients without intestinal inflammation). Moreover, micro-RNAs (miRNAs) are proposed to have a more active role in the pathogenesis of CD. miRNAs are short strands of noncoding RNA that post-transcriptionally regulate gene expression [123]. In the intestinal tract, miRNAs are involved in tissue homeostasis, intestinal cell differentiation, and maintenance of the intestinal barrier function, and they were proposed to be both possible biomarkers and therapeutic targets in IBD [124]. The innate immune response to bacterial infection is regulated by an intricate network of miRNA circuits that fine-tune the inflammatory response. Moreover, miRNAs appear to be involved in the dysregulation of autophagy and Th17 signaling in CD [125].

Given that environmental factors are known to influence epigenetic regulation, certain environmental risk factors for CD could mediate their negative action, at least to a certain extent, through epigenetics (Figure 2), with an imbalanced diet being one of those risk factors. To cite the most direct example, one-carbon metabolism is dependent on dietary food components (e.g., methionine, betaine, and folate), which participate in DNA methylation pathways and the supply of methyl groups [58]. Considering that the Western diet is often deficient in micronutrients, such as folate, it could provoke a dysregulation of DNA methylation and, consequently, an altered gene transcription profile. Moreover, the low intake of dietary fiber, which can lead to insufficient amounts of microbial-derived butyrate, can also provoke epigenetic dysregulation, which is due to the fact that butyrate is a natural histone deacetylase inhibitor and therefore has the potential to initiate and prolong gene activation. Consequently, butyrate insufficiency could be responsible, to a certain extent, for the excessive condensation of the chromatin structure and gene expression mediated by histone deacetylases [126]. In addition, the putative aforementioned role of butyrate upregulating *MUC2* expression is thought to be partly epigenetically mediated, via the acetylation/methylation of histones at the *MUC2* promoter [88]. In this case, a butyrate insufficiency caused by a poor diet could therefore affect the "normal"/ideal expression of *MUC2* in the intestinal mucosa via epigenetic dysregulation. In vivo studies with rodents showed that the microbiota regulates global histone acetylation and methylation in numerous host tissues in a diet-dependent manner. The consumption of a Western diet prevents many of the microbiota-dependent chromatin changes that occur in a polysaccharide-rich diet [127]. Another well-known environmental risk factor for CD that affects epigenetics is cigarette smoke. Active smoking is an established critical factor for epigenetic modification; alterations in DNA methylation were suggested as a possible mechanism for mediating cigarette smoke-induced diseases [124,125].

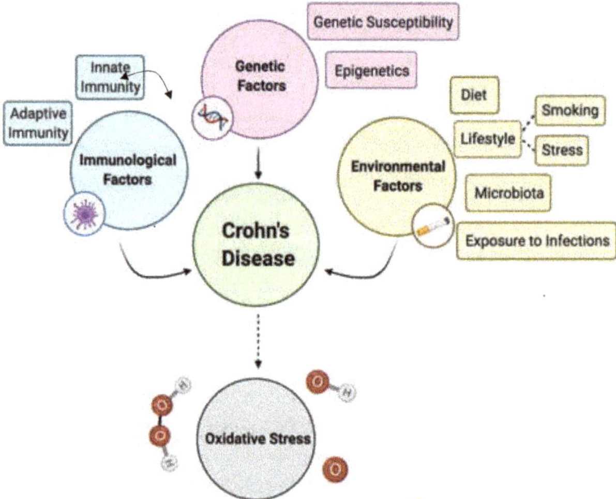

Figure 2. Factors linked to the etiopathogenesis of Crohn's disease and oxidative stress as a key effector mechanism underlying the pathogenesis. Immunological factors associated with CD comprise both impaired innate response (infiltration of activated neutrophils and macrophages into the affected gut mucosa) and adaptive response (accumulation of $CD4^+$ T cells in the lamina propria with a Th1/Th17 immune response). Environmental factors, especially diet, tobacco smoking, and the microbiota (the "in-vironment"), have an outstanding influence on the course and development of CD that appears to outweigh the influence of genetic factors (i.e., genetic susceptibility to the disease). Epigenetics provides a link between genetics and environmental factors and might constitute, at least to some extent, the mechanism through which some environmental factors mediate their impact on CD. Oxidative stress plays a central role in CD pathogenesis and was associated with the aforementioned factors.

Environmental factors and oxidative stress can have an impact on disease through epigenetics. Increasing evidence suggests that oxidative stress globally influences the chromatin structure, enzymatic, and nonenzymatic post-translational modifications of histones and the DNA-binding proteins. These chromatin alterations can therefore modulate gene expression, cell death, cell survival, and mutagenesis. Histones are extensively modified in an ROS-dependent and RNS-dependent manner and are glutathionylated in a redox-sensitive manner, which affects their ability to be post-translationally modified [128]. Oxidative stress not only alters global histone modification but also DNA methylation and can therefore have a modulating role in gene expression [129]. Nevertheless, there is a lack of studies addressing whether oxidative stress-induced epigenetic changes can have a further role in the pathogenesis of the diseases characterized by oxidative stress, such as CD, or just constitute collateral changes.

Epigenetic imprinting (not to be confused with genomic imprinting) could be defined as the mechanism through which environmental, external, and "in-vironmental" factors influence epigenetic changes, with potential consequences for health and disease [78]. A great example would be the recently recognized microbiota-sensitive epigenetic signature that predicts inflammation in CD [130]. As the authors indicate, their study defines the manner in which microbiota-derived signals can be integrated by the host via epigenetics, in priming the epithelium for overt clinical disease and with the subsequent disease-associated environmental triggers. Epigenetic imprinting can also act as a "disease memory" and can help explain the relapses, after resections in patients with CD, as well as explain why certain environmental factors appear to influence the intestinal mucosa and disease onset even when the environmental factor is long gone [131]. All in all, it is plausible to believe that the connection between external factors and the host DNA, mediated by epigenetic changes, has a key influence on the phenotypical expression of complex and multifactorial diseases such as CD [58]. However, we need to consider the difficulty in establishing causative relationships and distinguish between epigenetic changes with a possible role in the pathogenesis and those that are merely a consequence of the disease. Future studies should focus on establishing the epigenetic changes that can be derived from environmental risk factors, the microbiota, and even oxidative stress, and those that can contribute to the onset, progression, or relapse of CD.

4. Conclusions

Although the exact etiopathogenesis of CD remains unknown, the role of oxidative stress in its pathogenesis is widely recognized. We discussed how oxidative stress is present in CD, not only locally in the most-affected tissues but also at a systemic level. Oxidative stress is interconnected and feeds back into the impaired immune response and microbiota imbalance in CD.

External and environmental factors are known to have a large influence on the development and course of CD. The aforementioned primary risk factors are also related to oxidative stress, at least to a certain extent. Epigenetics provides a link between genetic and external factors and can provide greater insight into the pathogenesis of the disease. Further studies should seek to determine the environmental and oxidative stress-induced epigenetic changes that could have a role in the onset and development of CD, an area that has much to be explored.

Author Contributions: Writing—original draft preparation, E.A.-C. and E.S.-G.; writing—review and editing, E.A.-C. and E.S.-G.; visualization and review, I.M., B.M., and M.I.; supervision, J.S., B.B., and P.N. All authors have read and agreed to the published version of the manuscript.

Funding: This research received no external funding.

Conflicts of Interest: The authors declare no conflict of interest.

References

1. Bernstein, C.N. Review article: Changes in the epidemiology of inflammatory bowel disease-clues for aetiology. *Aliment. Pharmacol. Ther.* **2017**, *46*, 911–919. [CrossRef] [PubMed]
2. Patel, K.K.; Stappenbeck, T.S. Autophagy and intestinal homeostasis. *Annu. Rev. Physiol.* **2013**, *75*, 241–262. [CrossRef] [PubMed]

3. Moret, I.; Cerrillo, E.; Navarro-Puche, A.; Iborra, M.; Rausell, F.; Tortosa, L.; Beltrán, B. Oxidative stress in Crohn's disease. *Gastroenterol. Hepatol.* **2014**, *37*, 28–34. [CrossRef] [PubMed]
4. Iborra, M.; Moret, I.; Rausell, F.; Bastida, G.; Aguas, M.; Cerrillo, E.; Nos, P.; Beltrán, B. Role of oxidative stress and antioxidant enzymes in Crohn's disease. *Biochem. Soc. Trans.* **2011**, *39*, 1102–1106. [CrossRef] [PubMed]
5. Moret-Tatay, I.; Iborra, M.; Cerrillo, E.; Tortosa, L.; Nos, P.; Beltrán, B. Possible Biomarkers in Blood for Crohn's Disease: Oxidative Stress and MicroRNAs-Current Evidences and Further Aspects to Unravel. *Oxid. Med. Cell. Longev.* **2016**, *2016*, 2325162. [CrossRef]
6. Beltrán, B.; Nos, P.; Dasí, F.; Iborra, M.; Bastida, G.; Martínez, M.; O'Connor, J.-E.; Sáez, G.; Moret, I.; Ponce, J. Mitochondrial dysfunction, persistent oxidative damage, and catalase inhibition in immune cells of naïve and treated Crohn's disease. *Inflamm. Bowel Dis.* **2010**, *16*, 76–86. [CrossRef]
7. Bourgonje, A.R.; Feelisch, M.; Faber, K.N.; Pasch, A.; Dijkstra, G.; van Goor, H. Oxidative Stress and Redox-Modulating Therapeutics in Inflammatory Bowel Disease. *Trends Mol. Med.* **2020**, *26*, 1034–1046. [CrossRef]
8. Pizzino, G.; Irrera, N.; Cucinotta, M.; Pallio, G.; Mannino, F.; Arcoraci, V.; Squadrito, F.; Altavilla, D.; Bitto, A. Oxidative Stress: Harms and Benefits for Human Health. *Oxid. Med. Cell. Longev.* **2017**, *2017*, 1–13. [CrossRef]
9. Pereira, C.; Grácio, D.; Teixeira, J.P.; Magro, F. Oxidative Stress and DNA Damage: Implications in Inflammatory Bowel Disease. *Inflamm. Bowel Dis.* **2015**, *21*, 2403–2417. [CrossRef]
10. Zhu, H.; Li, Y.R. Oxidative stress and redox signaling mechanisms of inflammatory bowel disease: Updated experimental and clinical evidence. *Exp. Biol. Med.* **2012**, *237*, 474–480. [CrossRef]
11. Alzoghaibi, M.A. Concepts of oxidative stress and antioxidant defense in Crohn's disease. *World J. Gastroenterol.* **2013**, *19*, 6540. [CrossRef] [PubMed]
12. Circu, M.L.; Aw, T.Y. Redox biology of the intestine. *Free Radic. Res.* **2011**, *45*, 1245–1266. [CrossRef] [PubMed]
13. Kitahora, T.; Suzuki, K.; Asakura, H.; Yoshida, T.; Suematsu, M.; Watanabe, M.; Aiso, S.; Tsuchiya, M. Active oxygen species generated by monocytes and polymorphonuclear cells in Crohn's disease. *Dig. Dis. Sci.* **1988**, *33*, 951–955. [CrossRef] [PubMed]
14. Raddatz, D.; Bockemühl, M.; Ramadori, G. Quantitative measurement of cytokine mRNA in inflammatory bowel disease: Relation to clinical and endoscopic activity and outcome. *Eur. J. Gastroenterol. Hepatol.* **2005**, *17*, 547–557. [CrossRef]
15. Balmus, I.; Ciobica, A.; Trifan, A.; Stanciu, C. The implications of oxidative stress and antioxidant therapies in Inflammatory Bowel Disease: Clinical aspects and animal models. *Saudi J. Gastroenterol.* **2016**, *22*, 3. [CrossRef]
16. Kruidenier, L.; Kuiper, I.; van Duijn, W.; Mieremet-Ooms, M.A.; van Hogezand, R.A.; Lamers, C.B.; Verspaget, H.W. Imbalanced secondary mucosal antioxidant response in inflammatory bowel disease. *J. Pathol.* **2003**, *201*, 17–27. [CrossRef]
17. D'Odorico, A.; Bortolan, S.; Cardin, R.; D'Inca', R.; Martines, D.; Ferronato, A.; Sturniolo, G.C. Reduced plasma antioxidant concentrations and increased oxidative DNA damage in inflammatory bowel disease. *Scand. J. Gastroenterol.* **2001**, *36*, 1289–1294. [CrossRef]
18. Piechota-Polanczyk, A.; Fichna, J. Review article: The role of oxidative stress in pathogenesis and treatment of inflammatory bowel diseases. *Naunyn Schmiedebergs Arch. Pharmacol.* **2014**, *387*, 605–620. [CrossRef]
19. Rezaie, A.; Parker, R.D.; Abdollahi, M. Oxidative stress and pathogenesis of inflammatory bowel disease: An epiphenomenon or the cause? *Dig. Dis. Sci.* **2007**, *52*, 2015–2021. [CrossRef]
20. Yuksel, M.; Ates, I.; Kaplan, M.; Arikan, M.F.; Ozin, Y.O.; Kilic, Z.M.Y.; Topcuoglu, C.; Kayacetin, E. Is Oxidative Stress Associated with Activation and Pathogenesis of Inflammatory Bowel Disease? *J. Med. Biochem.* **2017**, *36*, 341–348. [CrossRef]
21. Sido, B.; Hack, V.; Hochlehnert, A.; Lipps, H.; Herfarth, C.; Dröge, W. Impairment of intestinal glutathione synthesis in patients with inflammatory bowel disease. *Gut* **1998**, *42*, 485–492. [CrossRef] [PubMed]
22. Iantomasi, T.; Marraccini, P.; Favilli, F.; Vincenzini, M.T.; Ferretti, P.; Tonelli, F. Glutathione Metabolism in Crohn's Disease. *Biochem. Med. Metab. Biol.* **1994**, *53*, 87–91. [CrossRef] [PubMed]
23. Kruidenier, L.; Kuiper, I.; Lamers, C.B.H.W.; Verspaget, H.W. Intestinal oxidative damage in inflammatory bowel disease: Semi-quantification, localization, and association with mucosal antioxidants. *J. Pathol.* **2003**, *201*, 28–36. [CrossRef] [PubMed]
24. Pinto, M.A.S.; Lopes, M.S.-M.S.; Bastos, S.T.; Reigada, C.L.; Dantas, R.F.; Neto, J.C.; Luna, A.S.; Madi, K.; Nunes, T.; Zaltman, C. Does active Crohn's disease have decreased intestinal antioxidant capacity? *J. Crohn's Colitis* **2013**, *7*, e358–e366. [CrossRef]
25. Bourgonje, A.R.; von Martels, J.Z.H.; Bulthuis, M.L.C.; van Londen, M.; Faber, K.N.; Dijkstra, G.; van Goor, H. Crohn's Disease in Clinical Remission Is Marked by Systemic Oxidative Stress. *Front. Physiol.* **2019**, *10*, 499. [CrossRef]
26. Cosnes, J.; Gower-Rousseau, C.; Seksik, P.; Cortot, A. Epidemiology and Natural History of Inflammatory Bowel Diseases. *Gastroenterology* **2011**, *140*, 1785–1794.e4. [CrossRef]
27. Chen, J.J.; Bertrand, H.; Yu, B.P. Inhibition of adenine nucleotide translocator by lipid peroxidation products. *Free Radic. Biol. Med.* **1995**, *19*, 583–590. [CrossRef]
28. Catalá, A. Lipid peroxidation of membrane phospholipids generates hydroxy-alkenals and oxidized phospholipids active in physiological and/or pathological conditions. *Chem. Phys. Lipids* **2009**, *157*, 1–11. [CrossRef]
29. Pelli, M.A.; Trovarelli, G.; Capodicasa, E.; De Medio, G.E.; Bassotti, G. Breath alkanes determination in ulcerative colitis and Crohn's disease. *Dis. Colon Rectum* **1999**, *42*, 71–76. [CrossRef]
30. Sampietro, G.M.; Cristaldi, M.; Cervato, G.; Maconi, G.; Danelli, P.; Cervellione, R.; Rovati, M.; Bianchi Porro, G.; Cestaro, B.; Taschieri, A.M. Oxidative stress, vitamin A and vitamin E behaviour in patients submitted to conservative surgery for complicated Crohn's disease. *Dig. Liver Dis.* **2002**, *34*, 696–701. [CrossRef]

31. Wang, Y.; Wang, W.; Yang, H.; Shao, D.; Zhao, X.; Zhang, G. Intraperitoneal injection of 4-hydroxynonenal (4-HNE), a lipid peroxidation product, exacerbates colonic inflammation through activation of Toll-like receptor 4 signaling. *Free Radic. Biol. Med.* **2019**, *131*, 237–242. [CrossRef]
32. Szczeklik, K.; Krzyściak, W.; Cibor, D.; Domagała-Rodacka, R.; Pytko-Polończyk, J.; Mach, T.; Owczarek, D. Markers of lipid peroxidation and antioxidant status in the serum and saliva of patients with active Crohn disease. *Pol. Arch. Intern. Med.* **2018**, *128*, 362–370. [CrossRef] [PubMed]
33. Sedelnikova, O.A.; Redon, C.E.; Dickey, J.S.; Nakamura, A.J.; Georgakilas, A.G.; Bonner, W.M. Role of oxidatively induced DNA lesions in human pathogenesis. *Mutat. Res.* **2010**, *704*, 152–159. [CrossRef] [PubMed]
34. Wiseman, H.; Halliwell, B. Damage to DNA by reactive oxygen and nitrogen species: Role in inflammatory disease and progression to cancer. *Biochem. J.* **1996**, *313 Pt 1*, 17–29. [CrossRef]
35. Dean, R.T.; Fu, S.; Stocker, R.; Davies, M.J. Biochemistry and pathology of radical-mediated protein oxidation. *Biochem. J.* **1997**, *324 Pt 1*, 1–18. [CrossRef]
36. Krzystek-Korpacka, M.; Neubauer, K.; Berdowska, I.; Boehm, D.; Zielinski, B.; Petryszyn, P.; Terlecki, G.; Paradowski, L.; Gamian, A. Enhanced formation of advanced oxidation protein products in IBD. *Inflamm. Bowel Dis.* **2008**, *14*, 794–802. [CrossRef] [PubMed]
37. Krzystek-Korpacka, M.; Kempiński, R.; Bromke, M.A.; Neubauer, K. Oxidative Stress Markers in Inflammatory Bowel Diseases: Systematic Review. *Diagnostics* **2020**, *10*, 601. [CrossRef]
38. Biddlestone, J.; Bandarra, D.; Rocha, S. The role of hypoxia in inflammatory disease (review). *Int. J. Mol. Med.* **2015**, *35*, 859–869. [CrossRef]
39. Cummins, E.P.; Crean, D. Hypoxia and inflammatory bowel disease. *Microbes Infect.* **2017**, *19*, 210–221. [CrossRef]
40. Tian, T.; Wang, Z.; Zhang, J. Pathomechanisms of Oxidative Stress in Inflammatory Bowel Disease and Potential Antioxidant Therapies. *Oxid. Med. Cell. Longev.* **2017**, *2017*, 1–18. [CrossRef]
41. Dryden, G.W.; Deaciuc, I.; Arteel, G.; McClain, C.J. Clinical implications of oxidative stress and antioxidant therapy. *Curr. Gastroenterol. Rep.* **2005**, *7*, 308–316. [CrossRef] [PubMed]
42. Luceri, C.; Bigagli, E.; Agostiniani, S.; Giudici, F.; Zambonin, D.; Scaringi, S.; Ficari, F.; Lodovici, M.; Malentacchi, C. Analysis of Oxidative Stress-Related Markers in Crohn's Disease Patients at Surgery and Correlations with Clinical Findings. *Antioxidants* **2019**, *8*, 378. [CrossRef] [PubMed]
43. Herszenyi, L.; Miheller, P.; Tulassay, Z. Carcinogenesis in inflammatory bowel disease. *Dig. Dis.* **2007**, *25*, 267–269. [CrossRef] [PubMed]
44. Nishida, A.; Inoue, R.; Inatomi, O.; Bamba, S.; Naito, Y.; Andoh, A. Gut microbiota in the pathogenesis of inflammatory bowel disease. *Clin. J. Gastroenterol.* **2018**, *11*, 1–10. [CrossRef]
45. O'Hara, A.M.; Shanahan, F. The gut flora as a forgotten organ. *EMBO Rep.* **2006**, *7*, 688–693. [CrossRef]
46. Round, J.L.; Mazmanian, S.K. The gut microbiota shapes intestinal immune responses during health and disease. *Nat. Rev. Immunol.* **2009**, *9*, 313–323. [CrossRef]
47. Clemente, J.C.; Ursell, L.K.; Parfrey, L.W.; Knight, R. The impact of the gut microbiota on human health: An integrative view. *Cell* **2012**, *148*, 1258–1270. [CrossRef]
48. Fiocchi, C. Genes and "in-vironment": How will our concepts on the pathophysiology of inflammatory bowel disease develop in the future? *Dig. Dis.* **2012**, *30* (Suppl. 3), 2–11. [CrossRef]
49. Biasi, F.; Leonarduzzi, G.; Oteiza, P.I.; Poli, G. Inflammatory Bowel Disease: Mechanisms, Redox Considerations, and Therapeutic Targets. *Antioxid. Redox Signal.* **2013**, *19*, 1711–1747. [CrossRef]
50. Xiong, Y.; Wang, G.-Z.; Zhou, J.-Q.; Xia, B.-Q.; Wang, X.-Y.; Jiang, B. Serum antibodies to microbial antigens for Crohn's disease progression: A meta-analysis. *Eur. J. Gastroenterol. Hepatol.* **2014**, *26*, 733–742. [CrossRef]
51. Sheehan, D.; Moran, C.; Shanahan, F. The microbiota in inflammatory bowel disease. *J. Gastroenterol.* **2015**, *50*, 495–507. [CrossRef] [PubMed]
52. Kostic, A.D.; Xavier, R.J.; Gevers, D. The microbiome in inflammatory bowel disease: Current status and the future ahead. *Gastroenterology* **2014**, *146*, 1489–1499. [CrossRef] [PubMed]
53. Ahmed, I.; Roy, B.C.; Khan, S.A.; Septer, S.; Umar, S. Microbiome, Metabolome and Inflammatory Bowel Disease. *Microorganisms* **2016**, *4*, 20. [CrossRef] [PubMed]
54. Tomasello, G.; Mazzola, M.; Leone, A.; Sinagra, E.; Zummo, G.; Farina, F.; Damiani, P.; Cappello, F.; Gerges Geagea, A.; Jurjus, A.; et al. Nutrition, oxidative stress and intestinal dysbiosis: Influence of diet on gut microbiota in inflammatory bowel diseases. *Biomed. Pap.* **2016**, *160*, 461–466. [CrossRef] [PubMed]
55. Hu, Y.; Chen, D.; Zheng, P.; Yu, J.; He, J.; Mao, X.; Yu, B. The Bidirectional Interactions between Resveratrol and Gut Microbiota: An Insight into Oxidative Stress and Inflammatory Bowel Disease Therapy. *BioMed Res. Int.* **2019**, *2019*, 5403761. [CrossRef]
56. Knights, D.; Lassen, K.G.; Xavier, R.J. Advances in inflammatory bowel disease pathogenesis: Linking host genetics and the microbiome. *Gut* **2013**, *62*, 1505–1510. [CrossRef]
57. Morgan, X.C.; Tickle, T.L.; Sokol, H.; Gevers, D.; Devaney, K.L.; Ward, D.V.; Reyes, J.A.; Shah, S.A.; LeLeiko, N.; Snapper, S.B.; et al. Dysfunction of the intestinal microbiome in inflammatory bowel disease and treatment. *Genome Biol.* **2012**, *13*, R79. [CrossRef]
58. Rapozo, D.C.M.; Bernardazzi, C.; de Souza, H.S.P. Diet and microbiota in inflammatory bowel disease: The gut in disharmony. *World J. Gastroenterol.* **2017**, *23*, 2124. [CrossRef]

59. Khanna, S.; Raffals, L.E. The Microbiome in Crohn's Disease. *Gastroenterol. Clin. N. Am.* **2017**, *46*, 481–492. [CrossRef]
60. The International IBD Genetics Consortium (IIBDGC); Jostins, L.; Ripke, S.; Weersma, R.K.; Duerr, R.H.; McGovern, D.P.; Hui, K.Y.; Lee, J.C.; Philip Schumm, L.; Sharma, Y.; et al. Host–microbe interactions have shaped the genetic architecture of inflammatory bowel disease. *Nature* **2012**, *491*, 119–124. [CrossRef]
61. Becker, C.; Neurath, M.F.; Wirtz, S. The Intestinal Microbiota in Inflammatory Bowel Disease. *ILAR J.* **2015**, *56*, 192–204. [CrossRef] [PubMed]
62. Hampe, J.; Franke, A.; Rosenstiel, P.; Till, A.; Teuber, M.; Huse, K.; Albrecht, M.; Mayr, G.; De La Vega, F.M.; Briggs, J.; et al. A genome-wide association scan of nonsynonymous SNPs identifies a susceptibility variant for Crohn disease in ATG16L1. *Nat. Genet.* **2007**, *39*, 207–211. [CrossRef] [PubMed]
63. Parkes, M.; Barrett, J.C.; Prescott, N.J.; Tremelling, M.; Anderson, C.A.; Fisher, S.A.; Roberts, R.G.; Nimmo, E.R.; Cummings, F.R.; Soars, D.; et al. Sequence variants in the autophagy gene IRGM and multiple other replicating loci contribute to Crohn's disease susceptibility. *Nat. Genet.* **2007**, *39*, 830–832. [CrossRef] [PubMed]
64. Ogura, Y.; Bonen, D.K.; Inohara, N.; Nicolae, D.L.; Chen, F.F.; Ramos, R.; Britton, H.; Moran, T.; Karaliuskas, R.; Duerr, R.H.; et al. A frameshift mutation in NOD2 associated with susceptibility to Crohn's disease. *Nature* **2001**, *411*, 603–606. [CrossRef] [PubMed]
65. Hugot, J.P.; Chamaillard, M.; Zouali, H.; Lesage, S.; Cézard, J.P.; Belaiche, J.; Almer, S.; Tysk, C.; O'Morain, C.A.; Gassull, M.; et al. Association of NOD2 leucine-rich repeat variants with susceptibility to Crohn's disease. *Nature* **2001**, *411*, 599–603. [CrossRef] [PubMed]
66. Wehkamp, J.; Harder, J.; Weichenthal, M.; Schwab, M.; Schäffeler, E.; Schlee, M.; Herrlinger, K.R.; Stallmach, A.; Noack, F.; Fritz, P.; et al. NOD2 (CARD15) mutations in Crohn's disease are associated with diminished mucosal alpha-defensin expression. *Gut* **2004**, *53*, 1658–1664. [CrossRef] [PubMed]
67. Frank, D.N.; Robertson, C.E.; Hamm, C.M.; Kpadeh, Z.; Zhang, T.; Chen, H.; Zhu, W.; Sartor, R.B.; Boedeker, E.C.; Harpaz, N.; et al. Disease phenotype and genotype are associated with shifts in intestinal-associated microbiota in inflammatory bowel diseases. *Inflamm. Bowel Dis.* **2011**, *17*, 179–184. [CrossRef]
68. Li, E.; Hamm, C.M.; Gulati, A.S.; Sartor, R.B.; Chen, H.; Wu, X.; Zhang, T.; Rohlf, F.J.; Zhu, W.; Gu, C.; et al. Inflammatory bowel diseases phenotype, C. difficile and NOD2 genotype are associated with shifts in human ileum associated microbial composition. *PLoS ONE* **2012**, *7*, e26284. [CrossRef]
69. Cooney, R.; Baker, J.; Brain, O.; Danis, B.; Pichulik, T.; Allan, P.; Ferguson, D.J.P.; Campbell, B.J.; Jewell, D.; Simmons, A. NOD2 stimulation induces autophagy in dendritic cells influencing bacterial handling and antigen presentation. *Nat. Med.* **2010**, *16*, 90–97. [CrossRef]
70. Shaw, M.H.; Kamada, N.; Warner, N.; Kim, Y.-G.; Nuñez, G. The ever-expanding function of NOD2: Autophagy, viral recognition, and T cell activation. *Trends Immunol.* **2011**, *32*, 73–79. [CrossRef]
71. Chu, H.; Khosravi, A.; Kusumawardhani, I.P.; Kwon, A.H.K.; Vasconcelos, A.C.; Cunha, L.D.; Mayer, A.E.; Shen, Y.; Wu, W.-L.; Kambal, A.; et al. Gene-microbiota interactions contribute to the pathogenesis of inflammatory bowel disease. *Science* **2016**, *352*, 1116–1120. [CrossRef] [PubMed]
72. Hedin, C.R.; McCarthy, N.E.; Louis, P.; Farquharson, F.M.; McCartney, S.; Taylor, K.; Prescott, N.J.; Murrells, T.; Stagg, A.J.; Whelan, K.; et al. Altered intestinal microbiota and blood T cell phenotype are shared by patients with Crohn's disease and their unaffected siblings. *Gut* **2014**, *63*, 1578–1586. [CrossRef] [PubMed]
73. Willing, B.P.; Dicksved, J.; Halfvarson, J.; Andersson, A.F.; Lucio, M.; Zheng, Z.; Järnerot, G.; Tysk, C.; Jansson, J.K.; Engstrand, L. A Pyrosequencing Study in Twins Shows That Gastrointestinal Microbial Profiles Vary with Inflammatory Bowel Disease Phenotypes. *Gastroenterology* **2010**, *139*, 1844–1854.e1. [CrossRef] [PubMed]
74. Zou, Y.; Wu, L.; Xu, W.; Zhou, X.; Ye, K.; Xiong, H.; Song, C.; Xie, Y. Correlation between antibiotic use in childhood and subsequent inflammatory bowel disease: A systematic review and meta-analysis. *Scand. J. Gastroenterol.* **2020**, *55*, 301–311. [CrossRef]
75. Dutta, A.K. Influence of environmental factors on the onset and course of inflammatory bowel disease. *World J. Gastroenterol.* **2016**, *22*, 1088. [CrossRef]
76. Kaplan, G.G.; Ng, S.C. Understanding and Preventing the Global Increase of Inflammatory Bowel Disease. *Gastroenterology* **2017**, *152*, 313–321.e2. [CrossRef]
77. Chen, Y.; Wang, Y.; Shen, J. Role of environmental factors in the pathogenesis of Crohn's disease: A critical review. *Int. J. Colorectal Dis.* **2019**, *34*, 2023–2034. [CrossRef]
78. Rogler, G.; Vavricka, S. Exposome in IBD: Recent Insights in Environmental Factors that Influence the Onset and Course of IBD. *Inflamm. Bowel Dis.* **2015**, *21*, 400–408. [CrossRef]
79. Hou, J.K.; Abraham, B.; El-Serag, H. Dietary Intake and Risk of Developing Inflammatory Bowel Disease: A Systematic Review of the Literature. *Am. J. Gastroenterol.* **2011**, *106*, 563–573. [CrossRef]
80. Ruemmele, F.M. Role of Diet in Inflammatory Bowel Disease. *Ann. Nutr. Metab.* **2016**, *68*, 32–41. [CrossRef]
81. Agus, A.; Denizot, J.; Thévenot, J.; Martinez-Medina, M.; Massier, S.; Sauvanet, P.; Bernalier-Donadille, A.; Denis, S.; Hofman, P.; Bonnet, R.; et al. Western diet induces a shift in microbiota composition enhancing susceptibility to Adherent-Invasive *E. coli* infection and intestinal inflammation. *Sci. Rep.* **2016**, *6*, 19032. [CrossRef] [PubMed]

82. Martinez-Medina, M.; Denizot, J.; Dreux, N.; Robin, F.; Billard, E.; Bonnet, R.; Darfeuille-Michaud, A.; Barnich, N. Western diet induces dysbiosis with increased *E. coli* in CEABAC10 mice, alters host barrier function favouring AIEC colonisation. *Gut* **2014**, *63*, 116–124. [CrossRef] [PubMed]
83. Cani, P.D.; Bibiloni, R.; Knauf, C.; Waget, A.; Neyrinck, A.M.; Delzenne, N.M.; Burcelin, R. Changes in Gut Microbiota Control Metabolic Endotoxemia-Induced Inflammation in High-Fat Diet-Induced Obesity and Diabetes in Mice. *Diabetes* **2008**, *57*, 1470–1481. [CrossRef]
84. Sáez-González, E.; Mateos, B.; López-Muñoz, P.; Iborra, M.; Moret, I.; Nos, P.; Beltrán, B. Bases for the Adequate Development of Nutritional Recommendations for Patients with Inflammatory Bowel Disease. *Nutrients* **2019**, *11*, 1062. [CrossRef] [PubMed]
85. Kelly, C.J.; Zheng, L.; Campbell, E.L.; Saeedi, B.; Scholz, C.C.; Bayless, A.J.; Wilson, K.E.; Glover, L.E.; Kominsky, D.J.; Magnuson, A.; et al. Crosstalk between Microbiota-Derived Short-Chain Fatty Acids and Intestinal Epithelial HIF Augments Tissue Barrier Function. *Cell Host Microbe* **2015**, *17*, 662–671. [CrossRef]
86. Saeedi, B.J.; Kao, D.J.; Kitzenberg, D.A.; Dobrinskikh, E.; Schwisow, K.D.; Masterson, J.C.; Kendrick, A.A.; Kelly, C.J.; Bayless, A.J.; Kominsky, D.J.; et al. HIF-dependent regulation of claudin-1 is central to intestinal epithelial tight junction integrity. *Mol. Biol. Cell* **2015**, *26*, 2252–2262. [CrossRef]
87. Peng, L.; Li, Z.-R.; Green, R.S.; Holzman, I.R.; Lin, J. Butyrate Enhances the Intestinal Barrier by Facilitating Tight Junction Assembly via Activation of AMP-Activated Protein Kinase in Caco-2 Cell Monolayers. *J. Nutr.* **2009**, *139*, 1619–1625. [CrossRef] [PubMed]
88. Burger-van Paassen, N.; Vincent, A.; Puiman, P.J.; van der Sluis, M.; Bouma, J.; Boehm, G.; van Goudoever, J.B.; van Seuningen, I.; Renes, I.B. The regulation of intestinal mucin MUC2 expression by short-chain fatty acids: Implications for epithelial protection. *Biochem. J.* **2009**, *420*, 211–219. [CrossRef]
89. Nielsen, D.S.G.; Jensen, B.B.; Theil, P.K.; Nielsen, T.S.; Knudsen, K.E.B.; Purup, S. Effect of butyrate and fermentation products on epithelial integrity in a mucus-secreting human colon cell line. *J. Funct. Foods* **2018**, *40*, 9–17. [CrossRef]
90. Bach Knudsen, K.E.; Lærke, H.N.; Hedemann, M.S.; Nielsen, T.S.; Ingerslev, A.K.; Gundelund Nielsen, D.S.; Theil, P.K.; Purup, S.; Hald, S.; Schioldan, A.G.; et al. Impact of Diet-Modulated Butyrate Production on Intestinal Barrier Function and Inflammation. *Nutrients* **2018**, *10*, 1499. [CrossRef]
91. Liu, H.; Wang, J.; He, T.; Becker, S.; Zhang, G.; Li, D.; Ma, X. Butyrate: A Double-Edged Sword for Health? *Adv. Nutr.* **2018**, *9*, 21–29. [CrossRef] [PubMed]
92. Furusawa, Y.; Obata, Y.; Fukuda, S.; Endo, T.A.; Nakato, G.; Takahashi, D.; Nakanishi, Y.; Uetake, C.; Kato, K.; Kato, T.; et al. Commensal microbe-derived butyrate induces the differentiation of colonic regulatory T cells. *Nature* **2013**, *504*, 446–450. [CrossRef] [PubMed]
93. Erickson, A.R.; Cantarel, B.L.; Lamendella, R.; Darzi, Y.; Mongodin, E.F.; Pan, C.; Shah, M.; Halfvarson, J.; Tysk, C.; Henrissat, B.; et al. Integrated Metagenomics/Metaproteomics Reveals Human Host-Microbiota Signatures of Crohn's Disease. *PLoS ONE* **2012**, *7*, e49138. [CrossRef] [PubMed]
94. Geirnaert, A.; Calatayud, M.; Grootaert, C.; Laukens, D.; Devriese, S.; Smagghe, G.; De Vos, M.; Boon, N.; Van de Wiele, T. Butyrate-producing bacteria supplemented in vitro to Crohn's disease patient microbiota increased butyrate production and enhanced intestinal epithelial barrier integrity. *Sci. Rep.* **2017**, *7*, 11450. [CrossRef]
95. Desai, M.S.; Seekatz, A.M.; Koropatkin, N.M.; Kamada, N.; Hickey, C.A.; Wolter, M.; Pudlo, N.A.; Kitamoto, S.; Terrapon, N.; Muller, A.; et al. A Dietary Fiber-Deprived Gut Microbiota Degrades the Colonic Mucus Barrier and Enhances Pathogen Susceptibility. *Cell* **2016**, *167*, 1339–1353.e21. [CrossRef]
96. Simopoulos, A.P. The Importance of the Omega-6/Omega-3 Fatty Acid Ratio in Cardiovascular Disease and Other Chronic Diseases. *Exp. Biol. Med.* **2008**, *233*, 674–688. [CrossRef]
97. Marion-Letellier, R.; Savoye, G.; Ghosh, S. Polyunsaturated fatty acids and inflammation: PUFA and Inflammation. *IUBMB Life* **2015**, *67*, 659–667. [CrossRef]
98. Shoda, R.; Matsueda, K.; Yamato, S.; Umeda, N. Epidemiologic analysis of Crohn disease in Japan: Increased dietary intake of n-6 polyunsaturated fatty acids and animal protein relates to the increased incidence of Crohn disease in Japan. *Am. J. Clin. Nutr.* **1996**, *63*, 741–745. [CrossRef]
99. Feagan, B.G.; Sandborn, W.J.; Mittmann, U.; Bar-Meir, S.; D'Haens, G.; Bradette, M.; Cohen, A.; Dallaire, C.; Ponich, T.P.; McDonald, J.W.D.; et al. Omega-3 Free Fatty Acids for the Maintenance of Remission in Crohn Disease: The EPIC Randomized Controlled Trials. *JAMA* **2008**, *299*, 1690. [CrossRef]
100. Lev-Tzion, R.; Griffiths, A.M.; Ledder, O.; Turner, D. Omega 3 fatty acids (fish oil) for maintenance of remission in Crohn's disease. *Cochrane Database Syst. Rev.* **2014**, CD006320. [CrossRef]
101. Chassaing, B.; Koren, O.; Goodrich, J.K.; Poole, A.C.; Srinivasan, S.; Ley, R.E.; Gewirtz, A.T. Dietary emulsifiers impact the mouse gut microbiota promoting colitis and metabolic syndrome. *Nature* **2015**, *519*, 92–96. [CrossRef] [PubMed]
102. Mu, W.; Wang, Y.; Huang, C.; Fu, Y.; Li, J.; Wang, H.; Jia, X.; Ba, Q. Effect of Long-Term Intake of Dietaryint Titanium Dioxide Nanoparticles on Intestine Inflammation in Mice. *J. Agric. Food Chem.* **2019**, *67*, 9382–9389. [CrossRef] [PubMed]
103. Reddavide, R.; Rotolo, O.; Caruso, M.G.; Stasi, E.; Notarnicola, M.; Miraglia, C.; Nouvenne, A.; Meschi, T.; De' Angelis, G.L.; Di Mario, F.; et al. The role of diet in the prevention and treatment of Inflammatory Bowel Diseases. *Acta Bio-Med. Atenei Parm.* **2018**, *89*, 60–75. [CrossRef]

104. D'Souza, S.; Levy, E.; Mack, D.; Israel, D.; Lambrette, P.; Ghadirian, P.; Deslandres, C.; Morgan, K.; Seidman, E.G.; Amre, D.K. Dietary patterns and risk for Crohn's disease in children. *Inflamm. Bowel Dis.* **2008**, *14*, 367–373. [CrossRef] [PubMed]
105. Marlow, G.; Ellett, S.; Ferguson, I.R.; Zhu, S.; Karunasinghe, N.; Jesuthasan, A.C.; Han, D.; Fraser, A.G.; Ferguson, L.R. Transcriptomics to study the effect of a Mediterranean-inspired diet on inflammation in Crohn's disease patients. *Hum. Genomics* **2013**, *7*, 24. [CrossRef]
106. Nagpal, R.; Shively, C.A.; Appt, S.A.; Register, T.C.; Michalson, K.T.; Vitolins, M.Z.; Yadav, H. Gut Microbiome Composition in Non-human Primates Consuming a Western or Mediterranean Diet. *Front. Nutr.* **2018**, *5*, 28. [CrossRef]
107. Oliveras-López, M.-J.; Berná, G.; Jurado-Ruiz, E.; López-García de la Serrana, H.; Martín, F. Consumption of extra-virgin olive oil rich in phenolic compounds has beneficial antioxidant effects in healthy human adults. *J. Funct. Foods* **2014**, *10*, 475–484. [CrossRef]
108. Slavin, J.L.; Lloyd, B. Health Benefits of Fruits and Vegetables. *Adv. Nutr.* **2012**, *3*, 506–516. [CrossRef]
109. Piovani, D.; Danese, S.; Peyrin-Biroulet, L.; Nikolopoulos, G.K.; Lytras, T.; Bonovas, S. Environmental Risk Factors for Inflammatory Bowel Diseases: An Umbrella Review of Meta-analyses. *Gastroenterology* **2019**, *157*, 647–659.e4. [CrossRef]
110. To, N.; Gracie, D.J.; Ford, A.C. Systematic review with meta-analysis: The adverse effects of tobacco smoking on the natural history of Crohn's disease. *Aliment. Pharmacol. Ther.* **2016**, *43*, 549–561. [CrossRef]
111. Benjamin, J.L.; Hedin, C.R.H.; Koutsoumpas, A.; Ng, S.C.; McCarthy, N.E.; Prescott, N.J.; Pessoa-Lopes, P.; Mathew, C.G.; Sanderson, J.; Hart, A.L.; et al. Smokers with active Crohn's disease have a clinically relevant dysbiosis of the gastrointestinal microbiota *. *Inflamm. Bowel Dis.* **2012**, *18*, 1092–1100. [CrossRef] [PubMed]
112. Bergeron, V.; Grondin, V.; Rajca, S.; Maubert, M.-A.; Pigneur, B.; Thomas, G.; Trugnan, G.; Beaugerie, L.; Cosnes, J.; Masliah, J.; et al. Current smoking differentially affects blood mononuclear cells from patients with crohn's disease and ulcerative colitis: Relevance to its adverse role in the disease. *Inflamm. Bowel Dis.* **2012**, *18*, 1101–1111. [CrossRef] [PubMed]
113. Berkowitz, L.; Schultz, B.M.; Salazar, G.A.; Pardo-Roa, C.; Sebastián, V.P.; Álvarez-Lobos, M.M.; Bueno, S.M. Impact of Cigarette Smoking on the Gastrointestinal Tract Inflammation: Opposing Effects in Crohn's Disease and Ulcerative Colitis. *Front. Immunol.* **2018**, *9*, 74. [CrossRef] [PubMed]
114. Yanbaeva, D.G.; Dentener, M.A.; Creutzberg, E.C.; Wesseling, G.; Wouters, E.F.M. Systemic Effects of Smoking. *Chest* **2007**, *131*, 1557–1566. [CrossRef]
115. Mateos, B.; Palanca-Ballester, C.; Saez-Gonzalez, E.; Moret, I.; Lopez, A.; Sandoval, J. Epigenetics of Inflammatory Bowel Disease: Unraveling Pathogenic Events. *Crohns Colitis 360* **2019**, *1*, otz017. [CrossRef]
116. Dawson, M.A.; Kouzarides, T. Cancer Epigenetics: From Mechanism to Therapy. *Cell* **2012**, *150*, 12–27. [CrossRef]
117. Scarpa, M.; Stylianou, E. Epigenetics: Concepts and relevance to IBD pathogenesis. *Inflamm. Bowel Dis.* **2012**, *18*, 1982–1996. [CrossRef]
118. Ventham, N.T.; Kennedy, N.A.; Nimmo, E.R.; Satsangi, J. Beyond Gene Discovery in Inflammatory Bowel Disease: The Emerging Role of Epigenetics. *Gastroenterology* **2013**, *145*, 293–308. [CrossRef]
119. Wawrzyniak, M.; Scharl, M. Genetics and epigenetics of inflammatory bowel disease. *Swiss Med. Wkly.* **2018**, *148*, w14671. [CrossRef]
120. Moret-Tatay, I.; Cerrillo, E.; Sáez-González, E.; Hervás, D.; Iborra, M.; Sandoval, J.; Busó, E.; Tortosa, L.; Nos, P.; Beltrán, B. Identification of Epigenetic Methylation Signatures With Clinical Value in Crohn's Disease. *Clin. Transl. Gastroenterol.* **2019**, *10*, e00083. [CrossRef]
121. Nimmo, E.R.; Prendergast, J.G.; Aldhous, M.C.; Kennedy, N.A.; Henderson, P.; Drummond, H.E.; Ramsahoye, B.H.; Wilson, D.C.; Semple, C.A.; Satsangi, J. Genome-wide methylation profiling in Crohn's disease identifies altered epigenetic regulation of key host defense mechanisms including the Th17 pathway. *Inflamm. Bowel Dis.* **2012**, *18*, 889–899. [CrossRef] [PubMed]
122. Somineni, H.K.; Venkateswaran, S.; Kilaru, V.; Marigorta, U.M.; Mo, A.; Okou, D.T.; Kellermayer, R.; Mondal, K.; Cobb, D.; Walters, T.D.; et al. Blood-Derived DNA Methylation Signatures of Crohn's Disease and Severity of Intestinal Inflammation. *Gastroenterology* **2019**, *156*, 2254–2265.e3. [CrossRef] [PubMed]
123. Filipowicz, W.; Bhattacharyya, S.N.; Sonenberg, N. Mechanisms of post-transcriptional regulation by microRNAs: Are the answers in sight? *Nat. Rev. Genet.* **2008**, *9*, 102–114. [CrossRef] [PubMed]
124. Loddo, I.; Romano, C. Inflammatory Bowel Disease: Genetics, Epigenetics, and Pathogenesis. *Front. Immunol.* **2015**, *6*, 551. [CrossRef] [PubMed]
125. Kalla, R.; Ventham, N.T.; Kennedy, N.A.; Quintana, J.F.; Nimmo, E.R.; Buck, A.H.; Satsangi, J. MicroRNAs: New players in IBD. *Gut* **2015**, *64*, 504–517. [CrossRef]
126. Silva, J.P.B.; Navegantes-Lima, K.C.; de Oliveira, A.L.B.; Rodrigues, D.V.S.; Gaspar, S.L.F.; Monteiro, V.V.S.; Moura, D.P.; Monteiro, M.C. Protective Mechanisms of Butyrate on Inflammatory Bowel Disease. *Curr. Pharm. Des.* **2019**, *24*, 4154–4166. [CrossRef]
127. Krautkramer, K.A.; Kreznar, J.H.; Romano, K.A.; Vivas, E.I.; Barrett-Wilt, G.A.; Rabaglia, M.E.; Keller, M.P.; Attie, A.D.; Rey, F.E.; Denu, J.M. Diet-Microbiota Interactions Mediate Global Epigenetic Programming in Multiple Host Tissues. *Mol. Cell* **2016**, *64*, 982–992. [CrossRef]
128. Kreuz, S.; Fischle, W. Oxidative stress signaling to chromatin in health and disease. *Epigenomics* **2016**, *8*, 843–862. [CrossRef]
129. Niu, Y.; DesMarais, T.L.; Tong, Z.; Yao, Y.; Costa, M. Oxidative stress alters global histone modification and DNA methylation. *Free Radic. Biol. Med.* **2015**, *82*, 22–28. [CrossRef]

130. Kelly, D.; Kotliar, M.; Woo, V.; Jagannathan, S.; Whitt, J.; Moncivaiz, J.; Aronow, B.J.; Dubinsky, M.C.; Hyams, J.S.; Markowitz, J.F.; et al. Microbiota-sensitive epigenetic signature predicts inflammation in Crohn's disease. *JCI Insight* **2018**, *3*. [CrossRef]
131. Rogler, G.; Biedermann, L.; Scharl, M. New insights into the pathophysiology of inflammatory bowel disease: Microbiota, epigenetics and common signalling pathways. *Swiss Med. Wkly.* **2018**, *148*, w14599. [CrossRef] [PubMed]

Article

Estrogen Receptor Beta (ERβ) Maintains Mitochondrial Network Regulating Invasiveness in an Obesity-Related Inflammation Condition in Breast Cancer

Toni Martinez-Bernabe [1,2,†], Jorge Sastre-Serra [1,2,3,†], Nicolae Ciobu [1], Jordi Oliver [1,2,3], Daniel Gabriel Pons [1,2,*] and Pilar Roca [1,2,3]

1 Grupo Multidisciplinar de Oncología Traslacional, Institut Universitari d'Investigació en Ciències de la Salut (IUNICS), Universitat de les Illes Balears, 07122 Palma de Mallorca, Illes Balears, Spain; toni.martinez@uib.es (T.M.-B.); jorge.sastre@uib.es (J.S.-S.); nicusorciobu@gmail.com (N.C.); jordi.oliver@uib.es (J.O.); pilar.roca@uib.es (P.R.)
2 Instituto de Investigación Sanitaria de las Islas Baleares (IdISBa), Hospital Universitario Son Espases, Edificio S, 07120 Palma de Mallorca, Illes Balears, Spain
3 Ciber Fisiopatología Obesidad y Nutrición (CB06/03), Instituto Salud Carlos III, 28029 Madrid, Madrid, Spain
* Correspondence: d.pons@uib.es; Tel.: +34-9711-73149
† These authors contributed equally to this work.

Citation: Martinez-Bernabe, T.; Sastre-Serra, J.; Ciobu, N.; Oliver, J.; Pons, D.G.; Roca, P. Estrogen Receptor Beta (ERβ) Maintains Mitochondrial Network Regulating Invasiveness in an Obesity-Related Inflammation Condition in Breast Cancer. *Antioxidants* **202**, *10*, 1371. https://doi.org/10.3390/antiox10091371

Academic Editors: Chiara Nediani and Monica Dinu

Received: 30 July 2021
Accepted: 24 August 2021
Published: 28 August 2021

Publisher's Note: MDPI stays neutral with regard to jurisdictional claims in published maps and institutional affiliations.

Abstract: Obesity, a physiological situation where different proinflammatory cytokines and hormones are secreted, is a major risk factor for breast cancer. Mitochondrial functionality exhibits a relevant role in the tumorigenic potential of a cancer cell. In the present study, it has been examined the influence of an obesity-related inflammation ELIT treatment (17β-estradiol, leptin, IL-6, and TNFα), which aims to stimulate the hormonal conditions of a postmenopausal obese woman on the mitochondrial functionality and invasiveness of MCF7 and T47D breast cancer cell lines, which display a different ratio of both estrogen receptor isoforms, ERα and ERβ. The results showed a decrease in mitochondrial functionality, with an increase in oxidative stress and invasiveness and motility, in the MCF7 cell line (high ERα/ERβ ratio) compared to a maintained status in the T47D cell line (low ERα/ERβ ratio) after ELIT treatment. In addition, breast cancer biopsies were analyzed, showing that breast tumors of obese patients present a high positive correlation between IL-6 receptor and ERβ and have an increased expression of cytokines, antioxidant enzymes, and mitochondrial biogenesis and dynamics genes. Altogether, giving special importance to ERβ in the pathology of obese patients with breast cancer is necessary, approaching to personalized medicine.

Keywords: obesity-related inflammation; oxidative stress; mitochondrial biogenesis; mitochondrial dynamics; epithelial-to-mesenchymal transition (EMT); estrogen receptor beta (ERβ); breast cancer

1. Introduction

Cancer, a pathology characterized by the excessive proliferation of tumor cells, is the second cause of death in the world, responsible for approximately 9.6 million deaths in 2018. In addition, breast cancer is a disease causing more than half a million deaths annually, projecting an increase in its incidence to a total of 3.2 million new cases per year in 2050 [1,2]. Breast cancer is a heterogeneous and multifactorial disease that involves both genetic predisposition, lifestyle, and environmental factors. In fact, it has been estimated that about 20% of breast cancer cases around the world are attributed to modifiable risk factors, which include obesity [1].

Obesity is a chronic metabolic disease characterized by excess fat accumulation in the body, and its prevalence has increased markedly in the last two decades in the most developed countries [3]. In obesity conditions, there is an alteration of the adipocyte secretome, which can lead to an imbalance of secreted adipokines, affecting processes as cell proliferation, invasive growth, apoptosis, angiogenesis, and metastasis in tumor cells of tissues such as the mammary gland [4–6]. Obesity is a risk factor for breast cancer due to increased circulating estrogens, such as 17β-estradiol. This is explained by the high conversion rate of the androgenic precursors produced by aromatase, an enzyme that increases its activity in the obese state [7].

The cellular effects of estrogens are mediated by their binding to estrogen receptors (ER) alpha (ERα), beta(ERβ), and GPER. Although ERα and ERβ show homology in the DNA and ligand-binding domains, their activity and affinity for 17β-estradiol are different, with ERα having a higher affinity for it. In addition, the expression pattern of the receptors also varies according to the cell type [8–10]. ERα is postulated as the main mediator of the carcinogenic effects of 17β-estradiol in breast cancer. In the case of ERβ, which is less studied, it is associated with an antiproliferative, cytostatic, and protective effect against tumor development, although other studies have observed contrary results, so its role in breast cancer remains to be clarified [10–12]. Secondly, GPER, also known as GPR30, is a G protein-coupled receptor that can be activated by estrogen to induce effects on proliferation, migration, and invasion, particularly in breast cancer [13].

Another adipokine synthesized and secreted by adipocytes is leptin, whose circulating levels are directly correlated with the individual's BMI [14]. Leptin has an important role as an independent predictor of risk and prognosis of breast cancer, which has been correlated with its circulating levels. Thus, women with breast cancer have higher plasma leptin levels and RNA expression in adipose tissue than healthy subjects [5,7]. Leptin exerts its biological role by binding to the leptin receptor, which is expressed in normal mammary epithelial cells and breast cancer cell lines, observing an effect on the stimulation of proliferation, cell division, invasion, and metastasis, through the JAK2/STAT3, MAPKs and PI3K/Akt signaling pathways. Furthermore, leptin is also capable of enhancing aromatase expression and activity [11], resulting in increased circulating levels of estrogens, which in turn have been shown to increase mRNA expression and leptin secretion from adipose tissue [14].

It is worth noting that obesity is considered an inducing factor of inflammation, where adipose tissue takes on the role of producer of cytokines and inflammatory proteins. However, the biochemical role between inflammation and cancer-inducing cellular modifications has not yet been elucidated [12]. In this way, the adipose tissue adjacent to the breast tissue is essential in the progression of cancer for the production of proinflammatory cytokines, leptin, and 17β-estradiol [15,16].

The expression of inflammatory cytokines, such as interleukin-6 (IL6) and Tumor necrosis factor alpha (TNFα), have been related to an increase in invasiveness and a poor prognosis in breast cancer [16]. IL6 is secreted by macrophages, fibroblasts, synovial cells, endothelial cells, and keratinocytes, but it is also synthesized by adipocytes [16,17]. Through the interaction with its receptor, IL6 would be able to induce aromatase, inducing a greater synthesis of 17β-estradiol [17]. On the other hand, TNFα is a key inflammatory cytokine produced by macrophages, T cells, B cells, and NK cells, in addition to tumor cells [14]. As a major cytokine in the tumor microenvironment, TNFα influences several of the hallmarks of cancer, such as stimulation of tumor growth, survival, invasion, metastasis, and angiogenesis [18,19]. In breast cancer cell lines, TNFα may have a role in both promoting and inhibiting cell growth, suggesting that the signaling pathway through which it contributes to proliferation is MAPK and PI3K/Akt [5].

Mitochondria is a cellular organelle that is highly influenced in response to estrogens, and it is the main source of reactive oxygen species (ROS), playing an important role in tumor processes, as well as in proliferation and apoptosis [20–22]. There is a balance between the production of ROS in the cell and its detoxification by antioxidant enzymes. If antioxidant mechanisms are diminished or ROS production undergoes a significant increase, an imbalance occurs in this system that leads to oxidative damage [23].

Mitochondrial function depends on mitochondria morphology, changes in shape, number, and location, which can translate into important functional modifications. Thus, mitochondrial biogenesis is the combination of both proliferation (an increase in the mitochondrial population) and differentiation (an improvement of the functional capabilities of pre-existing mitochondria) processes [24], while mitochondrial dynamics is a concept that describes the morphology and distribution of mitochondria in the cell [25].

Mitochondrial structure and function can be regulated by the activation of ERα and ERβ receptors by binding of 17β-estradiol [6,26]. Regarding the role of ERβ, some studies have implicated it as a tumor suppressor in breast cancer [6]. ERβ has been shown to colocalize in the mitochondria (mtERβ) and mediate estrogenic effects on the mitochondria, being able to increase mtDNA, increase respiratory capacity, increase antioxidant activity, and inhibit apoptosis [8].

Dysfunctional mitochondria play a relevant role in cancer and in the epithelial-to-mesenchymal transition (EMT) program in breast cancer [27]. As a consequence of mitochondrial defects, deteriorated OXPHOS has been shown to be involved in tumorigenesis [27]. In fact, alteration of mitochondrial functionality has been correlated with a specific mesenchymal phenotype. Moreover, previous studies have hypothesized a link between the regulation of mitochondrial genes, induction of EMT, and metastasis, which implies the worst clinical outcome for patients with cancer [28]. As mentioned above, obesity is an inflammatory disorder in which adipokines, as resistin or leptin, are secreted and play an important role in EMT. It has been described that obesity promotes the metastatic potential of breast cancer by inducing EMT [29]. Moreover, leptin promotes diverse biological events associated with tumorigenesis, as EMT. This process is induced by leptin through the expression of diverse transcription factors, which repress the epithelial markers and promote mesenchymal markers [30].

The aim of the present work was to study mitochondrial functionality and invasiveness, analyzing mitochondrial biogenesis and dynamics processes, as well as oxidative stress and inflammatory status and motility, in cell lines with different estrogen receptors ratio, exposed to a treatment consisting of 17β-estradiol, leptin, IL-6, and TNFα simulating circulating hormonal conditions in a postmenopausal obese woman [31–33] Likewise, the expression of the main antioxidant genes and genes related to inflammation and mitochondrial functionality in breast cancer tumors have been studied.

2. Materials and Methods

2.1. Reagents

Specific reagents 17β-estradiol, leptin, interleukin-6, and TNF-α were purchased from Sigma-Aldrich (St. Louis, MO, USA). Dulbecco's modified Eagle's medium without phenol red was purchased from GIBCO (Paisley, UK), Fetal Bovine Serum, and antibiotics solution (penicillin and streptomycin) from Biological Industries (Kibbutz Beit Haemek, Israel). Routine chemicals were supplied by Sigma-Aldrich (St. Louis, MO, USA), Panreac (Barcelona, Spain), and Bio-Rad Laboratories (Hercules, CA, USA).

2.2. Cell Culture and Treatments

MCF7 and T47D breast cancer cell lines were purchased from American Type Culture Collection ATCC (Manassas, VA, USA) and maintained in Dulbecco's modified Eagle's medium (DMEM) supplemented with 10% Foetal bovine serum (FBS) and 1% antibiotics (penicillin and streptomycin) at 37 °C with 5% CO_2. To avoid phenol red estrogenic effect, cells were seeded in 6-well or 96-well plates in phenol red-free DMEM containing 10% FBS and 1% antibiotics 24 h prior to treatment. Cells were treated, at 70–80% confluence, with vehicle (0.1% DMSO) or ELIT treatment (10 nM 17β-estradiol, 100 ng/mL leptin, 50 ng/mL interleukin-6 and 10 ng/mL TNFα) for 24 or 48 h.

2.3. Measurement of H_2O_2 Production and O_2^- Levels

Cells were seeded in 96-well plates (1.6×10^4 cells/well for MCF7 and 3.2×10^4 cells/well for T47D cell line) and treated with ELIT for 48 h. Hydrogen peroxide production was determined by Amplex® Red Hydrogen Peroxide/Peroxidase Assay Kit, following the manufacturer's protocol, as described previously [34]. Superoxide anion levels were determined by MitoSOX® Red reagent, following the manufacturer's protocol, as described previously [35]. The values obtained were normalized with the Hoechst 33342 fluorescence signal as previously described [36].

2.4. Cardiolipin Content

Cells were seeded in 96-well plates (1.6×10^4 cells/well for MCF7 and 3.2×10^4 cells/well for T47D cell line) and treated with ELIT for 48 h. Cells were stained with 250 nM of Nonyl Acridine Orange (NAO) to measure cardiolipin content as previously described [26]. The values obtained were normalized with the Hoechst 33342 fluorescence signal, as previously described [36].

2.5. RT-qPCR

Cells were seeded (4×10^5 cells/well for MCF7 and 6×10^5 cells/well for T47D cell line) in 6-well plates and treated with ELIT for 24 h. Total RNA from cultured cells or breast cancer human biopsies (25 mg) was isolated using TRI Reagent (Sigma-Aldrich, St. Louis, MO, USA), following the manufacturer's protocol, and then quantified using a BioSpec-nano spectrophotometer (Shimadzu Biotech, Kyoto, Japan) set at 260 nm and 280 nm, getting 260/280 nm ratio. Samples were retrotranscribed to cDNA, and PCR reactions were carried out as previously reported [35]. Genes, primers, and temperatures for the annealing step are specified in Table 1. Both GAPDH and 18S were used as housekeeping genes.

Table 1. Primers and conditions used for RT-qPCR.

Gene Accession Number	Forward Primer (5′-3′) Reverse Primer (5′-3′)	An. T° (°C)
ESR2 *NM_001437.3*	TAG TGG TCC ATC GCC AGT TAT GGG AGC CAC ACT TCA CCA T	60
NFE2L2 *NM_006164.5*	GCG ACG GAA AGA AGT ATG AGC GTT GGC AGA TCC ACT GGT TT	60
CAT *NM_001752.4*	CAT CGC CAC ATG AAT GGA TA CCA ACT GGG ATG AGA GGG TA	61
GSR *NM_000637.5*	TCA CGC AGT TAC CAA AAG GA CAC ACC CAA GTC CCC TGC AT	64
SOD1 *NM_000454.5*	TCA GGA GAC CAT TGC ATC ATT CGC TTT CCT GTC TTT GTA CTT TCT TC	64
SOD2 *NM_000636.4*	CGT GCT CCC ACA CAT CAA TC TGA ACG TCA CCG AGG AGA AG	64
UCP2 *NM_001381943.1*	GGT GGT CGG AGA TAC CAA CTC GGG CAA TGG TCT TGT	60
TNF *NM_000594.4*	AAG CCT GTA GCC CAT GTT GT GGA CCT GGG AGT AGA TGA GGT	58
PTGS2 *NM_000963.4*	CCC TTC TGC CTG ACA CCT TT TTC TGT ACT GCG GGT GGA AC	60
IL6 *NM_000600.5*	CAG GGG TGG TTA TTG CAT CT AGG AGA CTT GCC TGG TGA AA	60
CXCL8 *NM_000584.4*	GGC ACA AAC TTT CAG AGA CAG CAG GTT TCT TCC TGG CTC TTG TCC TAG	66
IL6R *NM_000565.4*	TGG GAG GTG GAG AAG AGA GA AGG ACC TCA GGT GAG AAG CA	60
CXCR8 *NM_000634.3*	AGT TCT TGG CAC GTC ATC GT CCC CTG AAG ACA CCA GTT CC	58
TGFB *NM_000660.7*	TCC TGG CGA TAC CTC AGC AA CGG TAG TGA ACC CGT TGA TG	60

Table 1. *Cont.*

Gene Accession Number	Forward Primer (5′-3′) Reverse Primer (5′-3′)	An. T° (°C)
NRF1 *NM_005011.5*	CCA CGT TAC AGG GAG GTG AG TGT AGC TCC CTG CTG CAT CT	60
SSBP1 *NM_001256510.1*	TGT GAA AAA GGG GTC TCG AA TGG CCA AAG AAG AAT CAT CC	60
PPARGC1A *NM_001330751.2*	TCA GTC CTC ACT GGT GGA CA TGC TTC GTC GTC AAA AAC AG	60
TFAM *NM_003201.3*	GTG GTT TTC ATC TGT CTT GGC ACT CCG CCC TAT AAG CAT CTT	60
TWNK *NM_021830.5*	GGG AGG AGG TGC TAG GAG AA TTC CTG GCT TGC TTT GGC T	61
MFN1 *NM_033540.3*	TTC GAT CAA GTT CCG GAT TC TTG GAG CGG AGA CTT AGC AT	51
MFN2 *NM_014874.4*	GCA GAA CTT TGT CCC AGA GC AGA GGC ATC AGT GAG GTG CT	56
OPA1 *NM_015560.3*	ACA ATG TCA GGC ACA ATC CA GGC CAG CAA GAT TAG CTA CG	51
OMA1 *NM_145243.5*	TTG GAT TGC TCT TTG TGG TG GGT ATC GGG CAT CTT TCT CA	51
DNM1L *NM_012062.5*	GTT CAC GGC ATG ACC TTT TT AAG AAC CAA CCA CAG GCA AC	51
FIS1 *NM_016068.3*	GCT GAA GGA CGA ATC TCA CTT GCT GTG TCC AAG TCC AA	55
SIRT1 *NM_012238.5*	GCA GAT TAG TAG GCG GCT TG TCT GGC ATG TCC CAC TAT CA	60
SIRT3 *NM_012239.6*	CGG CTC TAC ACG CAG AAC ATC CAG AGG CTC CCC AAA GAA CAC	56
GAPDH *NM 002046.7*	CCA CTC CTC CAC CTT TGA CG CTG GTG GTC CAG GGG TCT TA	60
18S *NR_146119.1*	GGACACGGACAGGATTGACA ACCCACGGAATCGAGAAAGA	60
STAT3 *NM_139276.3*	CTG GCC TTT GGT GTT GAA AT AAG GCA CCC ACA GAA ACA AC	61
SLC25A14 *NM_001282195.2*	CAA GCC GTT GGT CTC CTA AG CGT TTT CAA TGT CAC CCA TC	60
CDH1 *NM_004360.5*	GTCACTGACACCAACGATAATCCT TTTCAGTGTGGTGATTACGACGTTA	60
ESRRA *NM_004451.5*	TCG CTC CTC CTC TCA TCA TT TGG CCA AAC CCA AAA ATA AA	52
PPARG *NM_138712.5*	GAG CCC AAG TTT GAG TTT GC CTG TGA GGA CTC AGG GTG GT	61
ESR1 *NM_000125.4*	AAT TCA GAT AAT CGA CGC CAG GTG TTT CAA CAT TCT CCC TCC TG	61
MMP9 *NM_004994.3*	CGC AGA CAT CGT CAT CCA GT AAA CCG AGT TGG AAC CAC GA	60
GPER1 *NM_001505.3*	CAT CAT CGG CCT GTG CTA CT GAT GAA GAC CTT CTC CGG CA	60
GPX1 *NM_000581.4*	GCG GCG GCC CAG TCG GTG TA GAG CTT GGG GTC GGT CAT AA	61

2.6. Western Blot

After 48 h of ELIT treatment, cells were harvested as described previously by Torrens-Mas [35]. Protein content (supernatant) was determined with the bicinchoninic acid (BCA) protein assay kit (Thermo Fisher Scientific, Waltham, MA, USA). Ten micrograms of protein were resolved on a 12% SDS-PAGE gel and electrotransferred onto nitrocellulose membranes using the Trans-blot® Turbo™ transfer system (Bio-Rad, Hercules, CA, USA). Membranes were blocked in 5% non-fat powdered milk in TBS with 0.05% Tween for 1 h. Antisera against OXPHOS complexes (ab110411; Abcam, Bristol, UK), SOD-1 (574597, Calbiochem®, San Diego, CA, USA), SOD-2 (sc-30080; Santa Cruz Biotechnology, Santa Cruz, CA, USA), CAT (219010, Calbiochem®, San Diego, CA, USA), GRd (sc-133245; Santa Cruz Biotechnology, Santa Cruz, CA, USA), 4-HNE (HNE11-S, Alpha Diagnostic, San Antonio, TX, USA), COXIV (ab33985, Abcam, Bristol, UK), PGC1α (ab54481, Abcam, Bristol, UK), and GAPDH (sc-25778; Santa Cruz Biotechnology, Santa Cruz, CA, USA) were used as primary antibodies. Protein bands were visualized as described previously by [34].

2.7. Measurement of 4-HNE Adducts Levels

For 4-hydroxy-2-nonenal (4-HNE) adducts analysis, as lipid oxidative damage marker, 20 μg of total protein from cell lysate, processed as previously described by Pons et al. [37].

2.8. Wound Healing Assay

Cells were seeded in six-well plates at a density of 1×10^6 cells/well for T47D cell line and 8.5×10^5 cells/well for MCF7 cell line. Wound healing assay was performed as previously described by Torrens-Mas et al. [35]. The area of the scratch was measured using the MRI Wound Healing Tool macro for ImageJ software.

2.9. Confocal Microscopy

Cells were seeded on a glass coverslip inside 6-well plates at a density of 2×10^5 cells/well for MCF7 and 5×10^5 cells/well for T47D. After 24 h, cells were treated with ELIT treatment for 24 h. Then, cells were incubated with MitoTracker™ Green 0.5 μM (Invitrogen, M7514) for 1 h and LysoTracker™ Red 0.5 μM (L7528, Invitrogen, Waltham, MA, USA) for 20 min, both at 37 °C in the dark. For DNA staining, cells were incubated with 1/200 dilution of 1 μg/mL Hoechst 33342 (B2261, Sigma, St. Louis, MO, USA) for 5 min at 37 °C in the dark.

The fluorescence was monitored with a Leica TCS-SPE Confocal Microscope, using 63× immersion oil (147 N.A.) objective lens. Fluorescence excitation/emission was 490/516 nm for MitoTracker Green, 577/590 for LysoTracker Red, and 350/455 nm for Hoechst 33342.

Mitochondrial Roundness was analyzed in ImageJ software with Mito-Morphology Macro designed by Ruben K. Dagda at the University of Pittsburgh (2010). This macro is currently maintained and supported by grants NIH/NINDS R01NS105783-01 grant and by NIH/NIGMS R25 1R25-OD023795-01.

2.10. Seahorse Metabolic Analyzer

Real-time oxygen consumption rates (OCRs) were determined for MCF7 and T47D cells using the Seahorse Extracellular Flux (XFe96) analyzer (Seahorse Bioscience, North Billerica, MA, USA). Cells were seeded at a density of 4.8×10^3 cells/well for MCF7 and 9.600 cells/well for T47D into XFe96 well cell culture plates and incubated overnight to allow attachment at 37 °C in 5% CO_2. After 24 h, cells were incubated with vehicle or ELIT treatment. After 48 h of incubation, cells were maintained in 200 μL/well of XF assay media at 37 °C, in a non-CO_2 incubator for 1 h. During the incubation time, mitochondrial complex inhibitors (1 μM oligomycin, 2 μM FCCP, 0.5 μM rotenone, and 0.5 μM antimycin A) were preloaded for OCR measurements, in XF assay media into the injection ports in the XFe96 sensor cartridge.

2.11. Human Samples

Human breast cancer biopsies were obtained from 33 women, ages between 45–90 years. Samples of these patients were obtained from the Biological Specimen Bank of Son Llàtzer Hospital and as specified by and with the necessary permission granted from the Balearic Island Bioethics Committee. Tumor samples were collected immediately after tumor removal and were frozen in isopentane for analysis as described by Sastre-Serra et al. [38]. Written informed consent was obtained from the patients before surgery. All the patients presented an invasive ductal carcinoma (ER-positive, PR-negative, and HER2-negative, as determined by immunohistochemistry) and were classified in normal weight (nw), overweight (ow), and obese (o) by BMI (kg/m^2).

2.12. Statistical Analysis

The statistical analyses were performed with the Statistical Programme for the Social Sciences software for Windows (SPSS, version 27.0; SPSS Inc, Chicago, IL, USA). Data are presented as mean $\pm$ standard error of the mean (SEM). The statistical differences in cell lines between vehicle- and ELIT-treated cells were analyzed using a Student's *t*-test with statistical significance was set at $p < 0.05$ (*). The statistical differences in human samples were analyzed using Pearson's correlation with statistical significance was set at $p < 0.01$ (**) and $p < 0.05$ (*).

3. Results

3.1. Obesity-Related Inflammation Treatment Increases Inflammation-Related Genes Expression in Breast Cancer Cell Lines

To confirm the obesity-related inflammation treatment effectivity, mRNA expression of main inflammatory genes was determined. As shown in Figure 1, inflammatory genes expression (*IL6, IL6R, CXCL8, PTGS2, TNF*, and *STAT3*) showed a statistically significant increase in both MCF7 and T47D cell lines after 24 h of ELIT (17β-estradiol (10 nM), leptin (100 ng/mL), interleukin-6 (50 ng/mL), and TNFα (10 ng/mL)) treatment. In contrast, *PPARG* anti-inflammatory gene had decreased expression in the MCF7 cell line, and *TGFB* anti-inflammatory gene showed a statistically significant increase in the T47D cell line.

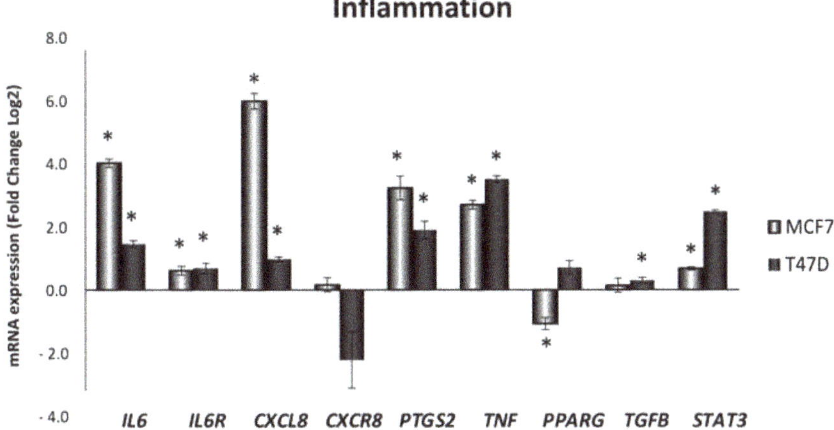

Figure 1. Obesity-related inflammation treatment increased the expression of genes related to inflammation in MCF7 and T47D breast cancer cell lines. *IL6*: interleukin-6; *IL6R*: interleukin-6 receptor; *CXCL8*: interleukin-8; *CXCR8*: interleukin-8 receptor; *PTGS2*: cyclooxygenase-2; *TNF*, tumor necrosis factor alpha; *PPARG*: peroxisome proliferator-activated receptor gamma; *TGFB*: transforming growth factor beta; *STAT3*: Signal transducer and activator of transcription 3. Breast cancer cells were incubated for 24 h with vehicle (DMSO) or ELIT. Data are represented as fold change (log2) of mRNA expression with respect to vehicle-treated cells, set at 0, of each cell line. Data represent means $\pm$ SEM (n = 6). * Statistically significant difference between ELIT treated and vehicle-treated cells (Student's *t*-test, $p < 0.05$).

3.2. *Obesity-Related Inflammation Treatment Increases Oxidative Stress in Breast Cancer Cell Lines with High ERα/ERβ Ratio*

As shown in Table 2, ELIT treatment, an increase ROS production in both MCF7 and T47D cell lines was seen. Superoxide anion levels increased more than 10 times in MCF7 cells (1177%) and 5 times in T47D cells (526%) after ELIT treatment, as shown in Table 2, and H_2O_2 production increased in both cell lines (+100% in MCF7 and +34% in T47D cells). However, cardiolipin content, as an indicator of mitochondrial inner membrane quantity, decreased in MCF7 treated cells. Moreover, as shown in Table 2, oxidative damage increased (+45%) in MCF7 cell line after ELIT treatment, but not in the T47D cell line. A representative blot of 4-HNE detection is shown in Supplementary Materials Figure S2.

Table 2. Oxidative stress in MCF7 and T47D breast cancer cell lines.

	MCF7		T47D	
	Control	**ELIT**	**Control**	**ELIT**
Superoxide anion levels (%)	100 ± 13	1177 ± 116 *	100 ± 16	526 ± 35 *
H_2O_2 production (%)	100 ± 1	200 ± 8 *	100 ± 2	134 ± 3 *
Cardiolipin content (%)	100 ± 2	82.7 ± 1.1 *	100 ± 1	101 ± 1
Oxidative damage (%)	100 ± 5	145 ± 8 *	100 ± 7	92.3 ± 11.7

Data represent the means ± SEM (n = 6). Values of control (DMSO-treated) cells were set at 100 in each cell line. * Significant difference between ELIT-treated and control cells (Student's test; $p < 0.05$).

The mRNA expression of antioxidant enzymes was analyzed (Figure 2) in both MCF7 and T47D cell lines after 24 h ELIT treatment. As shown in Figure 2, *SOD2* (mitochondrial superoxide dismutase) showed a high increase in both cell lines; in addition, a decrease in *SOD1* (copper/zinc superoxide dismutase) was observed. However, ELIT-treated MCF7 cells showed a general decrease in catalase (*CAT*) and glutathione reductase (*GSR*). Nevertheless, in the T47D cell line, ELIT treatment increased *GSR* and nuclear factor erythroid 2-related factor 2 (*NFE2L2*) expression, as shown in Figure 2.

Main antioxidant protein expression levels were determined (Table 3 and Supplementary Materials) after 48 h ELIT treatment and resulted in a high increase in SOD2 in both MCF7 and T47D cell lines (more than 10 times in MCF7 and 30 times in T47D). Moreover, SOD1 showed a statistically significant increase in the T47D cell line after treatment. As shown in Table 3, CAT and GSR protein levels decrease only in the MCF7 cell line after ELIT treatment (−30% and −51%, respectively). Representative detection bands are shown in Supplementary Materials Figures S3–S6.

Table 3. Antioxidant enzymes protein levels in MCF7 and T47D breast cancer cell lines.

	MCF7		T47D	
	Control	**ELIT**	**Control**	**ELIT**
SOD1 (%)	100 ± 12	89 ± 14	100 ± 25	156 ± 28 *
SOD2 (%)	100 ± 11	1631 ± 552 *	100 ± 15	3691 ± 350 *
CAT (%)	100 ± 6	70 ± 8 *	100 ± 14	88 ± 10
GSR (%)	100 ± 14	49 ± 6 *	100 ± 9	117 ± 8

Data represent the means ± SEM (n = 6). Values of control (DMSO-treated) cells were set at 100 in each cell line. * Significant difference between ELIT-treated and control cells (Student's test; $p < 0.05$).

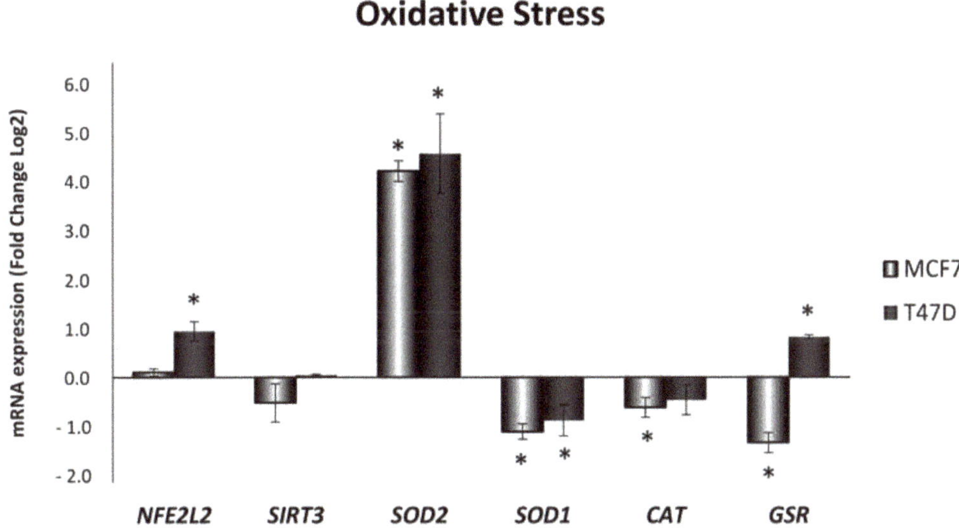

Figure 2. Obesity-related inflammation treatment modified expression of genes related to oxidative stress in MCF7 and T47D breast cancer cell lines. *NFE2L2*: nuclear factor erythroid 2-related factor 2; *SIRT3*: sirtuin 3; *SOD2*: manganese superoxide dismutase; *SOD1*: copper/zinc superoxide dismutase; *CAT*: catalase; *GSR*: glutathione reductase. Breast cancer cells were incubated for 24 h with vehicle (DMSO) or ELIT. Data are represented as fold change (log2) of mRNA expression with respect to vehicle-treated cells, set at 0, of each cell line. Data represent means ± SEM ($n = 6$). * Statistically significant difference between treated and vehicle-treated cells (Student's t-test, $p < 0.05$).

3.3. Mitochondrial Biogenesis and Functionality Are Reduced in Breast Cancer Cell Lines after ELIT Treatment with a High ERα/ERβ Ratio

Mitochondrial biogenesis genes were also checked, as shown in Figure 3. ELIT treatment decreased the expression of almost all mitochondrial biogenesis genes analyzed in MCF7 cell line. Nevertheless, in the T47D cell line, not only did the mRNA expression not decrease but also, especially high levels of estrogen-related receptor alpha (*ESRRA*) and Twinkle (*TWNK*) were found in ELIT-treated cells versus non-treated cells. Moreover, uncoupling proteins 2 and 5 (*UCP2* and *SLC25A14*, respectively) mRNA expression was analyzed, and a decrease in UCP2 expression in both cell lines was accompanied by a decrease in uncoupling protein 5 expression in the MCF7 cell line.

Mitochondrial biogenesis master regulator PPARGC1A protein levels increased after ELIT treatment in the MCF7 cell line. However, to check the effect of ELIT treatment on the mitochondrial respiratory chain (OXPHOS), protein levels of these OXPHOS complexes were analyzed. As seen in Table 4 and Supplementary Materials, ELIT treatment did not change protein levels in T47D, whereas, in MCF7 cell line OXPHOS complexes, protein levels decreased, except complex III (Q-cytochrome c oxidoreductase) after 48 h treatment. Representative detection bands are shown in Supplementary Materials in Figures S7 and S8.

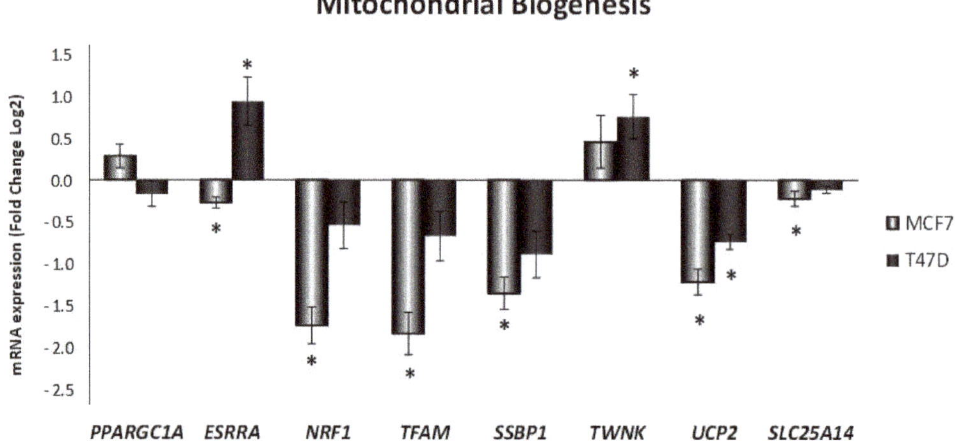

Figure 3. Obesity-related inflammation treatment modified expression of genes related to mitochondrial biogenesis in MCF7 and T47D breast cancer cell lines. *PPARGC1A*: peroxisome proliferator-activated receptor-gamma coactivator-1alpha; *ESRRA*: estrogen-related receptor alpha; *NRF1*: nuclear respiratory factor 1; *TFAM*: mitochondrial transcription factor A; *SSBP1*: mitochondrial single-strand DNA binding protein; *TWNK*: Twinkle mtDNA helicase; *UCP2*: uncoupling protein 2; *SLC25A14*: uncoupling protein 5. Breast cancer cells were incubated for 24 h with vehicle (DMSO) or ELIT. Data are represented as fold change (log2) of mRNA expression with respect to vehicle-treated cells, set at 0, of each cell line. Data represent means ± SEM ($n = 6$). * Statistically significant difference between treated and vehicle-treated cells (Student's t-test, $p < 0.05$).

Table 4. Mitochondrial-related protein levels in MCF7 and T47D breast cancer cell lines.

	MCF7		T47D	
	Control	**ELIT**	**Control**	**ELIT**
PPARGC1A (%)	100 ± 11	142 ± 2 *	100 ± 13	91 ± 16
Complex I (NDUFB8) (%)	100 ± 7	36 ± 2 *	100 ± 16	80 ± 21
Complex II (SDHB) (%)	100 ± 17	39 ± 4 *	100 ± 18	84 ± 10
Complex III (UQCRC2) (%)	100 ± 40	136 ± 24	100 ± 5	101 ± 8
Complex IV (COX II) (%)	100 ± 10	26 ± 4 *	100 ± 18	65 ± 13
(COX IV) (%)	100 ± 5	54 ± 6 *	100 ± 3	86 ± 13
Complex V (ATP5A) (%)	100 ± 9	56 ± 10 *	100 ± 14	117 ± 4

Data represent the means ± SEM ($n = 6$). Values of control (DMSO-treated) cells were set at 100 in each cell line. * Significant difference between ELIT-treated and control cells (Student's test; $p < 0.05$).

To further investigate mitochondrial function, the oxygen consumption rate (OCR) was determined. As shown in Figure 4a,c, basal OCR, Maximal respiratory capacity, ATP-linked respiration, and proton leak were statistically significantly lower in the MCF7 cell line after ELIT 48 h treatment. These changes were not observed in the T47D cell line (Figure 4b,d). Moreover, reserve capacity, calculated as basal minus maximal respiratory capacity rates, showing an increase in both cell lines, though this increase was higher in the T47D than in the MCF7 cell line.

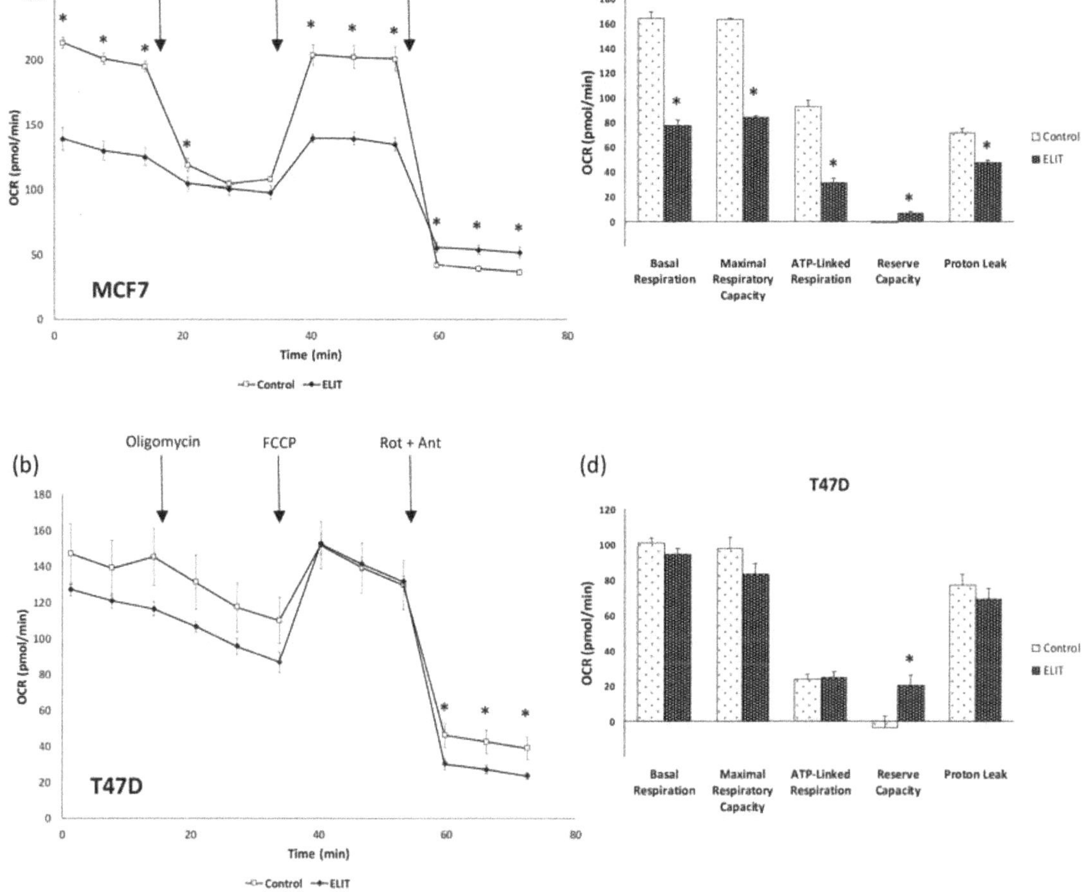

Figure 4. ELIT treatment reduced oxygen consumption ratio (OCR) in the MCF7 cell line. (**a**,**b**): OCR basal conditions and after oligomycin, FCCP, and antimycin A + rotenone addition. (**c**,**d**): calculated parameters from oxygen consumption analysis. Basal respiration: initial rate—antimycin A + rotenone rate; ATP-Linked Respiration: maximal respiratory capacity: FCCP rate—antimycin A+ rotenone rate; initial rate—oligomycin rate; Reserve capacity: FCCP rate—initial rate; proton leak: oligomycin rate—antimycin A+ rotenone rate. Values are expressed as means ± SEM (n = 5). * Statistically significant difference between treated and vehicle-treated cells (Student's t-test, $p < 0.05$).

3.4. Obesity-Related Inflammation Treatment Affects Mitochondrial Dynamics and Mitochondrial Network in Breast Cancer Cell Lines

As shown in Figure 5, mitochondrial dynamics genes expression was decreased in MCF7 cells after 24 h ELIT treatment, but in T47D treated cells, this situation was not observed. In fact, almost all the genes showed an increase after treatment in this cell line. The expression of mitochondrial fusion-related genes *MFN1*, *MFN2*, *OMA1*, *OPA1* showed decreased levels in the MCF7 cell line, whereas, in the T47D cell line, ELIT treatment increased expression of *MFN1*, *MFN2*, and *OMA1*. Fission-related mitochondrial genes presented a different pattern expression between cell lines, with a statistically significant increase in *FIS1* mRNA expression in the T47D cell line after ELIT treatment.

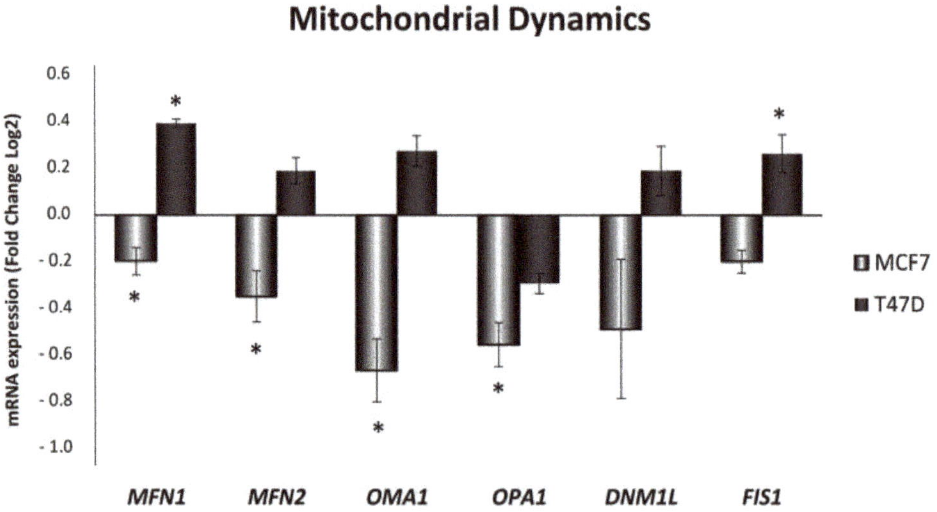

Figure 5. Obesity-related inflammation treatment modified expression of genes related to mitochondrial dynamics in MCF7 and T47D breast cancer cell lines. *MFN1*: mitofusin-1; *MFN2*: mitofusin-2; *OPA1*: mitochondrial dynamin-like GTPase; *OMA1*: zinc metallopeptidase; *DNM1L*: dynamin 1 like; *FIS1*: mitochondrial fission 1 protein. Breast cancer cells were incubated for 24 h with vehicle (DMSO) or ELIT. Data are represented as fold change (log2) of mRNA expression with respect to vehicle-treated cells, set at 0, of each cell line. Data represent means ± SEM (n = 6). * Statistically significant difference between treated and vehicle-treated cells (Student's t-test, * $p < 0.05$).

On the one hand, mitochondrial networking was modified after 24 h-ELIT treatment in both MCF7 and T47D cell lines. However, as shown in Figure 6a, the MTG signal was higher in the MCF7 cell line after treatment, but LTR intensity was also high. These changes were not observed in T47D cell lines after ELIT treatment. On the other hand, lysosomes distribution in MCF7 cell lines, whereas more equally distributed dispersed through the cytoplasm after 24 h-ELIT treatment, whereas in T47D cell line this distribution showed seemed to be more perinuclear. Index of elongation, calculated as the average circularity of mitochondria of confocal microscopy images, as shown in Figure 6b. As seen, the MCF7 cell line after 24 h ELIT treatment showed more circular mitochondria than in control cells. These morphological modifications were not seen in the T47D cell line with treatment.

3.5. Obesity-Related Inflammation Treatment Increase Invasiveness in Breast Cancer Cell Lines with a High ERα/ERβ Ratio

As shown in Figure 7a, Cadherin E (*CDH1*) expression was decreased in MCF7 cells after 24 h ELIT treatment, but Matrix Metalloproteinase 9 (*MMP9*) expression increased three times with respect to vehicle-treated cells. In contrast, CDH1 expression increased in the T47D cell line after treatment. In addition to these results, a wound-healing assay was performed. As shown in Figure 7b and in Supplementary Materials Figure S9, after 24 h, ELIT-treated MCF7 cells were able to better close the wound, leaving 80% of the initial scratch open, while the control cells were not able to close the wound. In the T47D cell line, there were no differences between control and ELIT-treated cells.

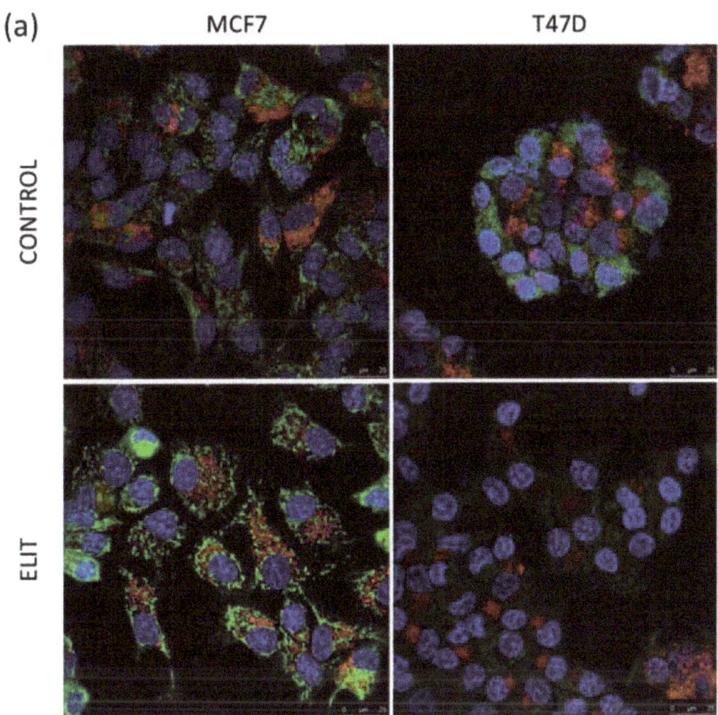

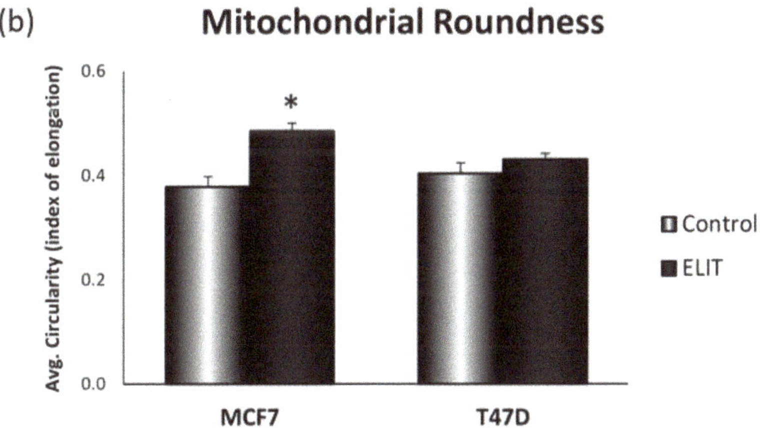

Figure 6. ELIT treatment induces mitochondrial networking changes in the MCF7 cell line. (**a**) Breast cancer cell lines MCF7 and T47D were treated with ELIT for 24 h followed by co-incubation with MTG, LTR, and Hoechst 33342. The fluorescence was monitored with a Leica Confocal Microscope using 63× oil 147 N.A. objective lens. Scale bar 25 µm. (**b**) Mitochondrial Roundness was analyzed with ImageJ Software. Breast cancer cells were incubated for 24 h with vehicle (DMSO) or ELIT. Data are represented as an index of elongation, set at one when a perfect circle. Data represent means ± SEM (n = 6). * Statistically significant difference between treated and vehicle-treated cells (Student's t-test, * $p < 0.05$).

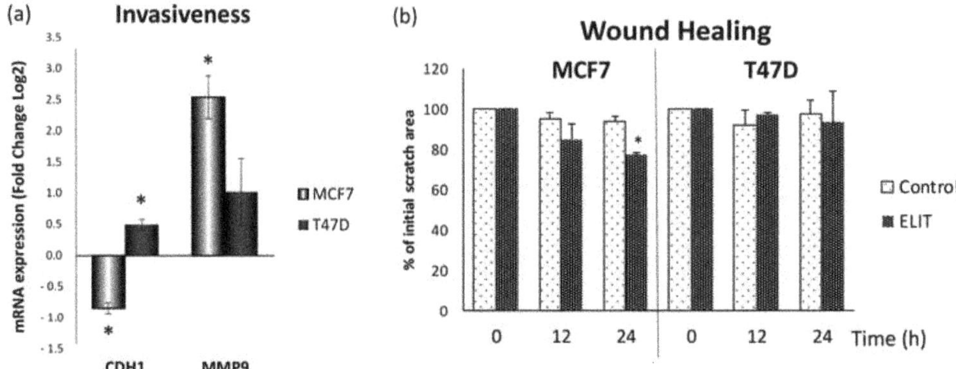

Figure 7. Invasiveness and cell motility were increased after ELIT treatment in the MCF7 cell line. (**a**) CDH1: cadherin-E; MMP9: Matrix Metalloproteinase 9. Breast cancer cells were incubated for 24 h with vehicle (DMSO) or ELIT. Data are represented as fold change (log2) of mRNA expression with respect to vehicle-treated cells, set at 0, of each cell line. Data represent means ± SEM (n = 6). (**b**) Area of the wound that remained open after 24 h with ELIT treatment. Data are represented as a percentage of the initial scratch area with respect to vehicle-treated cells of each well, set at 100, of each cell line. Data represent means ± SEM (n = 3). * Statistically significant difference between treated and vehicle-treated cells (Student's t-test, $p < 0.05$).

3.6. Estrogen Receptor Ratio Is Modified by ELIT Treatment in Breast Cancer Cell Lines

To observe the effects of obesity-related inflammation treatment over estrogen receptors alpha (ERα), beta (ERβ), and GPER in MCF7 and T47D breast cancer cell lines, mRNA expression was analyzed. As shown in Figure 8, estrogen receptor alpha (*ESR1*) mRNA expression was decreased in both cell lines after 24 h-ELIT treatment. However, estrogen receptor beta (*ESR2*) and GPER (*GPER1*) mRNA expression only shown a decrease in the MCF7 cell line, maintaining its expression in the T47D cell line.

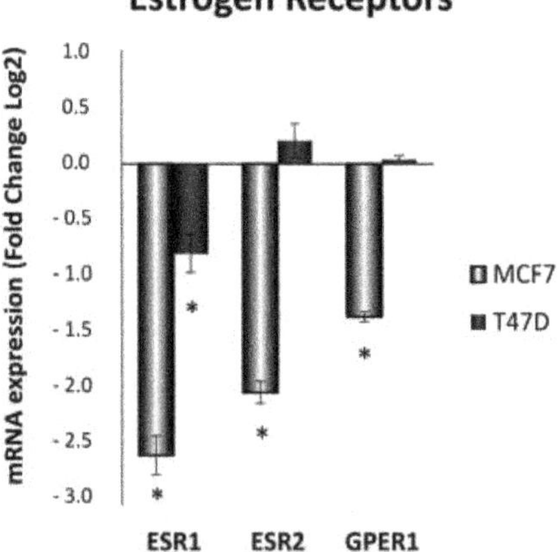

Figure 8. Obesity-related inflammation treatment modified expression of estrogen receptor alpha, beta, and GPER in MCF7 and T47D breast cancer cell lines. *ESR1*: estrogen receptor alpha; *ESR2*: estrogen receptor beta; GPER1: G-coupled protein estrogen receptor. Breast cancer cells were incubated for 24 h with vehicle (DMSO) or ELIT. Data are represented as fold change (log2) of mRNA expression with respect to vehicle-treated cells, set at 0, of each cell line. Data represent means ± SEM (n = 6). * Statistically significant difference between treated and vehicle-treated cells (Student's t-test, * $p < 0.05$).

3.7. IL6R in Breast Tumors Correlates with Inflammation, Mitochondrial Biogenesis, and Oxidative Stress Markers in Different BMI Situations

Table 5 shows the correlation with Pearson correlation values between the IL6R mRNA expression and studied genes in different BMI situations: normal weight (nw), overweight (ow), and obese (o). *CXCR8, CXCL8, TNF, SLC25A14, NRF1, NFE2L2, PPARGC1A,* and *FIS1* were significantly and positively correlated with IL6R expression in the nw group. *CXCR8, TNF, SLC25A14, NRF1, NFE2L2,* and *PPARGC1A* were significantly and positively correlated with IL6R expression in the ow group. *CXCR8, CXCL8, TNF, PTGS2, ESR2, GPX1, SOD1, SLC25A14, NRF1, NFE2L2, PPARGC1A, SIRT1, FIS1,* and *OMA1* were significantly and positively correlated with IL6R expression in the ow group.

Table 5. Correlation values of IL6R gene expression in breast tumors according to obesity status.

		nw	ow	o			nw	ow	o			nw	ow	o
IL6R	Pearson Correlation	1	1	1	CAT	Pearson Correlation	−0.205	0.421	0.164	SSBP1	Pearson Correlation	0.306	−0.218	0.229
	Sig.					Sig.	0.285	0.173	0.272		Sig.	0.195	0.319	0.197
ESR1	Pearson Correlation	−0.006	0.105	−0.337	GPX1	Pearson Correlation	0.358	0.406	0.704 **	NRF1	Pearson Correlation	0.870 **	0.878 **	0.936 **
	Sig.	0.494	0.412	0.101		Sig.	0.155	0.183	0.001		Sig.	0.001	0.005	0
ESR2	Pearson Correlation	0.396	0.703	0.605 *	GSR	Pearson Correlation	0.454	0.286	−0.009	PPARGC1 A	Pearson Correlation	0.656 *	0.807 *	0.830 **
	Sig.	0.19	0.059	0.024		Sig.	0.094	0.267	0.486		Sig.	0.02	0.014	0
CXCR8	Pearson Correlation	0.791 **	0.840 **	0.957 **	SOD1	Pearson Correlation	0.438	0.062	0.434 *	SIRT1	Pearson Correlation	0.228	0.261	0.604 **
	Sig.	0.003	0.009	0		Sig.	0.103	0.448	0.046		Sig.	0.263	0.286	0.007
IL6R	Pearson Correlation	0.421	0.113	0.304	SOD2	Pearson Correlation	0.289	−0.459	−0.106	TFAM	Pearson Correlation	0.279	−0.037	0.387
	Sig.	0.113	0.405	0.126		Sig.	0.209	0.15	0.348		Sig.	0.234	0.468	0.069
CXCL8	Pearson Correlation	0.732 **	0.662	0.911 **	SIRT3	Pearson Correlation	0.428	−0.11	0.3	FIS1	Pearson Correlation	0.609 *	0.326	0.883 **
	Sig.	0.008	0.053	0		Sig.	0.125	0.407	0.129		Sig.	0.041	0.237	0
TGFB	Pearson Correlation	0.296	−0.552	0.272	NFE2L2	Pearson Correlation	0.680 *	0.753 *	0.812 **	OMA1	Pearson Correlation	0.379	0.472	0.541 *
	Sig.	0.203	0.1	0.154		Sig.	0.015	0.025	0		Sig.	0.157	0.172	0.015
TNF	Pearson Correlation	0.604 *	0.684 *	0.806**	UCP2	Pearson Correlation	0.256	−0.266	0.332	OPA1	Pearson Correlation	−0.39	−0.45	−0.032
	Sig.	0.032	0.045	0		Sig.	0.238	0.282	0.104		Sig.	0.17	0.224	0.456
PTGS2	Pearson Correlation	0.07	−0.134	0.639 **	SLC25A 14	Pearson Correlation	0.874 **	0.901 **	0.940 **	* The correlation is significant at $p < 0.05$				
	Sig.	0.435	0.415	0.007		Sig.	0	0.003	0	** The correlation is significant at $p < 0.01$				

Data represent Pearson correlations and significance (unilateral) in normal weight ($n = 9$). overweight ($n = 8$) and obesity ($n = 16$) groups. * ($p < 0.05$) ** ($p < 0.01$) significant difference.

Moreover, *ESR1* and *ESR2* gene expression correlations are shown in Table 6a,b, respectively. On the one hand, *IL6, CXCL8, SOD1,* and *SSBP1* were significantly and negatively correlated with *ESR1* expression; instead, *SIRT3* was significantly and positively correlated. On the other hand, *IL6R, CXCR8, GPX1, SLC25A14, NRF1, PPARGC1A, SIRT1,* and *OMA1* were significantly and positively correlated with *ESR2* expression.

Table 6. Correlation values of estrogen receptors subtype alpha and beta gene expression in breast tumors. (**a**) *ESR1* correlation. (**b**) *ESR2* correlation.

(a)

Gene		Value	Gene		Value	Gene		Value
ESR1	Pearson Correlation	1.000	CAT	Pearson Correlation	0.089	SSBP1	Pearson Correlation	−0.461 **
	Sig.			Sig.	0.624		Sig.	0.007
IL6R	Pearson Correlation	−0.219	GPX1	Pearson Correlation	0.218	NRF1	Pearson Correlation	−0.154
	Sig.	0.221		Sig.	0.223		Sig.	0.393
ESR2	Pearson Correlation	0.010	GSR	Pearson Correlation	0.139	PPARGC1A	Pearson Correlation	−0.314
	Sig.	0.962		Sig.	0.440		Sig.	0.075
CXCR8	Pearson Correlation	−0.230	SOD1	Pearson Correlation	−0.391 *	SIRT1	Pearson Correlation	−0.012
	Sig.	0.205		Sig.	0.024		Sig.	0.945
IL6	Pearson Correlation	−0.364 *	SOD2	Pearson Correlation	0.177	TFAM	Pearson Correlation	−0.003
	Sig.	0.037		Sig.	0.323		Sig.	0.989
CXCL8	Pearson Correlation	−0.422 *	SIRT3	Pearson Correlation	0.571 **	FIS1	Pearson Correlation	−0.192
	Sig.	0.014		Sig.	0.001		Sig.	0.292
TGFB	Pearson Correlation	0.221	NFE2L2	Pearson Correlation	−0.122	OMA1	Pearson Correlation	0.110
	Sig.	0.216		Sig.	0.500		Sig.	0.555
TNF	Pearson Correlation	0.041	UCP2	Pearson Correlation	0.153	OPA1	Pearson Correlation	0.125
	Sig.	0.820		Sig.	0.396		Sig.	0.525
PTGS2	Pearson Correlation	−0.200	SLC25A14	Pearson Correlation	−0.163	* The correlation is significant at $p < 0.05$		
	Sig.	0.317		Sig.	0.364	** The correlation is significant at $p < 0.01$		

(b)

Gene		Value	Gene		Value	Gene		Value
ESR2	Pearson Correlation	1.000	CAT	Pearson Correlation		SSBP1	Pearson Correlation	0.260
	Sig.			Sig.			Sig.	0.220
IL6R	Pearson Correlation	0.639 **	GPX1	Pearson Correlation		NRF1	Pearson Correlation	0.483 *
	Sig.	0.001		Sig.			Sig.	0.017
ESR1	Pearson Correlation	0.010	GSR	Pearson Correlation		PPARGC1A	Pearson Correlation	0.409 *
	Sig.	0.962		Sig.			Sig.	0.047
CXCR8	Pearson Correlation	0.473 *	SOD1	Pearson Correlation		SIRT1	Pearson Correlation	0.560 **
	Sig.	0.023		Sig.			Sig.	0.004
IL6	Pearson Correlation	−0.060	SOD2	Pearson Correlation		TFAM	Pearson Correlation	0.326
	Sig.	0.781		Sig.			Sig.	0.120
CXCL8	Pearson Correlation	0.374	SIRT3	Pearson Correlation		FIS1	Pearson Correlation	0.294
	Sig.	0.071		Sig.			Sig.	0.163
TGFB	Pearson Correlation	0.169	NFE2L2	Pearson Correlation		OMA1	Pearson Correlation	0.446 *
	Sig.	0.430		Sig.			Sig.	0.029
TNF	Pearson Correlation	0.326	UCP2	Pearson Correlation		OPA1	Pearson Correlation	0.256
	Sig.	0.120		Sig.			Sig.	0.262
PTGS2	Pearson Correlation	−0.155	SLC25A14	Pearson Correlation		* The correlation is significant at $p < 0.05$		
	Sig.	0.502		Sig.		** The correlation is significant at $p < 0.01$		

4. Discussion

In this study, the effects of obesity related-inflammation on mitochondrial functionality in breast cancer cell lines and breast tumors, focusing on estrogen receptors ratio, were analyzed. Moreover, invasiveness in this situation was also analyzed. We demonstrated that ERβ, in an inflammatory and obesity condition, maintains mitochondrial functionality and avoids invasiveness in breast cancer cell lines. Moreover, we found a strong correlation between interleukin-6 receptor gene expression and inflammation, mitochondrial functionality, and oxidative stress markers, as well as with estrogen receptor beta, in breast cancer human samples in different BMI situations.

Obesity stimulates the adipose tissues to release inflammatory mediators such as tumor necrosis factor α and interleukin 6, predisposing them to a proinflammatory state and oxidative stress [14,39]. These signals also stimulate the release of inflammatory mediators by breast cancer cells, creating an autocrine feedback loop [40]. To start the study, we confirmed the effects of treatment on inflammatory genes expression. ELIT treatment (17β-estradiol (10 nM), Leptin (100 ng/mL), IL6 (50 ng/mL), and TNFα (10 ng/mL)), simulates circulating hormonal conditions in a postmenopausal obese woman and inducing a high increase in proinflammatory expression genes. Furthermore, it is worth noting that, in breast tumors in different BMI situations, inflammatory genes expression was positively correlated with interleukin-6 receptor gene expression. These results are in concordance with other studies where proinflammatory markers are studied in postmenopausal breast cancer patients [32].

Previous studies in our group and other research groups have shown the independent effects of 17β-estradiol or leptin on oxidative stress and mitochondrial biogenesis and dynamics [9,41–44], but never in an associated-inflammation situation. It is well known that ERα is the predominant estrogen receptor found in the MCF7 cell line, and it responds to estrogens by increasing proliferation, while, if ERβ is overexpressed in these cells, the proliferative effect of estrogens is inhibited [9]. Therefore, the response to estrogens in breast cancer not only depends on the concentration of estrogens in the cellular environment but also depends on the ERα/ERβ ratio presented by cells [9,43].

In this study, MCF7 and T47D cell lines have been treated with an ELIT inflammatory cocktail with the aim of generates an inflammation situation (IL6 and TNFα) in the presence of Leptin and 17β-estradiol levels, simulating the physiological condition of postmenopausal obese women. In this way, an increase in reactive oxygen species has been observed in both cell lines; however, in the T47D cell line, a lower increase in the production of hydrogen peroxide and levels of superoxide anion was observed. These data make more sense when studying the oxidative damage present in both cell lines, where a significant increase was only observed in the MCF7 cell line. On the other hand, it seems that the T47D treated cell line does not present differences compared to the control group, despite presenting higher levels of H_2O_2 and superoxide anion, as similarly described by some authors [41,45,46].

Oxidative stress is generated in breast cancer, and in many other pathologies, in two main ways. The first one, as observed in the high ERα/ERβ ratio MCF7 cell line after ELIt treatment, is a decrease in the expression of the antioxidant enzymes. It was observed that all antioxidant enzymes had decreased their gene or protein expression in the MCF7 cell line, except the SOD2 enzyme that had increased expression in both MCF7 and T47D cell lines. It has been described that SOD2 plays a role as a free radical detector, increasing gene and/or protein expression in order to alleviate oxidative damage [47]. Moreover, *NFE2L2*, a transcription factor that controls mainly antioxidant enzymes expression, and glutathione reductase, which recycles glutathione, was also increased, avoiding a high increase in free radical levels in low ERα/ERβ ratio T47D cell line. In addition, it should be noted that those tumors with a high correlation between *IL6R* and *ESR2* gene expression presented an increase in antioxidant enzymes, thus could palliate levels of free radicals, which can ultimately diminish oxidative damage.

The other pathway that increases oxidative stress is the poor maintenance of a functional mitochondrial pool [48]. To achieve good maintenance, two highly coordinated processes, such as mitochondrial biogenesis and dynamics, are very important [49,50]. In the MCF7 cell line, where ERα predominates, the protein levels of the OXPHOS system are decreased after ELIT treatment, as well as the amount of inner mitochondrial membrane (cardiolipin levels), required for the activity of complexes I, III, and IV, and plays an important role in mitochondrial biogenesis [51,52]. Likewise, both mitochondrial biogenesis and mitochondrial dynamics are diminished by ELIT treatment in the MCF7 cell line. Therefore, neither mitochondria are generated, nor are those that do not function properly eliminated; thus, the production of free radicals and, consequently, the oxidative damage is increased. Mitochondrial dysfunction has been postulated as one of the hallmarks of cancer, increasing free radicals and decrease energetic efficiency [19]. If this situation is accompanied by low levels of antioxidant enzymes, oxidative damage will be higher, as observed in the MCF7 cell line.

However, the T47D cell line, with high levels of ERβ, had mitochondrial dynamics elevated, maintaining a more functional pool of mitochondria. In the case of mitochondrial biogenesis, it should be noted that there are two genes that presented a very significant elevation, estrogen-related receptor alpha (*ESRRA*) and Twinkle (*TWNK*). ESRRA belongs to a superfamily of nuclear receptors independent of estrogen activation. Its expression is induced and activated by PGC1α, and the two factors together are capable of binding to response elements and promoting the initiation and elongation of the transcription of metabolic and mitochondrial target genes [53]. Twinkle is the mitochondrial helicase encoded by nuclear DNA that acts during mtDNA replication, and its overexpression has been shown to be related to increased mitochondrial biogenesis [54]. In fact, ESRRA should be considered as a possible factor responsible for the difference in the rate of mitochondrial biogenesis between both MCF7 and T47D cell lines in an obese-related inflammation situation. As shown in the results, the increased expression of the nuclear receptor ESRRA in the T47D cell line could be responsible for the maintenance of the mitochondrial pool in the cell. Furthermore, it has been observed that the STAT3 signaling pathway, activated by leptin and IL6, upregulates the expression of ESRRA [55].

Taking into account that patients with tumors with low ERα/ERβ ratios have a worse response to chemical agents oxidative damage inductors [56], we observed that breast tumors with a high correlation between IL6R and ESR2 in obese patients had genes related to mitochondrial biogenesis, and dynamics increased gene expression, which could be an attempt to increase the number of mitochondria due to the inflammatory state and oxidative stress generated over a long period of time, such happens in other situations like aging [57,58], but maintaining a functional mitochondrial pool. In addition, it is worthy to note that both interleukin 6 and 8 gene expression correlated negatively with estrogen receptor alpha gene expression, whereas both interleukin 6 and 8 receptors correlated positively with estrogen receptor beta gene expression. These results remark the importance of more studies are needed in order to better understand estrogen receptors subtypes' role in breast cancer.

In addition to mitochondrial biogenesis and dynamics, our study reveals an oxygen consumption rate decreased in the MCF7 cell line that supports all the results commented above. It is worth noting that ELIT treatment decreases basal respiration, maximal respiratory capacity, ATP-linked respiration as much as proton leak in the MCF7 cell line. Proton leak decrease is in concordance with low uncoupling proteins mRNA expression levels found in the MCF7 cell line after ELIT treatment. UCPs, which promote proton leak across the inner mitochondrial membrane, have emerged as essential regulators of mitochondrial membrane potential, respiratory activity, and ROS generation [59]. As seen in our results, the mitochondrial network was modified in the MCF7 cell line after treatment. Mitochondria appear more fragmented and circular than fused in the ELIT-treated MCF7 cell line but not in the T47D cell line. In fact, it seems that estrogen receptor beta could maintain mitochondrial network as well as oxygen consumption rate. Moreover, the T47D cell line showed an increase in the reserve capacity of mitochondria, a fact that supports good mitochondrial pool maintenance [48,60,61].

Oxidative stress, inflammation, and poor mitochondrial network observed in MCF7 cell lines after ELIT treatment leads to a worse situation, and some authors have described a relationship between this situation and metastasis [62]. Our results are in concordance with this idea due to increased motility and modification of invasiveness markers presented in the MCF7 cell line but not in the T47D cell line after treatment. In fact, the T47D cell line, with a better mitochondrial profile, including biogenesis, dynamics, and mitochondrial network, showed high levels of Cadherin-E that could simulate an epithelial-like phenotype, as suggested by some authors [63,64].

As mentioned above, the MCF7 cell line has a higher ERα/ERβ ratio, and the T47D cell line has a lower ratio. ELIT-treatment decreases estrogen receptor alpha mRNA expression in both MCF7 and T47D cell lines, as shown in results. However, estrogen receptor beta mRNA expression only decreases in the MCF7 cell line but not in the T47D cell line. This fact supports that all the results showed in our study could be through estrogen receptor beta T47D cell lines maintenance, giving to this receptor subtype the protective role that previously some studies have been described [9,39].

5. Conclusions

The presence of estrogen receptor beta allows maintaining a more functional mitochondrial pool, with active mitochondrial biogenesis and dynamics, which means less production of reactive oxygen species and better mitochondrial metabolism in an obesity-related inflammation condition. In addition, antioxidant enzymes are active, preventing oxidative damage and, at least in part, invasiveness.

This study could be important to remark the importance of estrogen receptor beta and mitochondria as an important organelle in the development and prognosis of breast cancer in obese patients. Likewise, more studies are necessary in order to clarify the estrogen receptor beta mechanism in breast cancer and establish it as a clinical biomarker, as is the estrogen receptor alpha.

Supplementary Materials: The following are available online at https://www.mdpi.com/article/10.3390/antiox10091371/s1, Figure S1: Representative bands and values of Inmunoblotting against GAPDH (approximately 37 kDa). Order of wells and Relative intensity values are above the blot image, control situation of each cell line set as 100; Figure S2: Representative bands and values of Inmunoblotting against 4HNE. Order of wells and Relative intensity values are above the blot image, control situation of each cell line set as 100; Figure S3: Representative bands and values of Inmunoblotting against SOD1 (Approximately 16 kDa). Order of wells and Relative intensity values are above the blot image, control situation of each cell line set as 100; Figure S4: Representative bands and values of Inmunoblotting against SOD2 (Approximately 25 kDa). Order of wells and Relative intensity values are above the blot image, control situation of each cell line set as 100; Figure S5: Representative bands and values of Inmunoblotting against GRd (Approximately 56 kDa). Order of wells and Relative intensity values are above the blot image, control situation of each cell line set as 100; Figure S6: Representative bands and values of Inmunoblotting against CAT (Approximately 60 kDa). Order of wells and Relative intensity values are above the blot image, control situation of each cell line set as 100; Figure S7: Representative bands and values of Inmunoblotting against OXPHOS (Cocktail antibody was used). Order of wells are above the blot image and Relative intensity values in table, control situation of each cell line set as 100; Figure S8: Representative bands and values of Inmunoblotting against COXIV; Figure S9: Illustrations of Wound Healing in both MCF7 and T47D cell lines after 0, 12 and 24 h of ELIT treatment.

Author Contributions: Conceptualization: D.G.P., J.O., P.R. and J.S.-S.; Methodology: T.M.-B., N.C. and J.S.-S.; Validation: D.G.P., T.M.-B., N.C., J.O., P.R. and J.S.-S.; Formal Analysis: D.G.P., J.O., P.R., T.M.-B. and J.S.-S.; Investigation: J.S.-S., D.G.P., T.M.-B. and P.R.; Data Curation: D.G.P., J.O., P.R., T.M.-B. and J.S.-S.; Writing—Original Draft Preparation: T.M.-B. and J.S.-S.; Writing—Review and Editing: D.G.P., J.O., P.R., T.M.-B. and J.S.-S.; Supervision: P.R. and J.S.-S.; Project Administration:

D.G.P., J.O., P.R. and J.S.-S.; Funding Acquisition: J.S.-S. and P.R. All authors have read and agreed to the published version of the manuscript.

Funding: This research was funded by PRIMUS program from the Balearic Islands Health Research Institute (IdISBa), grant number "PRI18/04", and the Fundraising Project: *Proyecto Investigación en Cáncer de Mama (InCaM), Fundació Universitat Empresa de les Illes Balears (FUEIB)—Oficina de Fundraising. Feim Camí per Viure—Santa Maria del Camí.* T. Martinez-Bernabe was funded by "Ayuda Formación Personal Investigador FPI 2020" grant from "Consejería de Educación, Universidad e Investigación del Gobierno de las Illes Balears".

Institutional Review Board Statement: The study was conducted according to the guidelines of the Declaration of Helsinki and approved by the "Comité de ética de la Investigación de las Islas Baleares (IB3888/18 PI, 10 June 2019).

Informed Consent Statement: Informed consent was obtained from all subjects involved in the study. Written informed consent has been obtained from the patient(s) to publish this paper.

Data Availability Statement: Data is contained within the article and Supplementary Materials.

Acknowledgments: We thank Laura Puigrós for her technician work inside "SOIB-Joves Qualificats" program. The authors thank Guillem Ramis from the coelomic unit (IUNICS—UIB) for assistance and the acquisition of the confocal microscopy images. This work was published thanks to funding from "LIBERI 2021" program from the Balearic Islands Health Research Institute (IdISBa).

Conflicts of Interest: The authors declare no conflict of interest.

References

1. Harbeck, N.; Penault-Llorca, F.; Cortés, J.; Gnant, M.; Houssami, N.; Poortmans, P.; Ruddy, K.; Tsang, J.; Cardoso, F. Breast cancer. *Nat. Rev. Dis. Primers* **2019**, *5*, 66. [CrossRef]
2. Tao, Z.Q.; Shi, A.; Lu, C.; Song, T.; Zhang, Z.; Zhao, J. Breast Cancer: Epidemiology and Etiology. *Cell Biochem. Biophys.* **2015**, *72*, 333–338. [CrossRef]
3. Calle, E.E.; Kaaks, R. Overweight, obesity and cancer: Epidemiological evidence and proposed mechanisms. *Nat. Rev. Cancer* **2004**, *4*, 579–591. [CrossRef]
4. KA, B. Metabolic pathways in obesity-related breast cancer. *Nat. Rev. Endocrinol.* **2021**, *17*, 350–363. [CrossRef]
5. Himbert, C.; Delphan, M.; Scherer, D.; Bowers, L.W.; Hursting, S.; Ulrich, C.M. Signals from the Adipose Microenvironment and the Obesity-Cancer Link—A Systematic Review. *Cancer Prev. Res.* **2017**, *10*, 494. [CrossRef] [PubMed]
6. Klinge, C.M. Estrogenic control of mitochondrial function. *Redox Biol.* **2020**, *31*, 101435. [CrossRef]
7. Roberts, D.L.; Dive, C.; Renehan, A.G. Biological mechanisms linking obesity and cancer risk: New perspectives. *Annu. Rev. Med.* **2010**, *61*, 301–316. [CrossRef] [PubMed]
8. Liao, T.L.; Tzeng, C.R.; Yu, C.L.; Wang, Y.P.; Kao, S.H. Estrogen receptor-β in mitochondria: Implications for mitochondrial bioenergetics and tumorigenesis. *Ann. N. Y. Acad. Sci.* **2015**, *1350*, 52–60. [CrossRef] [PubMed]
9. Sastre-Serra, J.; Nadal-Serrano, M.; Pons, D.G.; Roca, P.; Oliver, J. The over-expression of ERbeta modifies estradiol effects on mitochondrial dynamics in breast cancer cell line. *Int. J. Biochem. Cell Biol.* **2013**, *45*, 1509–1515. [CrossRef] [PubMed]
10. Simone, V.; D'Avenia, M.; Argentiero, A.; Felici, C.; Rizzo, F.M.; De Pergola, G.; Silvestris, F. Obesity and Breast Cancer: Molecular Interconnections and Potential Clinical Applications. *Oncologist* **2016**, *21*, 404. [CrossRef]
11. Shouman, S.; Wagih, M.; Kamel, M. Leptin influences estrogen metabolism and increases DNA adduct formation in breast cancer cells. *Cancer Biol. Med.* **2016**, *13*, 505. [CrossRef]
12. Postow, M.A.; Callahan, M.K.; Wolchok, J.D. Immune checkpoint blockade in cancer therapy. *J. Clin. Oncol.* **2015**, *33*, 1974–1982. [CrossRef] [PubMed]
13. Zhou, X.; Yang, S.; Wang, Z.; Feng, X.; Liu, P.; Lv, X.-B.; Li, F.; Yu, F.-X.; Sun, Y.; Yuan, H.; et al. Estrogen regulates Hippo signaling via GPER in breast cancer. *J. Clin. Investig.* **2015**, *125*, 2123–2135. [CrossRef] [PubMed]
14. Macciò, A.; Madeddu, C. Obesity, inflammation, and postmenopausal breast cancer: Therapeutic implications. *Sci. World J.* **2011**, *11*, 2020–2036. [CrossRef] [PubMed]
15. Bhat, H.K.; Calaf, G.; Hei, T.K.; Loya, T.; Vadgama, J.V. Critical role of oxidative stress in estrogen-induced carcinogenesis. *Proc. Natl. Acad. Sci. USA* **2003**, *100*, 3913–3918. [CrossRef]
16. Goldberg, J.E.; Schwertfeger, K.L. Proinflammatory Cytokines in Breast Cancer: Mechanisms of Action and Potential Targets for Therapeutics. *Curr. Drug Targets* **2010**, *11*, 1133–1146. [CrossRef]
17. Choi, J.; Cha, Y.J.; Koo, J.S. Adipocyte biology in breast cancer: From silent bystander to active facilitator. *Prog. Lipid Res.* **2018**, *69*, 11–20. [CrossRef]
18. Crusz, S.M.; Balkwill, F.R. Inflammation and cancer: Advances and new agents. *Nat. Rev. Clin. Oncol.* **2015**, *12*, 584–596. [CrossRef]
19. Hanahan, D.; Weinberg, R.A. Hallmarks of cancer: The next generation. *Cell* **2011**, *144*, 646–674. [CrossRef]

20. Pedram, A.; Razandi, M.; Wallace, D.C.; Levin, E.R. Functional estrogen receptors in the mitochondria of breast cancer cells. *Mol. Biol. Cell* **2006**, *17*, 2125–2137. [CrossRef]
21. Fariss, M.W.; Chan, C.B.; Patel, M.; Van Houten, B.; Orrenius, S. Role of mitochondria in toxic oxidative stress. *Mol. Interv.* **2005**, *5*, 94–111. [CrossRef]
22. Murphy, M.P. How mitochondria produce reactive oxygen species. *Biochem. J.* **2009**, *417*, 1–13. [CrossRef]
23. Moloney, J.N.; Cotter, T.G. ROS signalling in the biology of cancer. *Semin. Cell Dev. Biol.* **2018**, *80*, 50–64. [CrossRef]
24. Justo, R.; Boada, J.; Frontera, M.; Oliver, J.; Bermúdez, J.; Gianotti, M. Gender dimorphism in rat liver mitochondrial oxidative metabolism and biogenesis. *Am. J. Physiol. Physiol.* **2005**, *289*, C372–C378. [CrossRef] [PubMed]
25. Liesa, M.; Palacín, M.; Zorzano, A. Mitochondrial dynamics in mammalian health and disease. *Physiol. Rev.* **2009**, *89*, 799–845. [CrossRef] [PubMed]
26. Sastre-Serra, J.; Nadal-Serrano, M.; Pons, D.G.; Valle, A.; Oliver, J.; Roca, P. The effects of 17β-estradiol on mitochondrial biogenesis and function in breast cancer cell lines are dependent on the ERα/ERβ Ratio. *Cell. Physiol. Biochem.* **2012**, *29*, 261–268. [CrossRef] [PubMed]
27. Guha, M.; Srininvasan, S.; Ruthel, G.; Kashina, A.K.; Carstens, R.P.; Mendoza, A.; Khanna, C.; Van Winkle, T.; Avadhani, N.G. Mitochondrial retrograde signaling induces epithelial-mesenchymal transition and generates breast cancer stem cells. *Oncogene* **2014**, *33*, 5238–5250. [CrossRef] [PubMed]
28. Tsai, J.H.; Yang, J. Epithelial-mesenchymal plasticity in carcinoma metastasis. *Genes Dev.* **2013**, *27*, 2192–2206. [CrossRef] [PubMed]
29. Avtanski, D.; Garcia, A.; Caraballo, B.; Thangeswaran, P.; Marin, S.; Bianco, J.; Lavi, A.; Poretsky, L. Resistin induces breast cancer cells epithelial to mesenchymal transition (EMT) and stemness through both adenylyl cyclase-associated protein 1 (CAP1)-dependent and CAP1-independent mechanisms. *Cytokine* **2019**, *120*, 155–164. [CrossRef]
30. Olea-Flores, M.; Juárez-Cruz, J.C.; Mendoza-Catalán, M.A.; Padilla-Benavides, T.; Navarro-Tito, N. Signaling Pathways Induced by Leptin during Epithelial–Mesenchymal Transition in Breast Cancer. *Int. J. Mol. Sci.* **2018**, *19*, 3493. [CrossRef] [PubMed]
31. Valle, A.; Sastre-Serra, J.; Oliver, J.; Roca, P. Chronic leptin treatment sensitizes MCF-7 breast cancer cells to estrogen. *Cell. Physiol. Biochem.* **2011**, *28*, 823–832. [CrossRef] [PubMed]
32. Madeddu, C.; Gramignano, G.; Floris, C.; Murenu, G.; Sollai, G.; Macciò, A. Role of inflammation and oxidative stress in post-menopausal oestrogen-dependent breast cancer. *J. Cell. Mol. Med.* **2014**, *18*, 2519. [CrossRef] [PubMed]
33. Rose, D.P.; Vona-Davis, L. Biochemical and molecular mechanisms for the association between obesity, chronic Inflammation, and breast cancer. *BioFactors* **2014**, *40*, 1–12. [CrossRef]
34. Sastre-Serra, J.; Ahmiane, Y.; Roca, P.; Oliver, J.; Pons, D.G. Xanthohumol, a hop-derived prenylflavonoid present in beer, impairs mitochondrial functionality of SW620 colon cancer cells. *Int. J. Food Sci. Nutr.* **2019**, *70*, 396–404. [CrossRef] [PubMed]
35. Torrens-Mas, M.; Hernández-López, R.; Pons, D.G.; Roca, P.; Oliver, J.; Sastre-Serra, J. Sirtuin 3 silencing impairs mitochondrial biogenesis and metabolism in colon cancer cells. *Am. J. Physiol. Cell Physiol.* **2019**, *317*, C398–C404. [CrossRef]
36. Pons, D.G.; Vilanova-Llompart, J.; Gaya-Bover, A.; Alorda-Clara, M.; Oliver, J.; Roca, P.; Sastre-Serra, J. The phytoestrogen genistein affects inflammatory-related genes expression depending on the ERα/ERβ ratio in breast cancer cells. *Int. J. Food Sci. Nutr.* **2019**, *70*, 941–949. [CrossRef]
37. Pons, D.G.; Moran, C.; Alorda-Clara, M.; Oliver, J.; Roca, P.; Sastre-Serra, J. Micronutrients Selenomethionine and Selenocysteine Modulate the Redox Status of MCF-7 Breast Cancer Cells. *Nutrients* **2020**, *12*, 865. [CrossRef]
38. Nadal-Serrano, M.; Sastre-Serra, J.; Pons, D.G.; Miró, A.M.; Oliver, J.; Roca, P. The ERalpha/ERbeta ratio determines oxidative stress in breast cancer cell lines in response to 17Beta-estradiol. *J. Cell. Biochem.* **2012**, *113*, 3178–3182. [CrossRef]
39. Ellulu, M.S.; Patimah, I.; Khazaai, H.; Rahmat, A.; Abed, Y. Obesity & inflammation: The linking mechanism & the complications. *Arch. Med. Sci.* **2017**, *13*, 851–863. [CrossRef]
40. Hartman, Z.C.; Poage, G.M.; Den Hollander, P.; Tsimelzon, A.; Hill, J.; Panupinthu, N.; Zhang, Y.; Mazumdar, A.; Hilsenbeck, S.G.; Mills, G.B.; et al. Growth of triple-negative breast cancer cells relies upon coordinate autocrine expression of the proinflammatory cytokines IL-6 and IL-8. *Cancer Res.* **2013**, *73*, 3470–3480. [CrossRef]
41. Pons, D.G.; Torrens-Mas, M.; Nadal-Serrano, M.; Sastre-Serra, J.; Roca, P.; Oliver, J. The presence of Estrogen Receptor β modulates the response of breast cancer cells to therapeutic agents. *Int. J. Biochem. Cell Biol.* **2015**, *66*, 85–94. [CrossRef] [PubMed]
42. Sastre-Serra, J.; Valle, A.; Company, M.M.; Garau, I.; Oliver, J.; Roca, P. Estrogen down-regulates uncoupling proteins and increases oxidative stress in breast cancer. *Free Radic. Biol. Med.* **2010**, *48*, 506–512. [CrossRef] [PubMed]
43. Morani, A.; Warner, M.; Gustafsson, J.Å. Biological functions and clinical implications of oestrogen receptors alfa and beta in epithelial tissues. *J. Intern. Med.* **2008**, *264*, 128–142. [CrossRef] [PubMed]
44. Hartman, J.; Müller, P.; Foster, J.S.; Wimalasena, J.; Gustafsson, J.Å.; Ström, A. HES-1 inhibits 17β-estradiol and heregulin-β1-mediated upregulation of E2F-1. *Oncogene* **2004**, *23*, 8826–8833. [CrossRef] [PubMed]
45. Torrens-Mas, M.; Pons, D.G.; Sastre-Serra, J.; Oliver, J.; Roca, P. SIRT3 Silencing Sensitizes Breast Cancer Cells to Cytotoxic Treatments Through an Increment in ROS Production. *J. Cell. Biochem.* **2017**, *118*, 397–406. [CrossRef]
46. Abboud, M.M.; Al Awaida, W.; Alkhateeb, H.H.; Abu-Ayyad, A.N. Antitumor Action of Amygdalin on Human Breast Cancer Cells by Selective Sensitization to Oxidative Stress. *Nutr. Cancer* **2019**, *71*, 483–490. [CrossRef] [PubMed]
47. Hu, Y.; Rosen, D.G.; Zhou, Y.; Feng, L.; Yang, G.; Liu, J.; Huang, P. Mitochondrial manganese-superoxide dismutase expression in ovarian cancer: Role in cell proliferation and response to oxidative stress. *J. Biol. Chem.* **2005**, *280*, 39485–39492. [CrossRef]

48. Perron, N.R.; Beeson, C.; Rohrer, B. Early alterations in mitochondrial reserve capacity; a means to predict subsequent photoreceptor cell death. *J. Bioenerg. Biomembr.* **2013**, *45*, 101–109. [CrossRef]
49. Rodrigues, T.; Ferraz, L.S. Therapeutic potential of targeting mitochondrial dynamics in cancer. *Biochem. Pharmacol.* **2020**, *182*, 114282. [CrossRef]
50. Blanquer-Rossellõ, M.M.; Santandreu, F.M.; Oliver, J.; Roca, P.; Valle, A. Leptin Modulates Mitochondrial Function, Dynamics and Biogenesis in MCF-7 Cells. *J. Cell. Biochem.* **2015**, *116*, 2039–2048. [CrossRef]
51. Genova, M.L.; Lenaz, G. Functional role of mitochondrial respiratory supercomplexes. *Biochim. Biophys. Acta BBA Bioenerg.* **2014**, *1837*, 427–443. [CrossRef] [PubMed]
52. Paradies, G.; Paradies, V.; Ruggiero, F.M.; Petrosillo, G. Role of Cardiolipin in Mitochondrial Function and Dynamics in Health and Disease: Molecular and Pharmacological Aspects. *Cells* **2019**, *8*, 728. [CrossRef] [PubMed]
53. Sanchis-Gomar, F.; Garcia-Gimenez, J.; Gomez-Cabrera, M.; Pallardo, F. Mitochondrial Biogenesis in Health and Disease. Molecular and Therapeutic Approaches. *Curr. Pharm. Des.* **2014**, *20*, 5619–5633. [CrossRef]
54. Ikeda, M.; Ide, T.; Fujino, T.; Arai, S.; Saku, K.; Kakino, T.; Tyynismaa, H.; Yamasaki, T.; Yamada, K.I.; Kang, D.; et al. Overexpression of TFAM or twinkle increases mtDNA copy number and facilitates cardioprotection associated with limited mitochondrial oxidative stress. *PLoS ONE* **2015**, *10*, e0119687. [CrossRef] [PubMed]
55. Ma, J.H.; Qi, J.; Lin, S.Q.; Zhang, C.Y.; Liu, F.Y.; Xie, W.D.; Li, X. STAT3 targets ERR-α to promote epithelial-mesenchymal transition, migration, and invasion in triple-negative breast cancer cells. *Mol. Cancer Res.* **2019**, *17*, 2184–2195. [CrossRef]
56. Sastre-Serra, J.; Nadal-Serrano, M.; Pons, D.G.; Valle, A.; Garau, I.; García-Bonafé, M.; Oliver, J.; Roca, P. The oxidative stress in breast tumors of postmenopausal women is ERα/ERβ ratio dependent. *Free Radic. Biol. Med.* **2013**, *61*, 11–17. [CrossRef]
57. Guevara, R.; Santandreu, F.M.; Valle, A.; Gianotti, M.; Oliver, J.; Roca, P. Sex-dependent differences in aged rat brain mitochondrial function and oxidative stress. *Free Radic. Biol. Med.* **2009**, *46*, 169–175. [CrossRef]
58. Valle, A.; Guevara, R.; García-Palmer, F.J.; Roca, P.; Oliver, J. Caloric restriction retards the age-related decline in mitochondrial function of brown adipose tissue. *Rejuvenation Res.* **2008**, *11*, 597–604. [CrossRef]
59. Akhmedov, A.T.; Rybin, V.; Marín-García, J. Mitochondrial oxidative metabolism and uncoupling proteins in the failing heart. *Hear. Fail. Rev.* **2014**, *20*, 227–249. [CrossRef]
60. Rambold, A.S.; Kostelecky, B.; Lippincott-Schwartz, J. Together we are stronger: Fusion protects mitochondria from autophagosomal degradation. *Autophagy* **2011**, *7*, 1568. [CrossRef]
61. Rambold, A.S.; Kostelecky, B.; Elia, N.; Lippincott-Schwartz, J. Tubular network formation protects mitochondria from autophagosomal degradation during nutrient starvation. *Proc. Natl. Acad. Sci. USA* **2011**, *108*, 10190. [CrossRef] [PubMed]
62. Pani, G.; Galeotti, T.; Chiarugi, P. Metastasis: Cancer cell's escape from oxidative stress. *Cancer Metastasis Rev.* **2010**, *29*, 351–378. [CrossRef] [PubMed]
63. Adams, B.D.; Claffey, K.P.; White, B.A. Argonaute-2 Expression Is Regulated by Epidermal Growth Factor Receptor and Mitogen-Activated Protein Kinase Signaling and Correlates with a Transformed Phenotype in Breast Cancer Cells. *Endocrinology* **2009**, *150*, 14. [CrossRef] [PubMed]
64. Wang, Y.Y.; Attané, C.; Milhas, D.; Dirat, B.; Dauvillier, S.; Guerard, A.; Gilhodes, J.; Lazar, I.; Alet, N.; Laurent, V.; et al. Mammary adipocytes stimulate breast cancer invasion through metabolic remodeling of tumor cells. *JCI Insight* **2017**, *2*. [CrossRef] [PubMed]

Article

Di-Tyrosine Crosslinking and *NOX4* Expression as Oxidative Pathological Markers in the Lungs of Patients with Idiopathic Pulmonary Fibrosis

Sanja Blaskovic [1,2], Yves Donati [1,2], Isabelle Ruchonnet-Metrailler [1,2], Tamara Seredenina [2], Karl-Heinz Krause [2], Jean-Claude Pache [3], Dan Adler [4], Constance Barazzone-Argiroffo [1,2,*] and Vincent Jaquet [2,5]

1 Department of Pediatrics, Gynecology and Obstetrics, Children's Hospital, 1211 Geneva, Switzerland; Sanja.Blaskovic@unige.ch (S.B.); Yves.Donati@unige.ch (Y.D.); Isabelle.Ruchonnet-Metrailler@hcuge.ch (I.R.-M.)
2 Department of Pathology and Immunology, Medical School, University of Geneva, 1211 Geneva, Switzerland; tamara.seredenina@gmail.com (T.S.); Karl-Heinz.Krause@unige.ch (K.-H.K.); Vincent.Jaquet@unige.ch (V.J.)
3 Service of Clinical Pathology, Department of Pathology and Immunology, Medical School, University of Geneva, 1211 Geneva, Switzerland; Jean-Claude.Pache@hcuge.ch
4 Division of Pulmonary Diseases, Department of Medicine, Geneva University Hospitals, 1211 Geneva, Switzerland; Dan.Adler@hcuge.ch
5 READS Unit, Medical School, University of Geneva, 1211 Geneva, Switzerland
* Correspondence: Constance.Barazzone@hcuge.ch

Abstract: Idiopathic pulmonary fibrosis (IPF) is a noninflammatory progressive lung disease. Oxidative damage is a hallmark of IPF, but the sources and consequences of oxidant generation in the lungs are unclear. In this study, we addressed the link between the H_2O_2-generating enzyme NADPH oxidase 4 (*NOX4*) and di-tyrosine (DT), an oxidative post-translational modification in IPF lungs. We performed immunohistochemical staining for DT and *NOX4* in pulmonary tissue from patients with IPF and controls using validated antibodies. In the healthy lung, DT showed little or no staining and *NOX4* was mostly present in normal vascular endothelium. On the other hand, both markers were detected in several cell types in the IPF patients, including vascular smooth muscle cells and epithelium (bronchial cells and epithelial cells type II). The link between *NOX4* and DT was addressed in human fibroblasts deficient for *NOX4* activity (mutation in the *CYBA* gene). Induction of *NOX4* by Transforming growth factor beta 1 (TGFβ1) in fibroblasts led to moderate DT staining after the addition of a heme-containing peroxidase in control cells but not in the fibroblasts deficient for *NOX4* activity. Our data indicate that DT is a histological marker of IPF and that *NOX4* can generate a sufficient amount of H_2O_2 for DT formation in vitro.

Keywords: idiopathic pulmonary fibrosis; NADPH oxidase; *NOX4*; di-tyrosine; immunohistochemistry; human lung tissue

Citation: Blaskovic, S.; Donati, Y.; Ruchonnet-Metrailler, I.; Seredenina, T.; Krause, K.-H.; Pache, J.-C.; Adler, D.; Barazzone-Argiroffo, C.; Jaquet, V. Di-Tyrosine Crosslinking and *NOX4* Expression as Oxidative Pathological Markers in the Lungs of Patients with Idiopathic Pulmonary Fibrosis. *Antioxidants* **2021**, *10*, 1833. https://doi.org/10.3390/antiox10111833

Academic Editor: Stanley Omaye

Received: 18 October 2021
Accepted: 12 November 2021
Published: 18 November 2021

Publisher's Note: MDPI stays neutral with regard to jurisdictional claims in published maps and institutional affiliations.

1. Introduction

Idiopathic pulmonary fibrosis (IPF) is a progressive interstitial lung disease leading to lung tissue stiffness and fibrosis, decreased lung function and eventually respiratory failure. It is estimated that three million people are affected by IPF, worldwide. The median survival of IPF is estimated at 2–3 years after diagnosis [1,2]. IPF is characterized by the aberrant proliferation and activation of fibroblasts as well as the presence of myofibroblasts that secrete excessive collagen and extracellular matrix (ECM) leading to the scarring of lung tissue and destruction of the alveolar epithelium (reviewed in [3]). Lung remodeling is thought to occur following repetitive injuries leading to alveolar epithelial cells (AEC) apoptosis, proliferation of fibroblasts and Transforming growth factor beta 1 (TGFβ1)-induced myofibroblast differentiation as well as endothelial dysfunction. However, knowledge of the underlying molecular mechanisms leading to IPF is scarce and the clinically approved antifibrotic agents Nintedanib and Pirfenidone provide only limited benefit and are associated with adverse effects. Characterization of novel alternative pathogenic pathways is strongly needed to improve our knowledge on this devastating disease and develop novel classes of therapeutics.

A key hallmark of IPF pathogenesis is the presence of high levels of reactive oxygen species (ROS) in the lungs (reviewed in [4]). Although, the sources of ROS can be multiple, *NOX4* represents a key enzymatic source of ROS in IPF lungs and has been associated to fibrogenic properties in the lungs [5,6]. *NOX4* is a member of the family of nicotinamide adenine dinucleotide phosphate (NADPH) oxidases, whose sole function is the generation of ROS [7]. *NOX4* expression is strongly upregulated in fibroblasts from IPF patients and following treatment with the fibrogenic cytokine TGFβ1 leading to extracellular H_2O_2 production [5]. In addition, *NOX4*-deficient mice show decreased pulmonary fibrosis and AEC cell death in a mouse model of bleomycin-induced lung fibrosis [6]. *NOX4* inhibitors are currently investigated in clinical trials for IPF (NCT03865927). Nevertheless, the precise molecular mechanisms by which *NOX4*-derived ROS contribute to IPF pathology are not known. ROS are often considered as harmful molecules directly damaging cellular components, but ROS (and in particular H_2O_2) also exert a physiological function as signaling molecules [8]. In fact, H_2O_2 can serve as a substrate for enzymatic reaction catalyzed by peroxidases in various biological reactions, such as the generation of hypochlorous acid (HOCl) for germ killing by granulocytes and thyroid hormone synthesis [9]. One such specific reaction is called di-tyrosine (DT) crosslinking. In vitro, DT formation is catalyzed by a heme-containing peroxidase in the presence of H_2O_2 leading to the formation of a covalent carbon−carbon bond between two tyrosine residues of the same, or two adjacent proteins [10]. This phenomenon has physiological relevance and is well described in invertebrates. For instance, during egg fertilization of the sea urchin egg, the combined activity of a NADPH oxidase (Udx1) and a secreted ovoperoxidase is required for crosslinking of protein tyrosyl residues to create a stiffened ECM around the egg to avoid polyspermy [11]. However, DT is also present in different pathological conditions such as Alzheimer's disease [12], Parkinson's disease [13], atherosclerotic plaque formation [14], and IPF. A study by Pennathur and collaborators used mass spectrometric quantification of oxidative tyrosine modifications and documented a significant increase in the plasma of patients with interstitial lung disease and in the lungs of bleomycin-treated mice [15]. As of today, we are still unaware whether DT is formed in the lungs of IPF patients, and how its levels and localization contribute to IPF pathogenesis. A recent study has shown that the DT cross-linking of fibronectin, a key protein of the ECM, alters functional fibronectin characteristics in vitro [16]. In fact, ECM properties and glycogen crosslinking are known to be altered in IPF patients, thereby contributing to tissue stiffness [17]. In terms of mechanism, neither the H_2O_2 sources nor the potential peroxidases involved in DT formation in humans are known.

Thus, we aimed at addressing the localization and levels of DT as well as *NOX4* as the potential source of H_2O_2 involved in DT generation in the lungs of patients with IPF. We showed that normal lungs were virtually devoid of DT while lung tissue from IPF patients were characterized by a significant DT staining. *NOX4* was a sufficient source of H_2O_2 to form DT in vitro, but *NOX4* expression in IPF tissue did not clearly correlate with DT localization, especially in the smooth muscle cells of IPF lungs. Altogether, we conclude that DT and *NOX4* expression patterns are strongly affected in IPF while DT represents a key novel histopathological feature of IPF.

2. Materials and Methods

2.1. Antibodies, Enzymes and Reagents

The anti-di-tyrosine monoclonal antibody clone 1C3 was purchased from JaiCa, Fukuroi, Japan and the rabbit monoclonal antibody directed against the C-terminus of *NOX4* was a kind gift of Prof Jansen Dürr, University of Innsbruck (Innsbruck, Austria). Isotype negative control antibodies were the following: mouse IgG2a (DAKO, X0943) for

DT and recombinant rabbit IgG, monoclonal isotype control [EPR25A] (ACAM, ab172730) for *NOX4*. DAKO REAL Detection system peroxidase/DAB+, rabbit/mouse, K5001 was used for IHC on human tissues. Transforming Growth Factor-β1 human was from Sigma (T7039, Saint Louis, MO 63103, USA). Alexa Fluor 488 tyramide Reagent (TSA) was from Life Technology, (B40953, Eugene, Oregon, USA). Fetal calf serum was purchased from Chemie Brunschwig (K007310G). Fluorescent secondary antibody for DT immunofluorescence was goat anti-mouse IgG DyLight594 (Jackson ImmunoResearch, Cambridge, UK, 115-515-003). Panexin basic was purchased from Pan Biotech (P04-96090, Aidenbach, Germany). Phosphate-Buffered Saline, PBS (14190, Life Technologies, Paisley, Scotland, UK), Hepes 1M (15630, Life Technologies, Paisley, Scotland, UK) and Hanks' Balanced Salt Solution (HBSS), calcium, magnesium, HBSS++ (14025, Life Technologies, Paisley, Scotland, UK) were from GIBCO. L-tyrosine was from Carl Roth (1741.1, Karlsruhe, Germany). Normal goat serum (NGS) was from Invitrogen (10000C, Waltham, Massachusetts, USA), DAPI from Roche Diagnostics (10236276001, Basel, Switzerland), FluorSave Reagent (345789) and H_2O_2 (107209) were from Merck (Merck KGaA, Darmstadt, Germany).

2.2. Gene Expression in Fibroblasts

The human lung fibroblast cell line MRC5 (ATCC CCL-171) and primary human skin fibroblasts isolated from a healthy donor (hSFs) and a patient carrying mutation in *CYBA* gene (hSFp22 deficient) [18], were maintained in Dulbecco's modified Eagle's medium (DMEM) supplemented with fetal bovine serum (FBS, 10%), penicillin (100 U/mL), and streptomycin (100 μg/mL) at 37 °C in air with 5% CO_2. For differentiation experiments, the cells were kept in a serum-free medium for 24 h and treated with 2 ng TGF-β1 for 24 h and collected for RNA extraction.

RNA was extracted using RNeasy mini kit (Qiagen, Dusseldorf, Germany) according to manufacturer's protocol. Five hundred nanograms (500 ng) were used for cDNA synthesis using the PrimeScript RT reagent kit (Takara, Saint-Germain-en-Laye, France) following the manufacturer's instructions. Real-time PCR was performed using the SYBR green assay at the Genomics Platform, National Center of Competence in Research Frontiers in Genetics (Geneva, Switzerland), on a 7900HT SDS system (Applied Biosystems, Foster City, USA). The efficiency of each primer was assessed with serial dilutions of cDNA. Relative expression levels were calculated by normalization to the geometric mean of two housekeeping genes, β2-microglobulin and *GAPDH*, as previously described [19]. Normalized quantities are reported as mean ± SEM. The sequences of the primers used in this study are documented in Table 1.

Table 1. RT-qPCR primers used in the study.

Gene	Accession Number	Primers Forward 5′–3′	Primers Reverse 5′–3′
NOX4	NM_001143836.3	AACCGAACCAGCTCTCAGAA	TTGACCATTCGGATTTCCAT
ACTA2	NM_001141945.2	AATACTCTGTCTGGATCGGTGGCT	ACGAGTCAGAGCTTTGGCTAGGAA
FN1	NM_001306129.2	CGGTGGCTGTCAGTCAAAG	AAACCTCGGCTTCCTCCATAA
COL1A1	NM_000088.4	GAGGGCCAAGACGAAGACATC	CAGATCACGTCATCGCACAAC
GAPDH	NM_001256799.3	GCACAAGAGGAAGAGAGAGACC	AGGGGAGATTCAGTGTGGTG
B2M	NM_004048.4	TGCTCGCGCTACTCTCTCTTT	TCTGCTGGATGACGTGAGTAAAC

2.3. TSA Assay and DT Immunofluorescence

Fibroblasts were cultured on glass coverslips as described above. A starvation step was performed by adding new media with low FCS concentration (0.01% for HSF cells) or culture media supplemented with 10% panexin, and 0.1% FCS for MRC5 cells for 24 h. TGFβ1 was added in the starvation media (5 ng/mL) for 7 days with medium change every 2–3 days. On the day of the experiment a 100× dilution of TSA was added with or without HRP (5 U/mL) and with or without H_2O_2 (10 μM) in TRIO buffer (1 mM Hepes, 1 mM L-Tyrosine in HBSS++) for 1 h at RT. Cells were washed 2× with HBSS++ and fixed with 4% paraformaldehyde for 15 min at room temperature (RT) and washed with PBS.

Cells were blocked for 1 h at RT in blocking solution (PBS, 5% NGS, 0.3% TritonX-100). Cells were incubated with primary di-tyrosine antibody (dilution 1:400) at RT. After 1 h, cells were washed 3 times for 5 min with PBS at RT and a secondary anti-mouse antibody conjugated to D594 was added for 1 h (dilution 1:200). Cells were again washed 3 times for 5 min at RT and cell nuclei were stained with DAPI (1 μg/mL in PBS) 5 min at RT. Coverslips with cells were mounted with FluorSave and imaged with Axiocam Fluo (Carl Zeiss Microscopy GmbH, Jena, Germany).

2.4. Human Lung Samples

Human lung tissue samples used in this study (Table 2) were obtained from the department of Pathology, University Geneva hospital. Samples were collected either on biopsies or on necropsies. Clinical description was limited to definitive diagnosis made by the pathologist, the gender and the age of the patient. Samples were irreversibly anonymized, dated, processed and analyzed according to the Swiss Medical ethical guidelines and recommendations (Senate of the Swiss Academy of Medical Sciences, Basel, Switzerland, 23 May 2006). The study was approved by the regional research committee of the University Geneva Hospital (NAC10-052R; NAC 11-027R).

Table 2. Additional information of human lung tissue samples used in the study.

Patient Number	Pathology	Age at Diagnosis	Gender	Origin of Sample	Lung Region
1	IPF	77	M	Surgical biopsy	Right upper lobe
2	IPF (fibrotic hypersensitivity pneumoniae)	69	M	Surgical biopsy	Right upper lobe
3	IPF	74	M	Surgical biopsy	Left lower lobe
4	IPF	71	M	Surgical biopsy	Left lower lobe
5	IPF (pleural fibrosis)	75	N/A	Surgical biopsy	N/A
6	Papillary adenocarcinoma	62	M	Biopsy, adjacent tissue	Right middle lobe
7	Undifferentiated lung adenocarcinoma	69	M	Biopsy, adjacent tissue	Left upper lobe
8	Metastases from endometrial adenocarcinoma	56	F	Biopsy, adjacent tissue	Left lower lobe
9	Bullous emphysema	30	M	Biopsy, adjacent tissue	Right upper lobe
10	Acute bronchitis	85	M	Postmortem material	N/A

IPF: idiopathic pulmonary fibrosis; M: male; F: female; N/A: not available.

2.5. Immunohistochemical Staining

Paraffin-embedded sections (5 μm) of human control and IPF affected lungs were stained according to the cell signaling technology standard protocol [20] using DAKO Envision kit (K4011, Agilent Technologies, CA, USA) using DT and *NOX4* antibodies.

Briefly, paraffin-embedded samples were deparaffinized using xylene and 95–100% ethanol and subsequently hydrated in H_2O.

NOX4 labeling required pressure and heat-induced epitope retrieval (20 bar) in Tris–EDTA, pH 9.0 (10 mM/1 mM) buffer. DT did not require antigen retrieval. Endogenous peroxidases were blocked with DAKO peroxidase block solution. Both primary and secondary antibodies were diluted with DAKO antibody diluent. The anti-DT mouse monoclonal antibody and its mouse IgG2a isotype control were used at 0.25 μg/mL. The anti-*NOX4* rabbit monoclonal antibody and its rabbit monoclonal Ig isotype control were used at 2.5 μg/mL). We applied the primary antibodies for 1 h at RT. Finally, labeled polymer-horseradish peroxidase (HRP) anti-rabbit or anti-mouse (DAKO Envision system, Agilent Technologies, CA, USA) was used for 30 min at RT and the signal was visualized with diaminobenzidine (Envision system, Dako SA, Geneva, Switzerland). Sections were counterstained with hematoxylin (BioGnost, Zagreb, Croatia). Images were acquired using Axioscan Z1 microscope (Carl Zeiss Microscopy GmbH, Jena, Germany) and analyzed using the Definiens Developer XD™ 2.7 software (Definiens AG, Munich, Germany). Quantification of DT and *NOX4* was done in ImageJ (version 1.51, NIH, Rockville Pike, Bethesda, Maryland) on randomly selected 10 images/sample and statistical analysis was carried out with GraphPad PRISM (Prism 8.0.2, GraphPad Software, San Diego, CA, USA) using the Mann—Whitney nonparametric test. At least 10 adjacent sections were stained for each antibody. Control tissue consisted of adjacent healthy tissue for patients nos. 6–9 and from postmortem tissue for patient no. 10.

3. Results

Our study aimed at evaluating the pattern of expression of the H_2O_2-generating enzyme *NOX4* and the presence of oxidative post-translational modification DT in human lungs. The *NOX4* antibody used in this study was a rabbit monoclonal antibody directed against the intracellular C terminus of human *NOX4* produced in the laboratory of Professor Pidder Jansen-Dürr, Institute for Biomedical Ageing Research. University of Innsbruck. This antibody belongs to the small list of available validated *NOX4* antibodies, suitable for Western blot and IHC [21]. We previously documented the specificity of this antibody over other NOX isoforms [22] and its specificity for immunohistochemical detection in human tissue samples [23]. The mouse monoclonal DT antibody was generated using a DT conjugate as immunogen and selected for DT specificity [24]. In order to address the specificity of the DT antibody and the biological significance of *NOX4* in DT formation, we used an in vitro fibroblast experimental system inspired by the methodology described by Larios et al. [25].

3.1. Generation and Immunodetection of DT in Fibroblast Cell Culture

We used MRC5 lung fibroblasts treated with TGFβ1 to induce myofibroblast differentiation as evidenced by increased expression of *NOX4*, alpha smooth muscle actin (α-SMA) and the ECM proteins fibronectin and collagen 1a [16] (Figure 1A). The heme-containing peroxidase (HRP) and fluorescein-tyramide, a fluorescently labeled chemical analogue of tyrosine, were added to the fibroblast culture. Upon H_2O_2 addition, HRP converted fluorescein-tyramide into a highly-reactive tyramide radical that forms dimers with tyrosine residues allowing the detection of DT crosslinks formed with proteins of the ECM (Figure 1B). After washing unbound fluorescein tyramide, immunofluorescent staining using the anti-DT monoclonal antibody was performed. A striking colocalization of bound fluorescent tyramide and DT staining was observed (Figure 1B) confirming that the antibody recognized the DT formed between fluorescein tyramide and ECM proteins by the coordinated action of H_2O_2 and HRP.

3.2. NOX4-Derived H_2O_2 Is Sufficient for DT Generation in Skin Fibroblasts

In order to address whether *NOX4* was able to generate a sufficient amount of H_2O_2 upon activation by TGFβ1 to generate DT, we used primary skin fibroblasts isolated from a patient with a mutation in *CYBA*, the gene coding for p22phox [18], a necessary subunit for *NOX4* activity [26]. Similar to MRC5, *NOX4* was induced in both control and *CYBA*-deficient skin fibroblasts (Figure S1A). Interestingly, a weak overlapping staining was detected using both fluorescent tyramide and DT staining in the WT fibroblasts (Figure 1C, 1st and 2nd panel), which was completely absent in *CYBA*-deficient fibroblasts, whose *NOX4* is not active (Figure 1C, 3rd panel). This suggests that *NOX4* generates a sufficient amount of H_2O_2 for DT generation in differentiated skin myofibroblasts, provided that a heme containing peroxidase and tyrosine-rich proteins of the ECM are present.

3.3. Global DT Staining Is Significantly Increased in the Lungs of IPF Patients

In order to evaluate the presence of DT crosslinking and *NOX4* expression in the lungs of IPF patients, we performed immunohistochemistry (IHC) for both markers. Isotype control antibodies were used to confirm the specificity of the staining (Figure S2). We detected high levels of DT in the lungs of IPF patients, indicative of a wide dissemination of ECM proteins and oxidative events. On the other hand, DT staining was barely above background in control tissue (Figure 2A,B). *NOX4* staining was more localised than DT staining and *NOX4* levels varied widely between individuals. Altogether, no significant change of global *NOX4* levels was detected in the lungs of IPF patients compared to controls (Figure 2C,D).

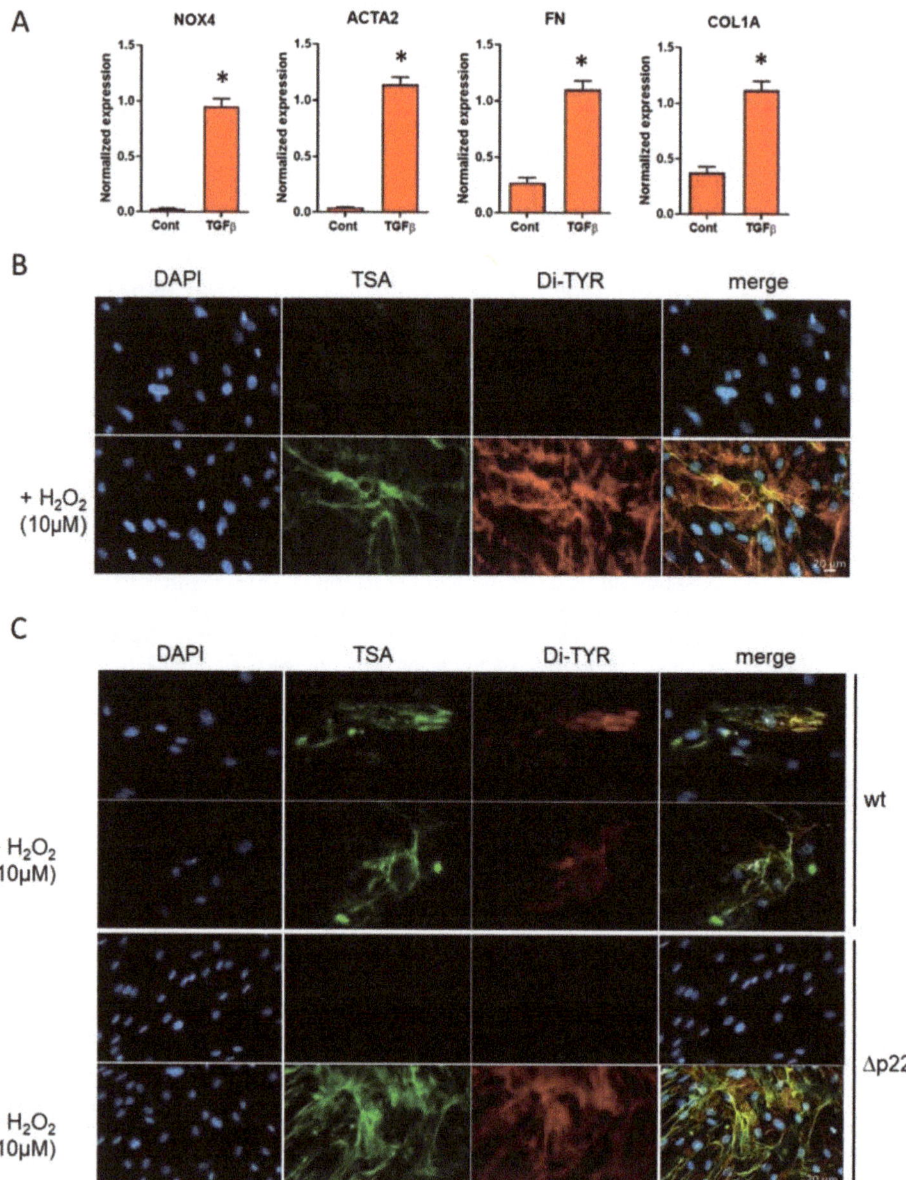

Figure 1. (**A**) TGFβ1-induced upregulation of the expression of *NOX4*, the myofibroblast marker alpha-SMA and proteins of the ECM in MRC5 human lung fibroblasts. $N = 4$, * $p < 0.05$ using Mann–Whitney nonparametric test. (**B**). Fluorescent microscopy of DT in TGF-β1-treated MRC5. Addition of H_2O_2 in presence of HRP cells induces crosslinking of the fluorescein tyramide (TSA, green) with the extracellular matrix network. The cross-links are detected by the DT antibody (red) and colocalize with crosslinked TSA (yellow). (**C**). Similar experiment with human skin fibroblasts from a healthy donor and a patient carrying a p22^{phox} mutation. Deletion of p22^{phox} completely abolishes the formation of DT in the absence of exogenous H_2O_2. Blue, DAPI staining of cell nuclei. *NOX4*: NADPH oxidase isoform 4; ACTA 2: Actin Alpha2, Smooth Muscle; FN: Fibronectin; COL1A: Collagen1A; DAPI: 4′,6-diamidino-2-phenylindole; TSA: Alexa Fluor 488 tyramide Reagent; Di-TYR: dityrosine; TGFβ: Transforming growth factor β; Cont: control; HRP: horseradish peroxidase.

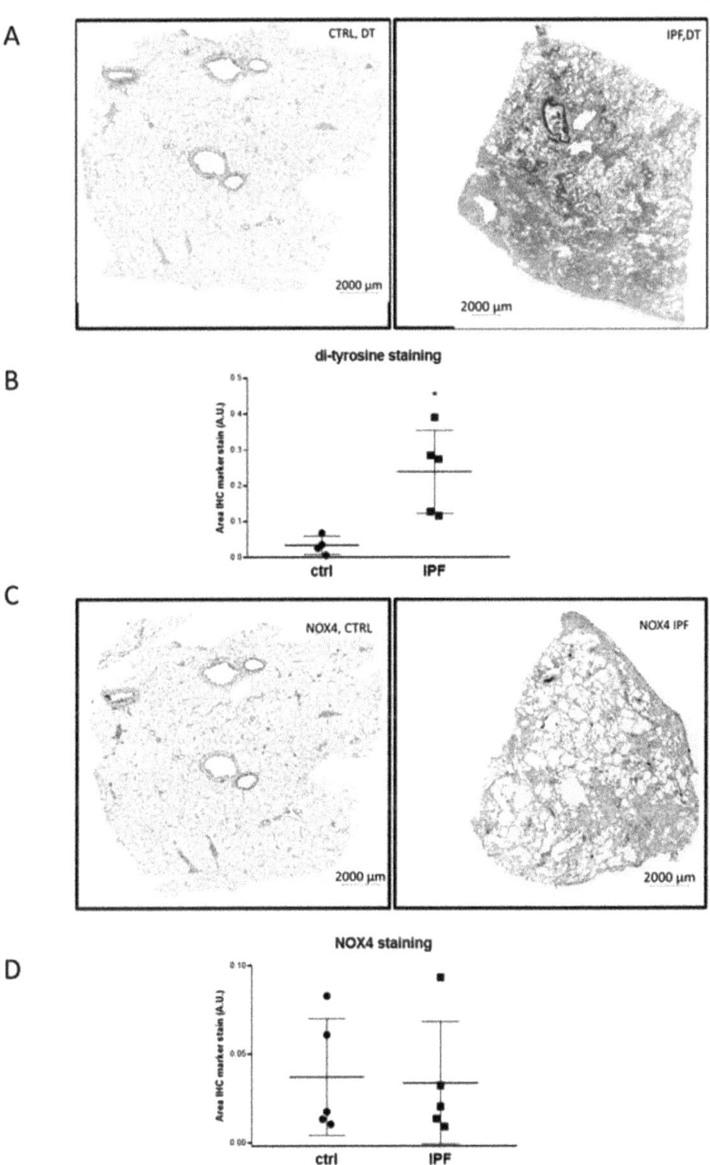

Figure 2. Representative images of DT (**A**) and *NOX4* (**C**) staining in the indicated patients are shown for lungs. Quantification of staining for DT (**B**) and *NOX4* (**D**). * $p < 0.05$ using Mann–Whitney nonparametric test. CTRL: control; IPF: idiopathic pulmonary fibrosis.

We further analyzed the lung material used in this study and focused on the cell types stained by *NOX4* and DT antibodies, namely smooth muscle cells, alveolar type II and bronchial epithelial cells and endothelial cells.

3.4. DT and NOX4 Are Present in Thickened Vessel Walls in the Lungs of IPF Patients

Vascular remodeling is a key pathological feature in the scarred areas of IPF lungs. It is characterized by the proliferation of pulmonary arterial smooth muscle cells in response to hypoxia leading to the thickening of the media, the smooth muscle layer around the blood vessels. Vascular smooth muscle cells in control lung tissues were negative for DT and *NOX4* (Figure 3A). However, fibrotic regions of IPF lungs showed abnormal vascular architecture and a strong staining for both DT and *NOX4* in pulmonary arterial smooth muscle cells (Figure 3B). At higher magnification, pulmonary arterial smooth muscle cells appeared highly positive for DT and to express high *NOX4* levels (insets in Figure 3B). While DT staining appeared cytoplasmic, *NOX4* staining was mainly found in perinuclear regions of smooth muscle cells. This indicates an upregulation of *NOX4* and DT in a specific vascular pathological modification of IPF and may suggest a potential role of *NOX4* in the generation of H_2O_2 needed for DT formation in proliferative pulmonary arterial smooth muscle cells.

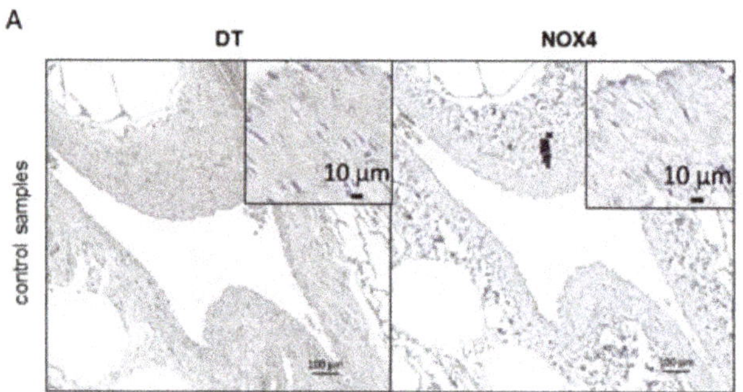

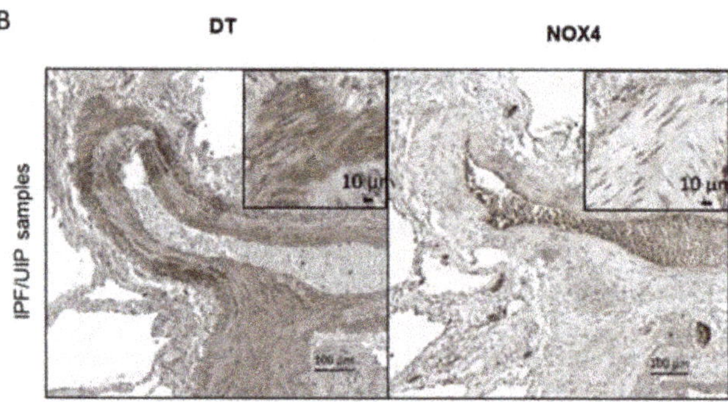

Figure 3. Representative images for DT ad *NOX4* staining of blood vessel smooth muscle cells in the control (**A**) and IPF samples (**B**) of human patients are shown. Inset shows a 10× higher magnification and depicts typical smooth muscle cell staining in these samples. DT and *NOX4* staining are shown in brown.

3.5. DT Staining and NOX4 Expression in Bronchi and Alveolar Epithelial Cells Type II of IPF Patients

The pulmonary alveolar epithelium is essential for the air–blood barrier function of the lungs. It is mainly composed of alveolar epithelial cells type I (AT1) and type II (AECII) cells. AT1 cells are abundant, large squamous cells involved in gas exchange, while AECII cells are smaller, cuboidal cells mostly involved in synthesis and secretion of the lung surfactant. Contrary to endothelial cells, smooth muscle cells, alveolar epithelial cells type II and macrophages which can be recognized by their tissue localization and morphology, it was impossible to correctly identify AT1 cells by IHC and hence we only looked at AECII cells.

AECII were not stained with DT and *NOX4* antibodies in control lung tissue. On the other hand, although the alveoli were severely damaged, both *NOX4* and DT antibodies decorated the remaining AECII cells in IPF (Figure 4B), consistent with a possible role of *NOX4* in DT formation in AECII.

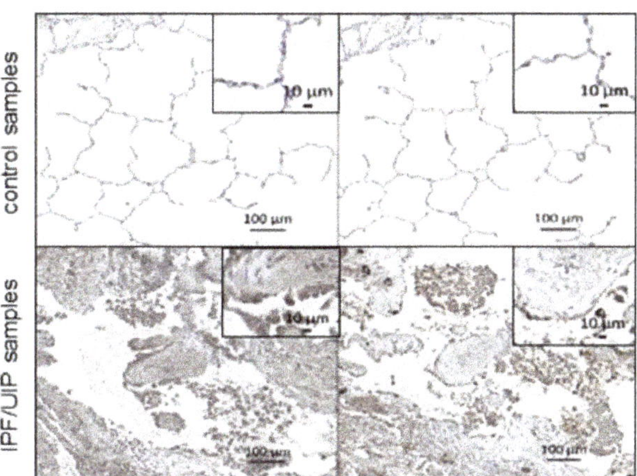

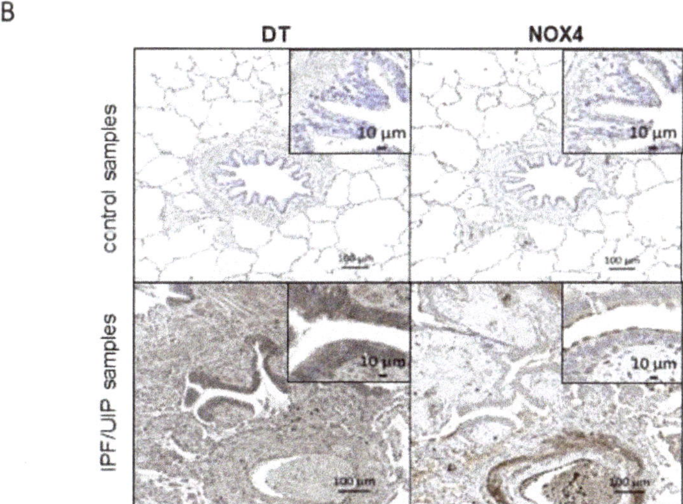

Figure 4. Representative images for DT and *NOX4* staining of AECII cells (**A**) and bronchi (**B**) in the control (upper panel) and IPF/UIP samples (lower panel) of human patients are shown. Insets shows a 10× higher magnification and depicts AECII (**A**) and ciliary bronchial cells (**B**) in these samples. DT and *NOX4* staining are shown in brown.

In normal AECII epithelial cells and bronchi, very little staining was observed for both markers (Figure 4A,B, upper panels). However, both AECII and bronchial cells were stained in IPF samples with *NOX4* and DT antibody with a different intensity, depending on the patient (Figure 4A,B, lower panels).

3.6. *Different Expression of DT and NOX4 in Vascular Endothelium in Normal and IPF Tissue*

Interestingly, while IPF tissue globally showed elevated levels of DT, only partial DT staining was observed in pulmonary endothelial cells (EC) (Figure 5). On the other hand, high *NOX4* levels were present in the endothelial cells of large vessels in both control and IPF tissue. *NOX4* staining was also detected in capillary endothelial cells in some samples (data not shown). Due to the high specificity of *NOX4* staining in endothelial cells and the fact that *NOX4* is generating H_2O_2 constitutively, the absence of DT staining in two out of four samples indicates that the conditions required for DT formation are not fulfilled in the intimal parts of the vessels.

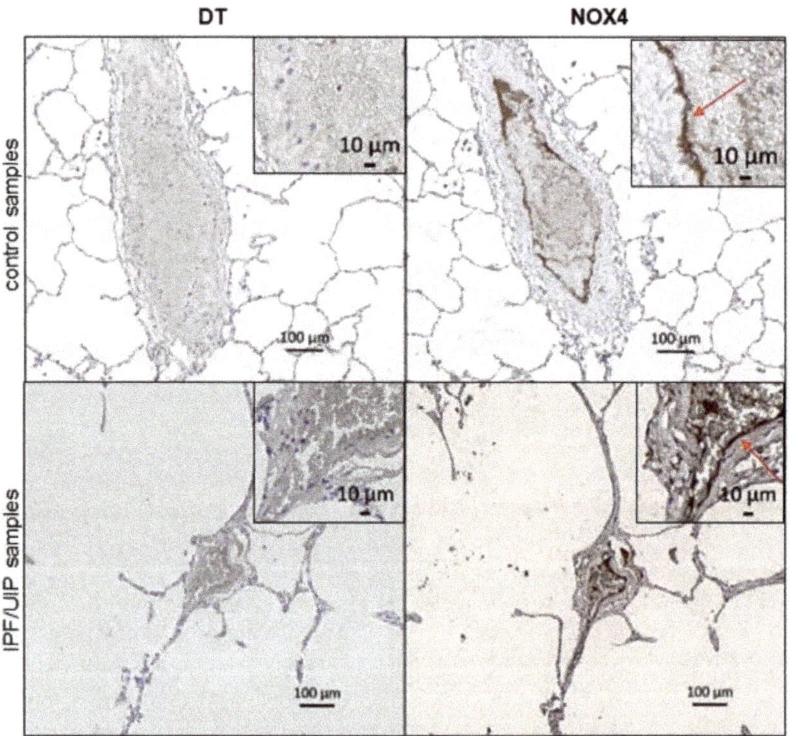

Figure 5. Representative images for DT and *NOX4* staining of blood vessel endothelial cells in the control (upper panels) and the IPF/UIP samples (lower panels) from human patients are shown. Inset shows a 10X higher magnification and depicts typical intimal endothelial region of these samples. DT and *NOX4* staining are shown in brown. The red arrows indicate the *NOX4* endothelial staining.

Altogether, DT staining was found in none of our control samples, but was present in SMC, AEC and AECII in all tested IPF patients (Figure 6A). *NOX4* staining was present in EC in both control and IPF samples, while AECII and SMC cells were mainly stained only in IPF patients (Figure 6B). Note that both antibodies stained macrophages in all samples (Figure 6A,B).

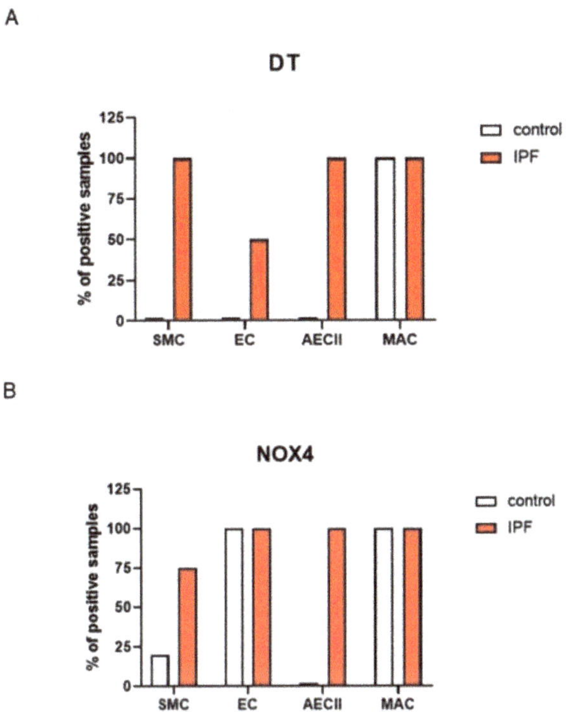

Figure 6. Histograms documenting the number of samples positive for DT and *NOX4* for each cellular subtype. All 4 IPF patients were positive for DT in SMC and AECII while none of the controls showed DT staining (**A**). All cell types (SMC, AECII and EC were stained for *NOX4* in IPF patients, while only EC were stained in the control (**B**). Macrophages were stained with both antibodies in both control and IPF (most likely not specific). SMC: smooth muscle cells, EC: endothelial cells, AECII: alveolar epithelial cells type 2, MAC macrophages.

4. Discussion

In this study, we showed that DT is a novel specific histopathological marker of IPF that can be detected with a specific antibody in lung tissue. We showed using differentiated myofibroblasts that *NOX4* generated a sufficient amount of H_2O_2 for DT formation in vitro and that *NOX4* and DT stained both proliferative pulmonary smooth muscle cells and epithelial bronchial and AECII cells. However, the endothelial cells of blood vessels were stained predominantly with *NOX4* and independently of the disease, though some IPF samples also showed the staining with DT.

Our data confirm that DT accumulation in the lungs is a pathological feature of IPF. Indeed, our data confirmed previous reports showing increased levels of proteins containing DT modifications, in particular the tyrosine-rich ECM protein, fibronectin, in the plasma of patients with fibrotic interstitial lung disease and the bleomycin-induced IPF mouse model [15,16].

Since DT bridges are covalent and most likely irreversible in vivo, a pharmacological strategy using small chemical molecules or anti-DT antibodies is unlikely to be successful. However, the well-described mechanism leading to DT formation provides a rational for future therapies for IPF. Such an approach would aim at inhibiting one or several of the components necessary for DT formation, namely, (i) abundant levels of proteins of the ECM, (ii) one or several sources of H_2O_2 and (iii) the presence of an active heme-containing peroxidase [16,24,25,27]. Existing therapeutic approaches in IPF that target the fundamental mechanisms of fibrosis leading to the deposition of ECM, including the proliferation and differentiation of fibroblasts are comprehensively reviewed in [28,29] and will not be developed here.

More than 40 enzymes are described as generating H_2O_2 when they are active [8]. Among them, NADPH oxidases are highly relevant sources of H_2O_2 in the context of DT formation in IPF lungs. NOX is known to coordinate their activity with specific heme-containing peroxidases for key biological functions in humans and other organisms, including host defense and thyroid hormone synthesis (reviewed in [9]). Several NOX isoforms are expressed in the lungs, but most studies point towards a contribution of *NOX4* in IPF. *NOX4* is upregulated in IPF lungs while *NOX4*-deficient mice are partially protected in a bleomycin mouse model of IPF [5,6]. *NOX4* expression is induced in fibroblasts concomitantly with ECM proteins following TGFβ1 treatment. Interestingly, high DT staining in dermal tissue observed during physiological wound healing is absent in *NOX4*-deficient mice [30]. The fact that p22^{phox}-deficient skin fibroblasts were not able to generate DT crosslinking with TSA, in spite of the presence of high levels of ECM and an active heme peroxidase, supports a role of *NOX4* in DT formation at least in the skin. No DT was observed in the sites with the highest *NOX4* levels in endothelial cells in healthy tissue, but we detected some DT staining in two out of four IPF samples. One possibility is that in endothelial cells, *NOX4* may generate H_2O_2 in the luminal side where ECM proteins or heme-containing peroxidases are absent. Alternatively, although the absence of colocalization does not exclude the diffusion of *NOX4*-derived H_2O_2 to other lung regions, *NOX4*-dependent DT formation might be limited to the fibrotic foci. Other sources of H_2O_2 might contribute to DT formation in IPF tissue. The dual oxidases DUOX1 and DUOX2 are expressed in airway epithelia where they generate H_2O_2 upon activation [31,32]. In particular DUOX1 and 2 may play a role in host defense in lung tissue through a coordinated function with the heme-containing lactoperoxidase [33]. DUOX1 was shown to be induced following lung injury, while DUOX1-deficient mice have an attenuated fibrotic phenotype in a bleomycin model of IPF [34]. However, to our knowledge, in spite of a strong rationale, the involvement of the DUOX1/lactoperoxidase coupling in DT formation in the lungs has not been studied yet.

5. Conclusions

In conclusion, our IHC study of postmortem and biopsy material describes elevated DT and specific DT and *NOX4* expression patterns as typical features of IPF human pathology. The validation of a specific antibody against DT allowing the detection of a novel histopathological marker of IPF pathology grants a great impact in future studies aiming at dissecting the pathogenic roles played by DT formation in IPF. Our study opens a number of important questions for further investigation. including the timing of DT formation in IPF lungs, the types of ECM proteins undergoing DT formation, the alternative sources of H_2O_2 and the necessary heme-containing peroxidases leading to high DT in IPF. Exploration of these questions might lead to the discovery of novel therapeutics for IPF and potentially other fibrotic diseases.

Supplementary Materials: The following are available online at https://www.mdpi.com/article/10.3390/antiox10111833/s1, Figure S1: *NOX4* expression in human skin fibroblasts. Figure S2: Isotype antibody controls.

Author Contributions: Conceptualization, C.B.-A., S.B., I.R.-M. and K.-H.K.; methodology, S.B., Y.D., T.S., J.-C.P. and D.A.; formal analysis, S.B., J.-C.P. and V.J.; writing—original draft preparation, V.J. and S.B.; writing—review and editing, C.B.-A. and I.R.-M.; funding acquisition, C.B.-A. All authors have read and agreed to the published version of the manuscript.

Funding: This work was supported by grants to C. Barazzone-Argiroffo from the Swiss National Science Foundation (Grant 310030-159500/1) and to K.H. Krause (Grant 31003A-179478) and by the Swiss Lung Liga and the OPO-Stiftung, Switzerland.

Institutional Review Board Statement: The study was conducted according to the guidelines of the Declaration of Helsinki, and approved by the regional research committee of the University of Geneva Hospital (Protocol no. 10-161R (NAC10-052R) and no. 11-087R (NAC 11-027R); date of approval 30 January 2011 and 10 June 2011).

Informed Consent Statement: Informed consent was obtained from all subjects involved in the study.

Data Availability Statement: Data is contained within the article or supplementary material.

Acknowledgments: The authors are grateful to the members of the Bioimaging Core Facility of the Faculty of Medicine, University of Geneva for their help in image analysis.

Conflicts of Interest: The authors declare no conflict of interest.

References

1. Wolters, P.J.; Collard, H.R.; Jones, K.D. Pathogenesis of Idiopathic Pulmonary Fibrosis. *Annu. Rev. Pathol. Mech. Dis.* **2014**, *9*, 157–179. [CrossRef]
2. Shumar, J.; Chandel, A.; King, C. Antifibrotic Therapies and Progressive Fibrosing Interstitial Lung Disease (PF-ILD): Building on INBUILD. *J. Clin. Med.* **2021**, *10*, 2285. [CrossRef]
3. Betensley, A.; Sharif, R.; Karamichos, D. A Systematic Review of the Role of Dysfunctional Wound Healing in the Patho-genesis and Treatment of Idiopathic Pulmonary Fibrosis. *J. Clin. Med.* **2016**, *6*, 2. [CrossRef] [PubMed]
4. Veith, C.; Boots, A.W.; Idris, M.; Van Schooten, F.-J.; Van Der Vliet, A. Redox Imbalance in Idiopathic Pulmonary Fibrosis: A Role for Oxidant Cross-Talk Between NADPH Oxidase Enzymes and Mitochondria. *Antioxid. Redox Signal.* **2019**, *31*, 1092–1115. [CrossRef]
5. Hecker, L.; Vittal, R.; Jones, T.; Jagirdar, R.; Luckhardt, T.R.; Horowitz, J.; Pennathur, S.; Martinez, F.J.; Thannickal, V.J. NADPH oxidase-4 mediates myofibroblast activation and fibrogenic responses to lung injury. *Nat. Med.* **2009**, *15*, 1077–1081. [CrossRef]
6. Carnesecchi, S.; Deffert, C.; Donati, Y.; Basset, O.; Hinz, B.; Preynat-Seauve, O.; Guichard, C.; Arbiser, J.L.; Banfi, B.; Pache, J.C.; et al. A key role for *NOX4* in epithelial cell death during development of lung fibrosis. *Antioxid. Redox Signal.* **2011**, *15*, 607–619. [CrossRef] [PubMed]
7. Bedard, K.; Krause, K.H. The NOX Family of ROS-Generating NADPH Oxidases: Physiology and Pathophysiology. *Physiol. Rev.* **2007**, *87*, 245–313. [CrossRef]
8. Sies, H.; Jones, D.P. Reactive oxygen species (ROS) as pleiotropic physiological signalling agents. *Nat. Rev. Mol. Cell Biol.* **2020**, *21*, 363–383. [CrossRef] [PubMed]
9. Sirokmány, G.; Geiszt, M. The Relationship of NADPH Oxidases and Heme Peroxidases: Fallin' in and Out. *Front. Immunol.* **2019**, *10*, 394. [CrossRef]
10. Aeschbach, R.; Amadoò, R.; Neukom, H. Formation of dityrosine cross-links in proteins by oxidation of tyrosine residues. *Biochim. Biophys. Acta Protein Struct.* **1976**, *439*, 292–301. [CrossRef]
11. Wong, J.L.; Creton, R.; Wessel, G.M. The Oxidative Burst at Fertilization Is Dependent upon Activation of the Dual Oxidase Udx1. *Dev. Cell* **2004**, *7*, 801–814. [CrossRef]
12. Hensley, K.; Maidt, M.L.; Yu, Z.; Sang, H.; Markesbery, W.R.; Floyd, R.A. Electrochemical analysis of protein nitrotyrosine and dityrosine in the Alzheimer brain indicates region-specific accumulation. *J. Neurosci.* **1998**, *18*, 8126–8132. [CrossRef]
13. Al-Hilaly, Y.K.; Biasetti, L.; Blakeman, B.J.; Pollack, S.J.; Zibaee, S.; Abdul-Sada, A.; Thorpe, J.R.; Xue, W.F.; Serpell, L.C. The involvement of dityrosine crosslinking in alpha-synuclein assembly and deposition in Lewy Bodies in Parkinson's disease. *Sci. Rep.* **2016**, *6*, 39171. [CrossRef] [PubMed]
14. Leeuwenburgh, C.; Rasmussen, J.E.; Hsu, F.F.; Mueller, D.M.; Pennathur, S.; Heinecke, J.W. Mass Spectrometric Quantification of Markers for Protein Oxidation by Tyrosyl Radical, Copper, and Hydroxyl Radical in Low Density Lipoprotein Isolated from Human Atherosclerotic Plaques. *J. Biol. Chem.* **1997**, *272*, 3520–3526. [CrossRef] [PubMed]
15. Pennathur, S.; Vivekanandan-Giri, A.; Locy, M.L.; Kulkarni, T.; Zhi, D.; Zeng, L.; Byun, J.; de Andrade, J.A.; Thannickal, V.J. Oxidative Modifications of Protein Tyrosyl Residues Are Increased in Plasma of Human Subjects with In-terstitial Lung Disease. *Am. J. Respir. Crit. Care Med.* **2016**, *193*, 861–868. [CrossRef] [PubMed]
16. Locy, M.L.; Rangarajan, S.; Yang, S.; Johnson, M.R.; Bernard, K.; Kurundkar, A.; Bone, N.B.; Zmijewski, J.W.; Byun, J.; Pennathur, S.; et al. Oxidative cross-linking of fibronectin confers protease resistance and inhibits cellular migration. *Sci. Signal.* **2020**, *13*, eaau2803. [CrossRef] [PubMed]
17. Wells, R.G. Tissue mechanics and fibrosis. *Biochim. Biophys. Acta Mol. Basis Dis.* **2013**, *1832*, 884–890. [CrossRef]
18. Brault, J.; Goutagny, E.; Telugu, N.; Shao, K.; Baquié, M.; Satre, V.; Coutton, C.; Grunwald, D.; Brion, J.-P.; Barlogis, V.; et al. Optimized Generation of Functional Neutrophils and Macrophages from Patient-Specific Induced Pluripotent Stem Cells:Ex VivoModels of X0-Linked, AR220- and AR470- Chronic Granulomatous Diseases. *BioRes. Open Access* **2014**, *3*, 311–326. [CrossRef]
19. Prior, K.-K.; Leisegang, M.S.; Josipovic, I.; Löwe, O.; Shah, A.M.; Weissmann, N.; Schröder, K.; Brandes, R.P. CRISPR/Cas9-mediated knockout of p22phox leads to loss of Nox1 and *Nox4*, but not Nox5 activity. *Redox Biol.* **2016**, *9*, 287–295. [CrossRef] [PubMed]

20. Vandesompele, J.; De Preter, K.; Pattyn, F.; Poppe, B.; Van Roy, N.; De Paepe, A.; Speleman, F. Accurate normalization of real-time quantitative RT-PCR data by geometric averaging of multiple in-ternal control genes. *Genome Biol.* **2002**, *3*, RESEARCH0034. [CrossRef] [PubMed]
21. Weinberg, R.A.; Comb, M.J. *Chapter 12: Immunohistochemistry (IHC) in CST Guide: Pathways & Protocols*, 1st ed.; Cell Signaling Technology, Inc.: Danvers, MA, USA, 2015; pp. 214–217.
22. Diebold, B.A.; Wilder, S.G.; De Deken, X.; Meitzler, J.L.; Doroshow, J.H.; McCoy, J.W.; Zhu, Y.; Lambeth, J.D. Guidelines for the Detection of NADPH Oxidases by Immunoblot and RT-qPCR. *Methods Mol. Biol.* **2019**, *1982*, 191–229.
23. Augsburger, F.; Filippova, A.; Rasti, D.; Seredenina, T.; Lam, M.; Maghzal, G.; Mahiout, Z.; Jansen-Durr, P.; Knaus, U.G.; Doroshow, J.; et al. Pharmacological characterization of the seven human NOX isoforms and their inhibitors. *Redox Biol.* **2019**, *26*, 101272. [CrossRef]
24. Rajaram, R.D.; Dissard, R.; Faivre, A.; Ino, F.; Delitsikou, V.; Jaquet, V.; Cagarelli, T.; Lindenmeyer, M.; Jansen-Duerr, P.; Cohen, C.; et al. Tubular *NOX4* expression decreases in chronic kidney disease but does not modify fibrosis evolution. *Redox Biol.* **2019**, *26*, 101234. [CrossRef]
25. Kato, Y.; Wub, X.; Naitoc, M.; Nomurac, H.; Kitamotoa, N.; Osawa, T. Immunochemical Detection of Protein Dityrosine in Atherosclerotic Lesion of Apo-E-Deficient Mice Using a Novel Monoclonal Antibody. *Biochem. Biophys. Res. Commun.* **2000**, *275*, 11–15. [CrossRef]
26. Larios, J.M.; Budhiraja, R.; Fanburg, B.L.; Thannickal, V.J. Oxidative protein cross-linking reactions involving L-tyrosine in transforming growth factor-beta1-stimulated fibroblasts. *J. Biol. Chem.* **2001**, *276*, 17437–17441. [CrossRef]
27. Woods, A.A.; Linton, S.M.; Davies, M. Detection of HOCl-mediated protein oxidation products in the extracellular matrix of human atherosclerotic plaques. *Biochem. J.* **2003**, *370*, 729–735. [CrossRef] [PubMed]
28. Maher, T.M.; Strek, M.E. Antifibrotic therapy for idiopathic pulmonary fibrosis: Time to treat. *Respir. Res.* **2019**, *20*, 1–9. [CrossRef]
29. Mora, A.L.; Rojas, M.; Pardo, A.; Selman, M. Emerging therapies for idiopathic pulmonary fibrosis, a progressive age-related disease. *Nat. Rev. Drug Discov.* **2017**, *16*, 755–772. [CrossRef] [PubMed]
30. Lévigne, D.; Modarressi, A.; Krause, K.-H.; Pittet-Cuénod, B. NADPH oxidase 4 deficiency leads to impaired wound repair and reduced dityrosine-crosslinking, but does not affect myofibroblast formation. *Free. Radic. Biol. Med.* **2016**, *96*, 374–384. [CrossRef]
31. Luxen, S.; Noack, D.; Frausto, M.; Davanture, S.; Torbett, B.E.; Knaus, U.G. Heterodimerization controls localization of Duox-DuoxA NADPH oxidases in airway cells. *J. Cell Sci.* **2009**, *122*, 1238–1247. [CrossRef] [PubMed]
32. Forteza, R.; Salathe, M.; Miot, F.; Forteza, R.; Conner, G.E. Regulated hydrogen peroxide production by Duox in human airway epithelial cells. *Am. J. Respir. Cell Mol. Biol.* **2005**, *32*, 462–469. [CrossRef] [PubMed]
33. Sarr, D.; Tóth, E.; Gingerich, A.; Rada, B. Antimicrobial actions of dual oxidases and lactoperoxidase. *J. Microbiol.* **2018**, *56*, 373–386. [CrossRef] [PubMed]
34. Louzada, R.A.; Corre, R.; Ameziane El Hassani, R.; Meziani, L.; Jaillet, M.; Cazes, A.; Crestani, B.; Deutsch, E.; Dupuy, C. NADPH oxidase DUOX1 sustains TGF-beta1 signalling and promotes lung fibrosis. *Eur. Respir. J.* **2021**, *57*, 1901949. [CrossRef] [PubMed]

 antioxidants

Article

Oxidative Stress in Patients with Advanced CKD and Renal Replacement Therapy: The Key Role of Peripheral Blood Leukocytes

Carmen Vida [1,2,*], Carlos Oliva [1], Claudia Yuste [2,3], Noemí Ceprián [1,2], Paula Jara Caro [2,3], Gemma Valera [1], Ignacio González de Pablos [2,3], Enrique Morales [2,3] and Julia Carracedo [1,2,*]

[1] Department of Genetics, Physiology and Microbiology (Unit of Animal Physiology), Faculty of Biology, University Complutense of Madrid, 28040 Madrid, Spain; coliva02@ucm.es (C.O.); nceprian@ucm.es (N.C.); gvalera@ucm.es (G.V.)

[2] Research Institute Hospital 12 de Octubre (imas12), 28041 Madrid, Spain; claudia.yuste@salud.madrid.org (C.Y.); paulajara.caro@salud.madrid.org (P.J.C.); ignacioangel.gonzalez@salud.madrid.org (I.G.d.P.); emoralesr@senefro.org (E.M.)

[3] Department of Nephrology, Hospital Universitario 12 de Octubre, 28041 Madrid, Spain

* Correspondence: mvida@ucm.es (C.V.); julcar01@ucm.es (J.C.); Tel.: +34-91-394-4939 (J.C.)

Abstract: Oxidative stress plays a key role in the pathophysiology of chronic kidney disease (CKD). Most studies have investigated peripheral redox state focus on plasma, but not in different immune cells. Our study analyzed several redox state markers in plasma and isolated peripheral polymorphonuclear (PMNs) and mononuclear (MNs) leukocytes from advanced-CKD patients, also evaluating differences of hemodialysis (HD) and peritoneal dialysis (PD) procedures. Antioxidant (superoxide dismutase (SOD), catalase (CAT), glutathione peroxidase (GPx), reduced glutathione (GSH)) and oxidant parameters (xanthine oxidase (XO), oxidized glutathione (GSSG), malondialdehyde (MDA)) were assessed in plasma, PMNs and MNs from non-dialysis-dependent-CKD (NDD-CKD), HD and PD patients and healthy controls. Increased oxidative stress and damage were observed in plasma, PMNs and MNs from NDD-CKD, HD and PD patients (increased XO, GSSG and MDA; decreased SOD, CAT, GPX and GSH; altered GSSG/GSH balance). Several oxidative alterations were more exacerbated in PMNs, whereas others were only observed in MNs. Dialysis procedures had a positive effect on preserving the GSSG/GSH balance in PMNs. Interestingly, PD patients showed greater oxidative stress than HD patients, especially in MNs. The assessment of redox state parameters in PMNs and MNs could have potential use as biomarkers of the CKD progression.

Keywords: chronic kidney disease; oxidative stress; polymorphonuclear leukocytes; mononuclear leukocytes; hemodialysis; peritoneal dialysis

Citation: Vida, C.; Oliva, C.; Yuste, C.; Ceprián, N.; Caro, P.J.; Valera, G.; González de Pablos, I.; Morales, E.; Carracedo, J. Oxidative Stress in Patients with Advanced CKD and Renal Replacement Therapy: The Key Role of Peripheral Blood Leukocytes. *Antioxidants* **2021**, *10*, 1155. https://doi.org/10.3390/antiox10071155

Academic Editors: Chiara Nediani and Monica Dinu

Received: 9 June 2021
Accepted: 15 July 2021
Published: 20 July 2021

Publisher's Note: MDPI stays neutral with regard to jurisdictional claims in published maps and institutional affiliations.

1. Introduction

Chronic kidney disease (CKD) is a high global systemic pathology that affects approximately between 11–13% of the population worldwide [1], with the most prevalence in elderly people [2]. Indeed, due to the aging population, CKD is exponentially increasing and is the main contributor to morbi-mortality from non-communicable diseases [3]. In this regard, several studies have shown that patients with CKD suffer accelerated aging [4,5], contributing to increased risk of adverse outcomes, such as cardiovascular diseases (CVDs), altered immune response, neurological complications, or premature death [5–7]. Finally, CKD may advance to end-stage renal disease (ESRD) requiring renal replacement therapy (RRT), such as hemodialysis (HD) or peritoneal dialysis (PD) [6,8].

A chronic oxidative stress situation (a progressive redox imbalance due to the excessive production and accumulation of oxidant compounds in cells and tissues that exceeds the scavenging capacity of antioxidant systems to detoxify these reactive products), together with a low grade of inflammation, are the basis of aging and several age-related diseases [9–11]. Thus, in addition to traditional risk factors (e.g., hypertension, diabetes mellitus, obesity, etc.), both chronic systemic inflammation and excessive oxidative stress also contribute to the onset, progression and pathogenesis of CKD and renal failure [5,12–16], being some of the major contributors to the associated-CVDs and elevated mortality in CKD patients [5,8]. It is also important to highlight that oxidative-inflammatory stress is increased even in the early stages of CKD and progresses in parallel to the deterioration of renal functions [5,12,14,15]. Moreover, the presence of other non-conventional risk factors, such as anemia [17,18], or several lifestyle factors, such as nutrition or dietary interventions [6,19], also have a great impact on the redox status of CKD patients, promoting oxidative-inflammatory stress, and consequently, aggravating progression to renal failure and CVDs in these patients [6,17–19]. In fact, advanced CKD patients may suffer from malnutrition or dietary restrictions, which, together with the loss of micronutrients during dialysis procedures, may further increase oxidative stress due to the reduction of extracellular antioxidant defenses [12–14]. In addition, an inadequate intake of oligoelements (e.g., copper, manganese, etc.) may draw on misleading antioxidant enzymes [19]. However, medical nutrition therapy (e.g., low-protein diet restriction) may be useful in decreasing uremic toxicity and increasing the efficiency of muscle and body metabolism in severe CKD patients [20].

Enhanced oxidative-inflammatory stress is not restricted to the kidney [8] but is also present in other peripheral locations, such as plasma or immune cells [21]. Thus, at peripheral levels, CKD patients exhibit increased oxidative stress and a chronic inflammation state [5,12–15], as well as an impaired immune response with/or activation of peripheral blood polymorphonuclear (PMNs) and mononuclear (MNs) leukocytes in the circulation [6,15,16,22,23]. Indeed, the leukocyte activation promotes an increased release of pro-inflammatory cytokines, such as interleukin-(IL)-6 or IL-1β, as well as overproduction of oxidant compounds, such as reactive oxygen species (ROS) [5,23]. Thus, the overproduction of ROS by the increased activity of several oxidant enzymatic systems (e.g., xanthine oxidase (XO), myeloperoxidase, etc.) or by the dysregulation of the mitochondrial respiratory chain, together with the impairment of the antioxidant systems, including enzymes (e.g., superoxide dismutase (SOD), catalase (CAT), glutathione peroxidase (GPx), etc.) and other non-enzymatic compounds (e.g., reduced glutathione (GSH), etc.) systems [5,24–27], further leading to oxidation of macromolecules (e.g., lipids, proteins and DNA) and, consequently, cellular and tissue damage [5,15,16,28]. Indeed, CKD patients show elevated levels of oxidized compounds (e.g., oxidized glutathione (GSSG), etc.), lipid peroxidation (e.g., malondialdehyde (MDA), etc.) and oxidized proteins and DNA [23,29], contributing to dysfunction in CKD patients [5,30] and especially in HD and PD patients [12–15,28,31,32]. Additionally, since oxidation and inflammation are interlinked processes [9], the accumulation of oxidant compounds triggers the activation of inflammatory mediators and, subsequently, amplify and perpetuate a vicious circle of oxidative damage [23], contributing to chronic kidney damage and systemic complications in CKD, such as CVDs [5,15,22,28]. However, most studies have been carried out on plasma, serum or monocytes of CKD patients but not in other types of peripheral blood leukocytes. In fact, human studies about the changes in redox state parameters in different types of leukocytes and at different stages of CKD are still scarce and preliminary.

The CKD patients undergoing RRT also exhibit higher levels of inflammation and oxidative stress compared with non-dialysis-dependent CKD patients (NDD-CKD) [12–14]. In fact, both HD and PD procedures are characterized by amplifying oxidative-inflammatory stress in dialysis patients by different underlying mechanisms [5,12,28,31,32]. On the one hand, the HD technique aggravates the accumulation of oxidative products due to several factors, such as duration of dialysis, the filtration of low molecular weight antioxidants, use of anticoagulants (e.g., heparin, etc.), or both the biocompatibility of dialyzer membrane (e.g., polisulfone, cellulose, etc.) and the type of dialysates (e.g., citrate, acetate, etc.) [8,12,14,33–35]. In fact, it is known that acetate dialysates induce complement and leukocytes activation, enhancing ROS production and the release of pro-inflammatory cytokines during HD, whereas citrate dialysates increase antioxidant enzymes (e.g., SOD, GSH) and reduce oxidized compounds and lipid damage (e.g., MDA, GSSG, etc.) [34–36]. Moreover, a few studies have also shown different patterns in the redox state and inflammation of HD patients depending on the type of dialyzer membrane (e.g., polisulfone, cellulose, etc.) and the HD modality (e.g., low- and high-flux HD, hemodiafiltration, etc.) [37–40]. However, these reports are not clear, showing increased, decreased or no changes in the oxidative-inflammatory stress parameters analyzed [37–40]. On the other hand, the composition of the PD dialysate solutions, together with chronic exposure of the peritoneal membranes, can trigger oxidative stress and inflammation in PD patients [12–14,32]. Thus, the increased glucose and lactate concentrations, low pH or hyperosmolality of the PD dialysates can promote ROS overproduction and the accumulation of oxidative damage products in the peritoneum, enhancing calcification and fibrosis [12]. Regarding PD modality, it is known that continuous ambulatory peritoneal dialysis (CAPD) also enhances both oxidative stress and lipid damage in plasma and red blood cells (e.g., MDA, GSSG, etc.) from PD patients compared with healthy subjects and NDD-CKD patients [12,13,41]. In addition, several studies also showed a strong linkage between residual renal function (RRF) and oxidative stress, correlating with CVDs and survival in PD patients [12]. Thus, a preserved RRF is associated with reduced oxidative stress and lipid damage in PD patients [12,42,43]. Nevertheless, the PD is considered a more biocompatible dialysis technique compared to the HD [12]. Although several studies have found a higher accumulation of oxidants compound and antioxidant depletion in HD patients compared with PD patients [12,13,32,44], the data are still scarce. Moreover, most research evidence has been evaluated in plasma/serum analysis but not in other blood cells from CKD patients. In this context, it is important to note that numerous peripheral blood cell types are involved in the maintenance of the redox state [10]. However, the most studied and commonly used in the clinical setting are MNs (mainly lymphocytes) but not other cells, such as isolated PMNs (phagocytes: mainly neutrophils). Since few studies have been suggested that phagocytes are the principal cells behind the chronic oxidative stress and damage associated with aging and age-related diseases [9–11], these cells could provide a helpful sample to clarify the molecular mechanisms underlying the increased oxidative stress throughout CKD progression. Furthermore, it is possible that the assessment of several markers of redox state in PMNs and MNs leukocytes may be useful biomarkers of the CKD progression. However, to our knowledge, no studies on this issue have been explored in HD or PD patients.

Therefore, the aim of this study was to analyze the changes in several oxidative stress and damage parameters in different types of peripheral blood leukocytes from advanced CKD patients (stages 4–5) and relative age-matched healthy controls. In addition, differences by RRT were also considered. Therefore, CKD patients were subdivided into NDD-CKD, HD and PD patients. To address this study, several pro-oxidant enzymes (XO) and oxidized compounds (GSSG), antioxidants enzyme activities and compounds (SOD, CAT, GPx and GSH), lipid oxidative damage (MDA) and inflammatory markers were investigated in both plasma and/or isolated PMNs and MNs cells from NDD-CKD, HD and PD patients and healthy subjects.

2. Materials and Methods

2.1. Research Participants

For this cross-sectional study, a total of 117 volunteers were finally enrolled selected and divided into four experimental groups: healthy controls (n = 17), advanced NDD-CKD (n = 39), HD (n = 39) and PD (n = 22) patients. All the patients were selected from a cohort of prevalent patients from the Nephrology Department, Hospital Universitario 12 de Octubre (Madrid, Spain). The study determinations were performed between February 2018 and December 2019. All procedures were performed according to good clinical practice guidelines, and all patients gave their written informed consent statements for study participation [45]. The study protocol was approved by the ethics in human research committee of Hospital Universitario 12 de Octubre (Ethical approval code: CEIC 17/407). NDD-CKD patients were selected by disease severity and impaired renal function (stages 4 and 5, estimated glomerular filtration rate, eGFR < 30 mL/min per 1.73 m^2). Regarding RRT, all the patients undergoing HD and PD in the Nephrology Department of Hospital Universitario 12 de Octubre. The only exclusion criterion was an onset of chronic HD or PD shorter than 30 days before the study. On the one hand, the prescription of HD sessions was based on a regimen of 3 sessions per week (4 h each). Pre-dialysis blood samples for measurement were drawn on mid-week treatment (Wednesday or Thursday). HD patients received online hemodiafiltration (75%) or conventional high-flux HD (25%) with polysulfone, asymmetric cellulose triacetate (CTA) and polyacrylonitrile (AN69) membrane dialyzers (95%, 2.5% and 2.5%, respectively) (Helixone® membrane, ©Fresenius Medical Care, Bad Homburg, Germany). Standard ultrafiltered citrate buffer (SelectBag Citrate®, Baxter, Madrid, Spain) was used as dialysate. Moreover, HD patients (90%) received heparin-based anticoagulants (95% sodium heparin and 5% low molecular weight). On the other hand, PD patients received automated (APD) or continuous ambulatory (CAPD) peritoneal dialysis (86% and 14%, respectively); citrate dialysates containing 1.5–2.5% glucose solution (©Fresenius Medical Care; 41% of PD patients) or 1.36% to 2.27% icodextrin content (Baxter; 59% of PD patients) were used. Regarding RRF, 16 HD patients (41%) had residual diuresis between 100–1300 mL/day, while 20 PD patients (91%) had RRF between 500–3000 mL/day.

Clinical and analytical variables selected for the study were recorded at the time of extraction of blood samples. These included age, gender, CKD etiology and comorbidities (arterial hypertension, hyperuricemia, diabetes mellitus, dyslipidemia, CVDs and tumors). Previous immunosuppression, treatment with erythropoietin stimulating agents or allopurinol and the type of vascular access were also recorded. The following analytical parameters were also analyzed in serum/plasma: albumin, creatinine, total proteins, hemoglobin (Hb), glycosylated hemoglobin (HbA1c), total cholesterol (TC), total triglycerides (TG), uric acid (UA) and C-reactive protein (CRP).

2.2. Collection of Blood Samples and Processing

Peripheral blood samples were collected using vein puncture and EDTA-buffered Vacutainer tubes (BD Diagnostic, Madrid, Spain). Blood extraction was performed between 9:00 to 10:00 a.m. to avoid circadian variations and after an overnight fast of 12 h. On the one hand, plasmas were obtained for biochemical analysis according to standard protocols. Plasma concentrations of albumin, creatinine, total proteins, Hb, TC, TG, HbA1c, UA and CRP were measured with an automated clinical biochemistry analyzer (Hitachi 7050, Hitachi Corp., Tokyo, Japan) and a hematology analyzer (KX21N, Sysmex, Kobe, Japan). On the other hand, whole blood (12 mL) was used for isolation of PMNs and MNs leukocytes following a previously described method [11]. Thus, PMNs and MNs cells were isolated using 1.119 and 1.077 density Hystopaque (Sigma-Aldrich, Madrid, Spain) separation, respectively. Collected cells were counted (95% of viability determined using trypan blue staining test) and adjusted to a specific number of leukocytes, depending on the parameter analyzed. For that, PMNs and MNs leukocytes were washed with cold PBS 0.05 M, pH 7.4 and centrifuged at 1200$\times$ g for 10 min at 4 °C, the supernatants were removed, and the aliquots of the pellet were stored at −80 °C until use. Moreover, plasma samples were also stored at −80 °C until use for the determination of inflammatory and oxidative stress parameters.

2.3. Peripheral Blood Lymphocytes Population Subtyping

Peripheral blood lymphocyte populations (PBLP): total lymphocytes and the following lymphocyte subtyping: CD3+ T-cell, CD4+ T-cell, CD8+ T-cell, CD19+ B-cells and CD3−CD56+CD16+ NK cells, were assessed following a previously described method [46]. Determination of PBLP was performed with a FACSCanto II flow cytometer, and the data were analyzed by FACSCanto clinical software (BD Biosciences, San Jose, CA, USA). Values were compared with the range of normality established by our laboratory using blood samples from healthy controls.

2.4. Superoxide Dismutase Activity

The total SOD activity was measured using a colorimetric assay kit (EnzyChromTM ESOD-100, BioAssay Systems, Hayward, CA, USA). In the assay, superoxide ($O_2^{\bullet-}$) is provided by XO catalysed reaction. $O_2^{\bullet-}$ reacts with a water-soluble tetrazolium (WST-1) dye to form a coloured product. SOD scavenges the $O_2^{\bullet-}$. Thus, less $O_2^{\bullet-}$ is available for the chromogenic reaction. Plasmas were diluted 1:5 according to the manufacturer's instructions. PMNs and MNs leukocytes (1×10^6 cells) were resuspended in cold lysis buffer (500 μL), sonicated and centrifuged (12,000× *g*, 5 min; 4 °C). Clear supernatants and plasmas (20 μL) were incubated with XO and WST-1 for 1 h, and absorbance was measured at 440 nm. The results were expressed as units (U) SOD/mg protein or U SOD/mL.

2.5. Catalase Activity

The CAT activity was determined using a fluorimetric kit (A-22180 Amplex® Red Catalase Assay Kit, Molecular Probes, Paisley, UK). In the assay, CAT reacts with hydrogen peroxide (H_2O_2) to produce water and oxygen. The Amplex Red reagent containing horseradish peroxidase (HRP) reacts with any unreacted H_2O_2 (1:1 stoichiometry) to produce the fluorescent resorufin. Plasmas were assayed directly according to the manufacturer´s protocol. PMNs and MNs leukocytes (1×10^6 cells) were resuspended in lysis solution (100 μL), sonicated and centrifuged (3200× *g*, 20 min; 4 °C). Clear supernatants and plasmas (25 μL) were mixed with H_2O_2 40 μM for 30 min at room temperature. After incubation (30 min at 37 °C) with the working solution, fluorescence was measured at 530–590 nm excitation-emission detection. The results were expressed as U CAT/mg protein or U/mL.

2.6. Gluthathione Peroxidase Activity

The GPx activity was assessed using a colorimetric assay kit (EnzyChromTM EGPX-100, BioAssay Systems, USA), which quantifies the consumption of nicotinamide adenine dinucleotide phosphate (NADPH) in the GPx-coupled reactions by the decline in absorbance at 340 nm (directly proportional to the GPx activity). Plasmas were assayed directly according to manufacturer´s instructions. PMNs and MNs leukocytes (1×10^6 cells) were resuspended in cold lysis buffer, sonicated and centrifuged (14,000× *g*, 10 min; 4 °C). Clear supernatants and plasmas (10 μL) were mixed with the working reagent assay buffer and substrate solution. Absorbance was measured at 340 nm immediately (time zero) and again at 4 min. The results were expressed as U GPx/mg protein or U/L.

2.7. Xanthine Oxidase Activity

The XO activity was quantified using a commercial kit (A-22182 Amplex Red Xanthine/Xanthine Oxidase Assay Kit, Molecular Probes). In the assay, XO catalyzes the oxidation of xanthine/hypoxanthine to UA and $O_2^{\bullet-}$, which spontaneously degrades to H_2O_2. Amplex Red reagent containing HRP reacts with H_2O_2 to generate the resorufin. The XO assay was evaluated in plasma (50 μL) and isolated PMNs and MNs leukocytes (3×10^6 cells) as previously described [10]. Red-fluorescent resofurin production was measured at 530–590 nm excitation-emission detection. Results were expressed as milliunits (mU) XO/mg protein or mU/mL.

2.8. Glutathione Content Assay

Both reduced (GSH) and oxidized (GSSG) glutathione were quantified using a fluorimetric assay [47], which is based on the reaction of GSSG and GSH with o-phthalaldehyde (OPT, Sigma-Aldrich, Spain), at pH 12 and pH 8, respectively, resulting in the formation of a fluorescent product measured at 420 nm. The assay was evaluated in plasma (50 μL) and isolated PMNs and MNs leukocytes (3×10^6 cells) as previously described [11]. Results were expressed as nmol/mg protein or nmol/mL. Furthermore, the GSSG/GSH coefficient was also calculated.

2.9. Lipid Peroxidation Assay

Malondialdehyde (MDA) content, the most commonly used maker of lipid peroxidation, were assessed using the commercial kit "MDA Assay Kit" (Biovision, Milpitas, CA, USA), which measures the reaction of MDA with thiobarbituric acid (TBA) and the MDA-TBA adduct formation measured at 532 nm. The assay was evaluated in plasma (300 μL) and isolated PMNs and MNs leukocytes (1×10^6 cells) as previously described [11]. Results were expressed as nmol MDA/mg protein or MDA/mL.

2.10. Protein Content Assay

The protein contents of PMNs and MNs were determined using the bicinchoninic acid protein (BCA) assay kit protocol (Sigma-Aldrich, Madrid, Spain) according to the manufacturer´s instructions.

2.11. Cytokine IL-1β Measurement

The basal release of IL-1β was measured in plasma using a multiplex luminometry detection system (MILLIPLEX human high sensitivity T cell magnetic bead panel, HSTCMAG-28SK, EMD Millipore, MA, USA), with minimum detectable doses of IL-1β under 0.14 pg/mL. The results were expressed as pg/mL.

2.12. Statistical Analysis

Statistical analysis was performed in SPSS Statistics 21.0 (IBM, Chicago, IL, USA). All tests were two-tailed, with a significant level of $\alpha = 0.05$. Data were presented with a percentage and mean ± standard deviation (SD) for categorical and continuous variables, respectively. To compare the clinical features and comorbidities of the different groups, Pearson χ^2 test and one-way analysis of variance (ANOVA) were used. Normality and homogeneity of the variances were tested by the Kolmogorov-Smirnov test and Levene test, respectively. To determine the differences between groups, we used the Student´s t-test for independent samples as well as one-way ANOVA followed by non-parametric Kruskal–Wallis test or post hoc comparison analysis (Tukey test and Games–Howell test for homogeneous or unequal variances, respectively). Linear correlations between biochemical, inflammatory and oxidative stress parameters were explored using bivariate Pearson´s correlation coefficients (r). GraphPad Prism 6 Software (LLC, San Diego, CA, USA) was used to perform the figures.

3. Results

3.1. Demographical, Clinical and Biochemical Characteristics

The demographic and clinical characteristics of the participants are shown in Table 1. No differences in the mean age were found between the different study groups. A similar percentage of males and females was observed in the control and HD groups, whereas more male than female patients were included in NDD-CKD and HD groups.

Regarding clinical comorbidities (Table 1), 35 NDD-CKD (89.7%), 33 HD (84.6%) and 20 PD (90.9%) patients had hypertension, whereas 17 NDD-CKD (43.6%), 7 HD (17.9%) and 5 (22.7%) PD patients had diabetes mellitus. Almost half of the NDD-CKD, HD and DP patients had dyslipidemia, and the other half had hyperuricemia and various CVDs. HD and PD had a lower % of diabetes, dyslipidemia and hyperuricemia than NDD-CDK patients, whereas PD patients showed a higher % of diabetes and hyperuricemia and lower % of peripheral vascular disease than HD patients (see p-values in Table 1). CKD etiologies were similar among the HD and PD groups, with glomerulonephritis being the predominant cause of renal disease (28% and 36%, respectively), whereas in NDD-CKD patients, the main cause was diabetes (31%). Furthermore, almost half of the NDD-CKD, HD and DP patients used medications (e.g., antiplatelet agents, statins, allopurinol and erythropoietin). NDD-CKD group included significantly more patients receiving statins and less patients being treated with antiplatelet compared to dialysis patients. HD group had a significantly higher and lower % of patients receiving erythropoietin and allopurinol, respectively, than the NDD-CKD and PD groups. Regarding RRT, 56.41% of HD patients had an AVF as vascular access. In PD, 75% and 25% of PD patients underwent APD and CAPD, respectively. The RRF was higher in the PD group ($p < 0.01$) than in the HD group.

Table 1. The demographic and clinical characteristics of the study population.

Characteristics	Control	NDD-CKD	HD	PD
Demographic data	$n = 17$	$n = 39$	$n = 39$	$n = 22$
Male, *n* (%)	8 (47.1)	25 (64.1)	27 (69.2)	11 (50)
Female, *n* (%)	9 (52.9)	14 (35.9)	12 (30.8)	11 (50)
Age (years; mean ± SD)	52.17 ± 15.33	60.41 ± 17.26	57.46 ± 14.75	54.63 ± 15.82
BMI (kg/m^2; mean ± SD)	24.54 ± 3.55	25.19 ± 9.17	24.29 ± 4.05	24.54 ± 4.17
eGFR (mL/min per 1.73 m^2)	-	15.84 ± 2.74	-	-
Comorbidity				
Hypertension, *n* (%)	1 (5.9)	35 (89.7)	33 (84.6)	20 (90.9)
Diabetes, *n* (%)	2 (11.8)	17 (43.6)	7 (17.9) ^{bb}	5 (22.7) ^{b}
Dyslipidemia, *n* (%)	0 (0)	31 (79.5)	23 (59.0) ^{b}	14 (63.6) ^{b}
Hyperuricemia, *n* (%)	0 (0)	27 (69.2)	10 (25.6) ^{bb}	13 (59.1) ^{c}
CVD, *n* (%)	0 (0)	19 (48.7)	19 (48.7)	7 (31.8)
Peripheral vascular disease, *n* (%)	0 (0)	1 (2.6)	9 (23.1) ^{b}	2 (9.1) ^{c}
ACVA, *n* (%)	0 (0)	5 (12.8)	2 (5.1)	3 (13.6)
Tumors, *n* (%)	0 (0)	1 (2.6)	1 (2.6)	1 (4.5)
Etiology of the nephropathy				
Glomerulonephritis, *n* (%)	0 (0)	6 (15)	11 (28)	8 (36)
Hypertension/vascular, *n* (%)	0 (0)	7 (18)	6 (15)	4 (18)
Diabetes, *n* (%)	0 (0)	12 (31)	4 (10)	4 (18)
Polycystic kidneys	0 (0)	4 (10)	1 (3)	1 (5)
Tubulointersticial nephritis, *n* (%)	0 (0)	6 (15)	8 (21)	2 (5)
Others, *n* (%)	0 (0)	4 (10)	9 (23)	3 (14)
Dialysis data				
AVF (%)	-	-	56.41	-
OL-HDF/HFD (%)	-	-	75/25	-
APD/CAPD (%)	-	-	-	86/14
Kt/v (mean ± SD)	-	-	1.67 ± 0.25	2.34 ± 0.51
RRF (mL/day; median)	-	-	400 (100–1300)	1275 (500–3000) ^{cc}
Treatment				
Statins	0 (0)	30 (76.9)	16 (41) ^{b}	14 (63.6) ^{c}
Allopurinol	0 (0)	23 (59)	8 (20.5) ^{b}	14 (63.6) ^{c}
Antiplatelet	0 (0)	8 (20.5)	16 (41) ^{b}	7 (31.8)
Erythropoietin	0 (0)	18 (46.2)	34 (87.2) ^{bb}	16 (72.7) ^{bb}

Abbreviations: ACVA, cerebral vascular accident; APD, automated peritoneal dialysis; AVF, arteriovenous fistula; BMI, body mass index; CAPD, continuous ambulatory peritoneal dialysis; CVD, cardiovascular disease; eGFR, estimated glomerular filtration rate; HD, hemodialysis; HFD, high-flux dialysis; Kt/v, standardized dialysis adequacy. NDD-CKD, non-dialysis-dependent chronic kidney disease; OL-HDF, online hemodiafiltration; PD, peritoneal dialysis; SD, standard deviation. ^{b} $p < 0.05$ and ^{bb} $p < 0.01$ versus NDD-CKD patients. ^{c} $p < 0.05$ and ^{cc} $p < 0.01$ versus HD patients. (-): data not included.

Biochemical characteristics, inflammatory markers and lymphocytes populations of the study groups are also displayed in Table 2. Regarding biochemical variables, the NDD-CKD, HD and PD patients showed lower albumin and Hb content and higher serum creatinine ($p < 0.001$) levels than healthy controls. Moreover, NDD-CKD patients had also higher TG ($p < 0.01$) and UA ($p < 0.05$) levels than controls. Interestingly, dialysis patients had lower TG ($p < 0.01$ in HD; $p < 0.05$ in PD) and higher serum creatinine ($p < 0.001$) levels than NDD-CKD patients. Interestingly, HD patients had lower TC and HbA1c contents

($p < 0.05$) than NDD-CKD patients, whereas the PD group showed lower albumin content ($p < 0.05$) than the NDD-CKD group. No significant differences were observed for the other parameters.

Table 2. Biochemical and inflammatory parameters and lymphocytes population of the study participants.

Variables	CONTROL	NDD-CKD	HD	PD
Biochemical parameters	$n = 17$	$n = 39$	$n = 39$	$n = 22$
Serum creatinine (mg/dL)	0.82 ± 0.16	4.17 ± 1.04 [aaa]	7.79 ± 1.91 [aaa,bbb]	7.46 ± 2.72 [aaa,bbb]
Albumin (g/dL)	4.65 ± 0.28	4.25 ± 0.36 [aa]	4.11 ± 0.35 [aa]	3.82 ± 0.43 [aaa,bb]
TG (mmol/L)	96.94 ± 43.44	171.02 ± 98.36 [aa]	117.23 ± 44.70 [bb]	132.68 ± 61.64 [b]
TC (mmol/L)	178.41 ± 27.64	168.41 ± 52.30	147.15 ± 30.96 [b]	159.45 ± 37.39
HDL-C (mmol/L)	55.17 ± 8.88	46.02 ± 13.43	48.72 ± 11.21	47.41 ± 13.80
LDL-C (mmol/L)	103.82 ± 26.33	89.85 ± 44.32	77.82 ± 31.30	90.45 ± 28.03
UA (mg/dL)	5.10 ± 1.10	6.32 ± 1.70 [a]	5.78 ± 1.22	5.76 ± 1.29
Hb (g/dL)	13.61 ± 2.30	9.98 ± 2.82 [a]	8.79 ± 2.85 [aa]	10.21 ± 2.52 [a]
HbA1c (mg/dL)	5.43 ± 0.43	5.95 ± 1.20	5.14 ± 0.70 [b]	42 ± 0.49
Inflammatory markers				
CRP (mg/dL)	0.29 ± 0.09	0.45 ± 0.07 [a]	0.86 ± 0.13 [aa,b]	0.83 ± 0.16 [aa,b]
IL-1β (pg/mL)	4247 ± 1218	5200 ± 2450 [a]	5156 ± 2697	5280 ± 3031
Lymphocyte populations (cells/µL)				
Total lymphocytes	1798 ± 449	1585 ± 548	1083 ± 509 [aaa,bbb]	1271 ± 293 [aa,b]
$CD3^+$ T-cells	1272 ± 350	1187 ± 440	761 ± 425 [aaa,bbb]	916 ± 537 [a]
$CD4^+$ T-cells	831 ± 308	731 ± 295	430 ± 280 [aaa,bbb]	603 ± 224 [a]
$CD8^+$ T-cells	410 ± 143	429 ± 250	301 ± 190	387 ± 136
$CD4^+/CD8^+$ ratio	2.23 ± 0.22	2.20 ± 0.17	1.49 ± 0.09 [a,bbb]	1.71 ± 0.21 [b]
$CD19^+$ B-cells	194 ± 22	130 ± 18	118 ± 21 [a]	81 ± 11 [aa,b]
$CD3^-CD56^+CD16^+$ NK	243 ± 23	204 ± 22	173 ± 24 [a,b]	228 ± 13

Abbreviations: CRP, C-reactive protein; Hb, hemoglobin; HbA1c, glycosylated hemoglobin; HDL-C, high-density lipoprotein cholesterol; HD, hemodialysis; IL-1β, interleukine-1β; LDL-C, low-density lipoprotein cholesterol; NK, natural killer cells; NDD-CKD, non-dialysis-dependent chronic kidney disease; PD, peritoneal dialysis; TC, total cholesterol; TG, triglycerides; UA, uric acid. Data are shown as mean ± standard deviation. [a] $p < 0.05$, [aa] $p < 0.01$ and [aaa] $p < 0.001$ versus control; [b] $p < 0.05$, [bb] $p < 0.01$ and [bbb] $p < 0.001$ versus NDD-CKD patients.

Regarding the inflammatory markers, we assessed plasma CRP levels, together with extracellular basal IL-1β release under resting conditions (an indicator of sterile inflammation). NDD-CKD patients showed higher basal IL-1β and CRP levels ($p < 0.05$) than healthy subjects. In addition, HD and PD patients also had higher CRP levels than control ($p < 0.01$) and NDD-CKD ($p < 0.05$) groups. Regarding RRT, no differences were observed in IL-1β and CRP levels between HD and PD groups.

We also investigated the changes in lymphocytes populations, measuring the surface-marker profile by flow cytometry (Table 2). The lymphocytes subsets were divided between T-cells (CD3+; isotype CD4+ or CD8+), B-cells and NK cells. No significant differences were observed between NDD-CKD and control groups. However, dialysis patients showed a significant decrease in the total number of lymphocytes, as well as a lower number of CD3+ and CD4+ T-lymphocytes, B-lymphocytes and NK cells, and the CD4+/CD8+ ratio than controls (see p-values in Table 2). Interestingly, HD patients had a marked reduction in the total number of lymphocytes, CD3+ and CD4+ T-lymphocytes and NK cells, as well as a lower CD4+/CD8+ ratio than NDD-CKD patients. Moreover, PD patients had a lower total number of lymphocytes, B-cells and CD4+/CD8+ compared with NDD-CKD patients (see p-values in Table 1). No significant differences between HD and PD-group were found.

3.2. Changes in Oxidative Stress and Lipid Damage Parameters in Plasma

The results regarding oxidative stress and lipid damage parameters in plasma from healthy subjects and NDD-CKD, HD and PD patients are shown in Figure 1.

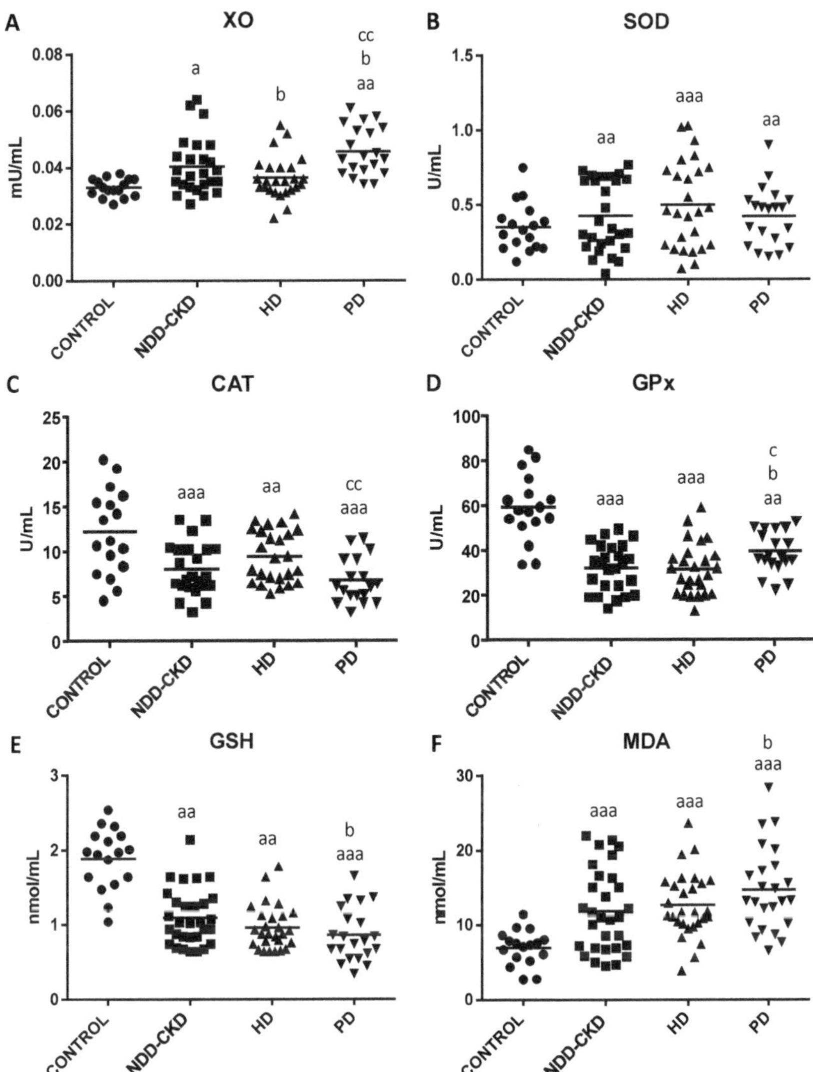

Figure 1. Oxidative stress and lipid damage parameters in plasma from healthy subjects (control), non-dialysis-dependent chronic kidney disease (NDD-CKD), hemodialysis (HD) and peritoneal dialysis (PD) patients. (**A**) Xanthine oxidase (XO; mU/mL), (**B**) superoxide dismutase (SOD; U/mL), (**C**) catalase (CAT; U/mL) and (**D**) glutathione peroxidase (GPx; U/mL) activities; (**E**) reduced glutathione (GSH; nmol/mL) and (**F**) malondialdehyde (MDA; nmol/mL) contents. Data are shown as the mean (horizontal bar) of 17–39 values corresponding to the number of subjects analyzed in each group. Each value is the mean of duplicate assays. ^{a} $p < 0.05$, ^{aa} $p < 0.01$ and ^{aaa} $p < 0.001$ versus control; ^{b} $p < 0.05$ versus NDD-CKD patients; ^{c} $p < 0.05$ and ^{cc} $p < 0.01$ versus HD patients.

The pro-oxidant XO activity (Figure 1A) was significantly increased in NDD-CKD ($p < 0.05$) and PD ($p < 0.01$) patients in comparison with healthy controls. Moreover, HD and PD patients showed lower and higher XO activity ($p < 0.05$), respectively, compared with NDD-CKD patients. Interestingly, PD patients had significantly higher XO activity ($p < 0.01$) than HD patients.

Furthermore, we also observed impairment of the circulating antioxidant enzymes SOD, CAT and GPx (Figure 1B–D), as well as of the GSH content (Figure 1E). In fact, NDD-CKD, HD and PD patients showed lower CAT and GPx activities and GSH concentration, as well as higher SOD activities than healthy controls (see p-values in Figure 1). In addition, PD patients presented higher GPx activity and lower GSH content ($p < 0.05$) than NDD-CKD patients. Interestingly, PD patients also exhibited lower and higher CAT and GPx activities compared with HD patients.

Finally, we also found a marked increase in lipid peroxidation in NDD-CKD, HD and PD patients (Figure 1F), showing higher MDA content ($p < 0.001$) than healthy subjects. Moreover, PD patients also showed higher MDA levels ($p < 0.05$) than NDD-CKD patients.

3.3. *Changes in Oxidative Stress and Lipid Damage Parameters in Isolated Peripheral Blood Leukocytes*

The results regarding oxidative stress and lipid damage parameters in isolated PMNs and MNs leukocytes from healthy subjects and NDD-CKD, HD and PD patients are shown in Figures 2 and 3.

First, regarding both oxidant (XO) and antioxidant (CAT, GPx and SOD) enzymes (Figure 2), we found significant differences in PMNs and MNs from NDD-CKD patients relative to healthy controls. Thus, in both leukocytes populations, NDD-CKD patients exhibited higher XO activity (Figure 2A,B) and lower CAT (Figure 2C,D) and GPx (Figure 2E,F) activities than controls (see p-values in Figure 2). Interestingly, a different pattern was observed in the SOD activity, with NDD-CKD patients showing an increased SOD in PMNs ($p < 0.05$; Figure 2G) and a decreased SOD in MNs ($p < 0.001$; Figure 2H). Similar results were observed in HD and PD patients as compared with healthy controls. However, these redox state alterations were much more marked in PD patients, showing in both leukocytes population an increased XO activity and decreased CAT, GPx and SOD activities (see p-values in Figure 2). Furthermore, it highlights the redox state differences between NDD-CKD patients and dialysis patients. Thus, in both leukocytes populations, HD patients showed lower XO (Figure 2A,B) and CAT (Figure 2C,D) activities than NDD-CKD patients, also accompanied by decreased GPx activity in PMNs (Figure 2E). By contrast, an increased GPx and SOD activities (Figure 2F,H) were found in MNs from HD patients compared to NDD-CKD patients. Interestingly, in PD patients, these redox balance alterations were only observed in PMNs, showing higher CAT and lower SOD activities (Figure 2C,G) than NDD-CKD patients (see p-values in Figure 2). In regard to RRT, the PD group presented higher XO activity (Figure 2A,B) and lower SOD activity (Figure 2G,H) in PMNs and MNs compared with the HD group. In addition, PD patients also showed increased CAT and GPx activities in PMNs (Figure 2C,E), whereas in MNs, the GPx activity (Figure 2F) was significantly decreased in comparison to HD patients.

Secondly, regarding the glutathione balance, we observed significant differences in both GSH and GSSG contents and GSSG/GSH ratios in leukocytes from NDD-CKD, HD and PD patients relative to healthy controls (Figure 3). Interestingly, the MNs showed a greater alteration of glutathione than PMNs in dialysis patients. In general, NDD-CKD patients showed in both leukocytes populations significantly lower GSH contents (Figure 3A,B), as well as higher GSSG amounts (Figure 3C) and GSSG/GSH ratios (Figure 3E,F), than healthy subjects. However, HD and PD patients presented a marked increase of GSSG content and GSSG/GSH ratios in MNs (Figure 3D,F) compared with controls, also accompanied by a significant decrease of GSH content (Figure 3B) (see p-values in Figure 3). By contrast, no significant differences were observed in the glutathione parameters in PMNs between dialysis patients and healthy controls. In addition, HD and PD patients presented in PMNs a higher GSH content (Figure 3A) and lower GSSG levels and GSSG/GSH ratios (Figure 3C,E) than NDD-CKD patients. By contrast, these patients also presented a marked increase of GSSG content in MNs (Figure 3D) in comparison to NDD-CKD patients (see p-values in Figure 3). Regarding RRT, PD patients showed in MNs higher GSSG levels and GSSG/GSH ratio ($p < 0.05$; Figure 3D,F) compared with HD patients, whereas no differences were observed in PMNs.

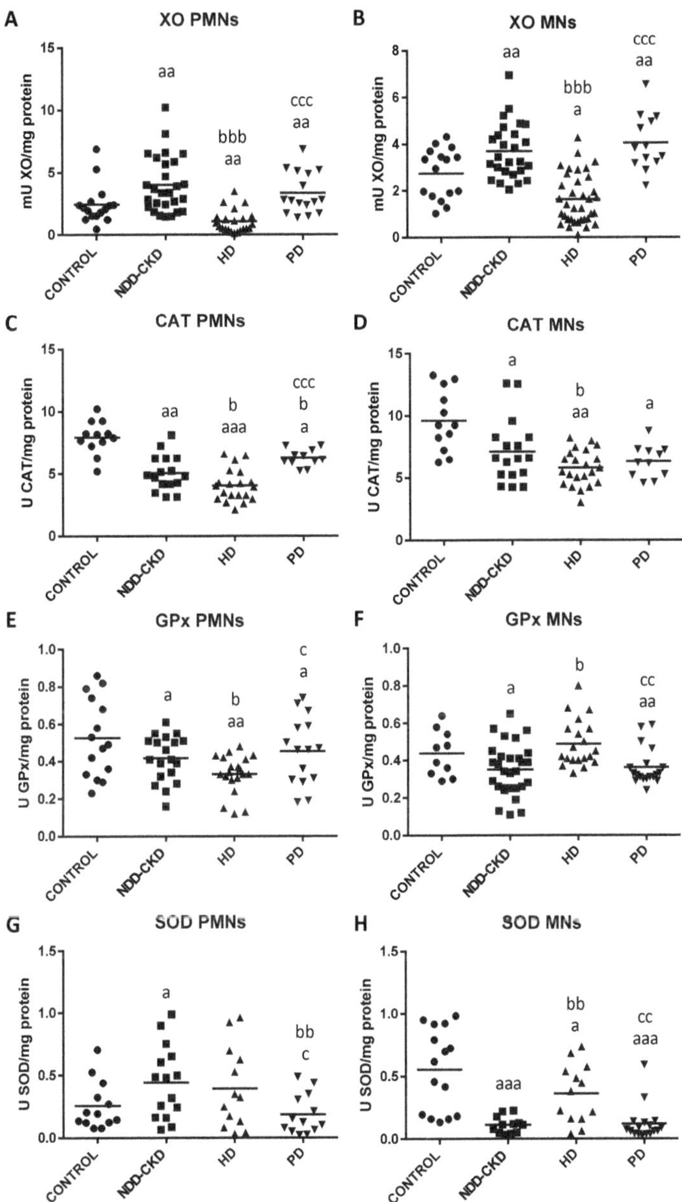

Figure 2. Oxidative stress parameters in isolated peripheral blood polymorphonuclear (PMNs) and mononuclear (MNs) leukocytes from healthy subjects (control), non-dialysis-dependent chronic kidney disease (NDD-CKD), hemodialysis (HD) and peritoneal dialysis (PD) patients. (**A**) PMNs and (**B**) MNs xanthine oxidase activity (XO; mU/mg protein); (**C**) PMNs and (**D**) MNs catalase activity (CAT; U/mg protein); (**E**) PMNs and (**F**) MNs glutathione peroxidase activity (GPx; U/mg protein); (**G**) PMNs and (**H**) MNs superoxide dismutase activity (SOD; U/mg protein). Data are shown as the mean (horizontal bar) of 10–32 values corresponding to the number of subjects analyzed in each group. Each value is the mean of duplicate assays. ^{a} $p < 0.05$, ^{aa} $p < 0.01$ and ^{aaa} $p < 0.001$ versus control; ^{b} $p < 0.05$, ^{bb} $p < 0.01$ and ^{bbb} $p < 0.001$ versus NDD-CKD patients; ^{c} $p < 0.05$, ^{cc} $p < 0.01$ and ^{ccc} $p < 0.001$ versus HD patients.

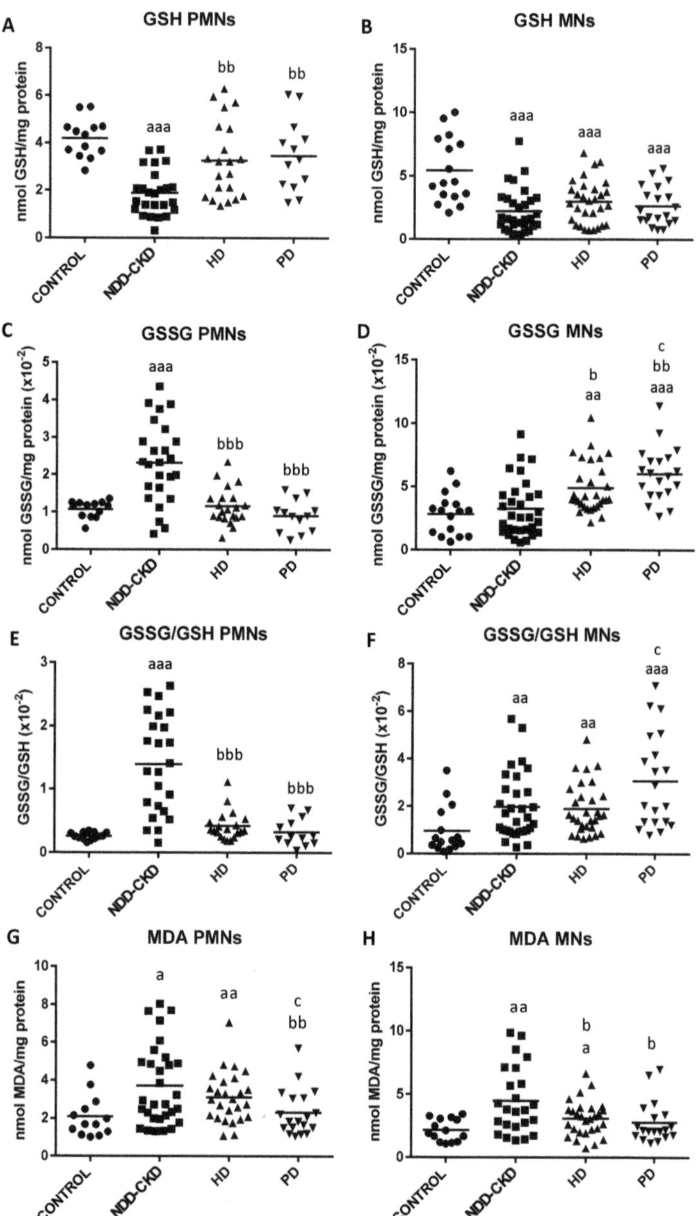

Figure 3. Intracellular glutathione balance and lipid damage in isolated peripheral blood polymorphonuclear (PMNs) and mononuclear (MNs) leukocytes from healthy subjects (control), non-dialysis-dependent chronic kidney disease (NDD-CKD), hemodialysis (HD) and peritoneal dialysis (PD) patients. (**A**) PMNs and (**B**) MNs reduced glutathione content (GSH; nmol/mg protein); (**C**) PMNs and (**D**) MNs oxidized glutathione content (GSSG; nmol/mg protein); (**E**) PMNs and (**F**) MNs GSSG/GSH ratio. (**G**) PMNs and (**H**) MNs malondialdehyde content (MDA; nmol/mg protein). Data are shown as the mean (horizontal bar) of 13–39 values corresponding to the number of subjects analyzed in each group. Each value is the mean of duplicate assays. ^{a} $p < 0.05$, ^{aa} $p < 0.01$ and ^{aaa} $p < 0.001$ versus control; ^{b} $p < 0.05$, ^{bb} $p < 0.01$ and ^{bbb} $p < 0.001$ versus NDD-CKD patients; ^{c} $p < 0.05$ versus HD patients.

Finally, regarding lipid damage (Figure 3G,H), NDD-CKD patients exhibited significantly higher MDA levels in PMNs and MNs ($p < 0.01$) than healthy subjects. Similar results were also observed in leukocytes from HD patients (MDA, $p < 0.05$) but not in PD patients. Interestingly, in both leukocyte populations, the PD group showed lower MDA content ($p < 0.05$) than NDD-CKD patients, whereas the HD group only showed this decreased MDA content in MNs ($p < 0.01$). Regarding RRT, PD patients had in PMNs significantly lower MDA content than HD patients ($p < 0.05$), whereas no differences were observed in MNs.

3.4. Correlations Between Biochemical, Inflammatory and Oxidative Stress Parameters in Plasma

The simple linear correlations between selected biochemical, inflammatory and oxidative stress parameters assessed in plasma from NDD-CKD, HD and PD patients are shown in Table 3.

Table 3. Pearson´s correlation coefficients (*r*) between selected biochemical, inflammatory and oxidative stress parameters in plasma of study participants.

Group	Parameters	*r*	*p*-Value
NDD-CKD	XO; IL-1β	0.333	0.048 *
	XO; CRP	0.776	0.039 *
	IL-1β; MDA	0.511	0.036 *
	SOD; GPx	−0.579	0.002 **
	SOD; LDL-C	−0.689	0.040 *
	GPx; TG	0.810	0.020 *
	GPx; LDL-C	0.775	0.041 *
	GSH; HDL-C	0.651	0.022 *
	MDA; HbA1c	0.676	0.010 **
HD	BMI; HbA1c	0.357	0.028 *
	IL-1β; HbA1c	0.553	0.006 **
	XO; GPx	0.598	0.003 **
	CAT; HbA1c	−0.434	0.043 *
	UA; SOD	0.478	0.022 *
PD	UA; HDL-C	−0.451	0.035 *

Abbreviations: BMI, body mass index; CAT, catalase activity; CRP, C-reactive protein; GPx, glutathione peroxidase activity; GSH, reduced glutathione levels; HbA1c, glycosylated hemoglobin; HDL-C, high-density lipoprotein cholesterol; HD, hemodialysis; IL-1β, interleukine-1β release; LDL-C, low-density lipoprotein cholesterol; MDA, malondialdehyde levels; NDD-CKD, non-dialysis-dependent chronic kidney disease; PD, peritoneal dialysis; SOD, superoxide dismutase activity; TG, triglycerides; UA, uric acid; XO, xanthine oxidase activity. * $p < 0.05$ and ** $p < 0.01$.

In the NDD-CKD group, XO activity was positively associated with the inflammatory markers, such as IL-1β and CRP levels ($p < 0.05$). Remarkably, lipid damage (MDA content) was also positively related with the IL-1β release ($p < 0.05$) and HbA1c levels ($p < 0.01$). In addition, GPx activity was negatively associated with both SOD activity ($p < 0.01$), whereas a direct relationship was observed with both TG and LDL-C levels ($p < 0.05$). Interestingly, SOD activity was significantly negatively correlated with the LDL-C (p < 0.05), meanwhile GSH content directly correlated with the HDL-C ($p < 0.05$).

When the correlations were analyzed in the RRT groups, in HD patients, BMI was significantly positively correlated with the HbA1C. Interestingly, HbA1C content was also positively and negatively associated with the IL-1β and CAT ($p < 0.05$), respectively. Moreover, a positive correlation was also observed between XO and GPx activities ($p < 0.01$), and UA content and SOD activity ($p < 0.05$). Finally, the UA was positively correlated with the HDL-C ($p < 0.05$) in PD patients. No significant differences were found between the other variables analyzed (data not shown).

4. Discussion

To our knowledge, this is the first study that analyzed the changes in several redox state parameters in different types of isolated peripheral blood leukocytes, as well as in plasma, from advanced NDD-CKD patients, that also analyzed the effect of HD and PD procedures on redox status. Our results clearly revealed an altered redox state with a marked increase of oxidative stress and damage in plasma and isolated PMNs and MNs from NDD-CKD patients, and interestingly, is more exacerbated in HD and PD. Furthermore, we also found a different pattern of increased oxidative stress and damage depending on the localization (plasma or leukocytes) and cell type (PMNs and MNs). Thus, several of these redox state alterations were greatly impaired in PMNs, whereas others were only observed in MNs. Interestingly, our results also demonstrated differences between HD and PD procedures, the PD patients showing greater oxidative stress and damage than the HD patients, especially in MNs. This altered redox balance could be mediated by the higher inflammation and/or the changes in lymphocytes populations also observed in NDD-CKD, HD and PD patients.

A chronic oxidative-inflammatory state plays a crucial role in the onset and progression of CKD [5,12–16]. Indeed, numerous studies have demonstrated a significantly increased peripheral oxidative-inflammatory state in ESRD patients, being exacerbated by dialysis treatments [12,13,15,31,48]. Nevertheless, most research has been focused on the study of redox makers in serum or plasma, but not in immune cells. Since nutrition and diet significantly influence the extracellular redox status (plasma), and most enzymatic and non-enzymatic antioxidant mechanisms are intracellular [49], we analyzed the changes in several redox state parameters, not only in plasma but also in different types of peripheral leukocytes from advanced CKD patients. Furthermore, an optimal redox status of immune cells is essential for adequate homeostasis in the physiological systems [49]. In CKD patients, most studies have focused on lymphocyte populations, but little is known about the changes in the redox status of PMNs leukocytes along CKD progression. Since phagocytes have been proposed as the main cells that contribute to the oxidative stress associated with age-related diseases [9–11], we also assessed several redox markers in PMNs and MNs leukocytes in order to elucidate possible differences between them.

Regarding the plasma, our study demonstrated increased circulating levels of oxidative stress markers (higher XO activity and MDA content) and an impairment of the antioxidant systems (lower CAT and GPx activities and GSH concentration) in NDD-CKD, HD and PD patients, compared to healthy subjects. Our results are in consonance with other studies that have revealed increased oxidative stress markers associated with CKD [5,8,15,16,44]. This research demonstrated a high production of oxidant compounds (ROS, GSSG, etc.), lipid and protein oxidative damage markers (MDA, AOPPs, etc.), or altered levels of antioxidant (CAT, GPx, SOD, GSH, etc.) in plasma or serum of ESRD patients [5,8,15,16], as well as in HD and PD patients [12–15,28,31,32,48,50]. In this pathophysiological context, the possible malnutrition, dietary restrictions and the hypoalbuminemia state, together with the higher accumulation of non-dialyzable uremic toxins (e.g., indoxyl-sulphate, etc.) that CKD patients suffer, could contribute to the low availability of GSH and other antioxidants compounds and also promote higher oxidative-inflammatory stress in these subjects [5,48]. Indeed, the uremic toxins promote leukocyte activation, stimulating both ROS overproduction (mainly via XO and NADPH oxidase) and pro-inflammatory cytokines release [30,48]. The XO is a pro-oxidant enzyme involved in purine metabolism responsible for UA and ROS production [51]. On the one hand, circulating XO activity may be dangerous because once in circulation, it has the ability to activated phagocytic cells and produce $O_2^{\bullet-}$ and H_2O_2, and it can be distributed to remote tissues [51,52], as well as internalized into vascular and other cells; therefore, it may further initiate oxidative damage exerting pathological effects in CKD patients [51,53]. On the other hand, although in optimal concentrations, UA has strong antioxidant properties, under conditions of oxidative stress, high UA concentrations may function as a powerful oxidizing compound, mainly when antioxidant defenses are diminished [8,27]. Moreover, hyperuricemia has been associated with CVDs and morbi-mortality in CKD patients, possibly through oxidative stress and endothelial dysfunction [48]. In our study, NDD-CKD patients showed higher UA levels than controls. Thus, these high UA concentrations, together with the increased XO activity and impaired antioxidant systems (GSH, GPx, CAT), could explain the higher plasma lipid damage (MDA content) also observed in NDD-CKD patients. MDA is a powerful oxidant compound, which can interact with others oxidized compounds, being potentially mutagenic and atherogenic [44]. Interestingly, it has been found that HD patients with CVDs had marked elevation of serum MDA compared with those without CVDs [33,54]. Moreover, Goundoin et al. found a positive correlation between plasma XO activity and MDA content in CKD and HD patients and proposed that, regardless of UA levels, plasma XO activity is an important predictor of CVDs in these patients [27]. Indeep, an increase in the circulating XO content could accelerate vascular oxidation via ROS generation, contributing to endothelial dysfunctions in CKD patients [55]. In our study, we also found higher percentages of CVDs in NDD-CKD and HD populations. However, the higher XO activity in plasma and isolated leukocytes were observed in NDD-CKD and PD patients, but not in HD patients, as will be discussed later.

Because oxidation and inflammation pathways can promote and exacerbate one another [9–11], we also analyzed the basal inflammation, measuring the IL-1β release and CRP levels in plasma, which provide higher specificity than other biomarkers (e.g., albumin) [2,56]. We observed increased CRP and IL-1β levels in NDD-CKD, HD and DP patients, which may contribute to triggering substantial collateral cellular oxidative damage in these patients. In fact, we also found a positive and significant association between IL-1β release and MDA content in NDD-CKD patients, as well as between IL-1β and HbA1c levels in HD patients. Although the studies are controversial, a strong upregulation of pro-inflammatory cytokines (e.g., IL-1β, IL-6, etc.) has been observed in CKD patients but also in dialyzed patients [5,15,50]. The increase of IL-1β has been also associated with the development of insulin resistance and modifications of lipid components, lipoproteins and proteins, resulting in lipid metabolism disorders in CKD [8]. In our study, a high percentage of NDD-CKD, HD and PD patients (79.5%, 59.0% and 63.6 %, respectively) suffered dyslipidemia, whereas NDD-CKD patients also showed marked hypertriglyceridemia. Because CRP is a powerful predictor of CVDs-morbidity and mortality in dialysis patients [6], the high plasma CRP levels observed in HD and PD patients may also contribute to CVDs in these patients. Moreover, because the outcome of local inflammation may directly build on the antioxidant ability [10], the impaired antioxidant systems observed in plasma from NDD-CKD, HD and PD patients may contribute to oxidative tissue damage, and consequently, to a systemic inflammatory response. Interestingly, although we did not find a relationship between inflammatory markers (IL-1β and CRP) and lipometabolism parameters in plasma from CKD patients, we observed a direct correlation between the GPx activity and both TG and LDL-C levels in NDD-CKD patients and between the GSH content and HLD-C levels. By contrast, SOD activity was significantly negatively correlated with LDL-C.

Regarding the isolated peripheral leukocytes, our study revealed enhanced oxidative stress and damage in PMNs and MNs from NDD-CKD, HD and PD patients, showing a marked increase of XO activity and lipid damage (MDA), an impairment of the antioxidant enzymes (higher SOD and lower CAT and GPx activities) and an altered redox GSSG/GSH balance compared to healthy controls. Interestingly, we observed notable differences in the redox status between PMNs and MNs in NDD-CKD, as well as between HD and PD patients. Concerning the pro-oxidant XO enzyme, our results revealed a higher XO activity in PMNs and MNs from NDD-CKD and PD patients, which can promote ROS overproduction, as well as exert cytotoxicity on the kidney and the vascular endothelium during CKD progression [16]. Since XO and its products are directly and indirectly implicated in immunomodulatory activities [52], this excessive increase in XO activity could induce the activation of PMNs, promoting phagocytic killing, the expression of other pro-oxidant enzymes (NADPH oxidase, etc.) or the nuclear factor-κB activation and translocation, and consequently, a pro-inflammatory cytokines overproduction [51–53]. Moreover, because the XO gene expression is markedly upregulated by pro-inflammatory cytokines (e.g., IL-1β) [51], the higher inflammation observed in NDD-CKD patients could also mediate a higher expression of XO, explaining the increased XO activity and the positive correlation observed between XO activity and inflammatory markers (IL-1β and CRP levels) in these patients. Interestingly, HD patients showed a marked reduction of XO activity in plasma, PMNs and MNs compared to PD and NDD-CKD patients. Although the research regarding XO in dialyzed patients is not clear, one study revealed that XO activity was markedly elevated in HD and PD patients independently of dialysis modality [57], whereas other authors proposed that the extracellular XO activity differ depending on the nutritional status of the HD patients [53]. Indeep, serum XO activity decreased during HD procedure in patients with high nutritional risk index and vice versa [53]. Although the decreased XO activity in HD patients appears a priori to be beneficial for these subjects, it could also be indicative of a basically higher content of endothelium-bound XO. Further, circulating XO has the ability to reversibly bind to the endothelial cell surface via glycosaminoglycan rich receptors, leading to ROS overproduction [51–53], especially at locations exposed to mechanical forces [53,58]. Thus, we propose that HD patients could have high circulating XO binding to the extracellular matrix of endothelial cells. This can promote increased lipid damage and the inflammatory cascade activation, and therefore, cause vascular dysfunction in HD patients [53]. In fact, increased XO levels have been documented in the arteries of HD patients [53]. Further studies are needed to elucidate the endothelium-bound XO and its effects on oxidative damage in chronic HD and PD patients.

Concerning the SOD, CAT and GPx antioxidant enzymes, which are essential to prevent oxidative damage throughout CKD progression [5,16], we found striking differences. The SOD catalyzes the conversion of $O_2^{\bullet-}$ into H_2O_2 and oxygen, while CAT and GPx mediate the conversion of toxic H_2O_2 into water. Under normal conditions, H_2O_2 is principally degraded by GPx (due to its higher affinity for H_2O_2 than CAT), whereas under severe oxidative stress conditions, H_2O_2 is mainly degraded by CAT [49]. On the one hand, NDD-CKD patients showed lower SOD activity in MNs and higher SOD in PMNs and plasma in comparison with healthy subjects. This increased SOD activity may be a balancing mechanism to neutralize the excessive $O_2^{\bullet-}$ levels, as has also been observed in both plasma and PMNs of CKD patients [8,11–13]. Paradoxically, PD patients had a decreased SOD activity in PMNs compared with NDD-CKD patients, whereas HD patients presented an enhanced SOD activity in MNs. Most studies have shown higher SOD activity in the serum of HD patients, being associated with the impaired renal functions [16]. Other studies have found high SOD gene expression in leukocytes from ESRD patients, which also correlated with high plasma CRP and $O_2^{\bullet-}$ levels [6,59]. Interestingly, we found a positive correlation between SOD activity and UA content in HD patients. On the other hand, we also observed a significant down-regulation of CAT and GPx in PMNs and MNs from NDD-CKD, HD and PD patients. This is in agreement with several studies that have shown a gradual decrease in the plasma GPx and CAT activities along with the CKD progression [8,12,13,16]. Interestingly, HD patients presented the most marked alteration of these antioxidant enzymes, especially in PMNs. Indeep, HD patients showed lower CAT and GPx activities in PMNs than NDD-CKD patients, whereas increased GPx activity was found in MNs, which could be a compensatory mechanism to protect against excessive oxidative stress. In fact, an increased GPx activity in erythrocytes protects against Hb oxidation and hemolysis, preventing against CVDs in CKD patients [16]. Interestingly, we found a negative correlation between GPx and SOD activities in NDD-CKD patients. Meanwhile, in HD patients, GPx activity was positively associated with XO activity, whereas CAT activity was negatively correlated with HbA1c content. Furthermore, an excess of SOD, CAT and GPx activities may promote detrimental outcomes as a consequence of a deficit of vital cell oxidants (optimal ROS concentrations), causing "reductive stress," which could contribute to weakening cell dysfunctions [16,49]. GPx not only detoxifies H_2O_2 but also catalyzes the reduction of lipid peroxides, thus preventing lipid peroxidation [49].

Therefore, the altered GPx activity observed in leukocytes from NDD-CKD patients may promote the accumulation of lipoperoxides, explaining the higher MDA content also observed in both PMNs and MNs of these patients. Our results are in agreement with other studies that also found elevated MDA levels in CKD patients undergoing HD [25,54]. MDA is a low molecular weight and water-soluble molecule, so it can undergo renal clearance or be dialyzed as a possible renal elimination pathway [25]. Interestingly, HD and PD patients had lower MDA content in their leukocytes than NDD-CKD patients. Although we have not found any study that has evaluated the differences in lipid peroxidation between PMNs and MNs in both NDD-CKD and dialysis patients, it has been proposed that MDA markedly decreased during the HD procedure [12–14,31]. Due to the absence of blood contact with artificial surfaces, PD patients are a particularly interesting population to study the effect of uremia on lipid peroxidation [12,31,38]. Interestingly, in our study, PD patients showed lower MDA levels in their leukocytes compared to HD patients. In this context, it has been described that PD promotes protein oxidation, whereas HD more intensely impacts lipid peroxidation [50].

The glutathione cycle is one of the main intracellular mechanisms to preserve an adequate intracellular redox state [11,60]. In fact, GSH is the major antioxidant involving in cell-signaling regulation, whereas GSSG is highly toxic to cells [60]. Therefore, an optimal GSSG/GSH balance is essential for the preservation of oxidative damage and the maintenance of immune functions [11,60]. In our study, NDD-CKD patients showed in PMNs and MNs lower GSH content, together with higher GSSG and GSSG/GSH ratios, compared with controls. Interestingly, this altered GSSG/GSH balance was more exacerbated in PMNs than in MNs leukocytes. Since GSH is essential to neutralize ROS, its reduction could contribute to higher accumulation of $O_2^{\bullet -}$ and H_2O_2 (mainly via XO or NADPH oxidase production), and together with the increased GSSG content, could explain the marked increase of lipid damage (MDA) also observed in PMNs from NDD-CKD patients. Interestingly, RRT had a different influence on the GSSG/GSH balance depending on the cell typed analyzed. Moreover, HD and PD procedures had a positive effect on the GSSG/GSH status in PMNs, with dialysis patients showing higher GSH content and lower GSSG levels and GSSG/GSH ratio than NDD-CKD patients. This could be a compensatory response in PMNs to cope with high exposure to oxidant compounds. By contrast, lower GSH and higher GSSG content and GSSG/GSH ratios were observed in MNs from HD and PD patients compared with NDD-CKD patients. These findings are in line with previous research showing decreased GSH and increased GSSG and GSSG/GSH in lymphocytes or whole-blood from HD or PD patients [61,62].

In regards to RRT, it is known that HD and PD patients presented both ROS overproduction and mostly compromised antioxidant mechanisms [12–14,31,38]. In fact, HD therapy per se seems to contribute to the oxidant/antioxidant imbalance due to several factors (e.g., the biocompatibility of dialysis membrane and dialysates, etc.) [8,14,31,38], whereas PD does not seem to have a strong benefit over HD procedure, even though the biocompatibility advantage [38]. Indeed, several studies have indicated that increased ROS production in peripheral/peritoneal phagocytes is activated during PD, mainly due to the composition of PD dialysates [12,32]. Interestingly, in our group of PD patients, we found oxidative stress values significantly higher in plasma and isolated leukocytes compared to HD patients, especially in MNs. Thus, PD patients showed a higher XO activity and lower SOD activity in PMNs and MNs than HD patients. However, in MNs, PD patients had a significant reduction of SOD and GPx activities, together with increased GSSG and GSSG/GSH ratio, which may contribute to the accumulation of ROS and lipid damage in these patients. By contrast, PD patients had in PMNs higher CAT and GPx activities and a preserved GSSG/GSH ratio compared with HD patients, suggesting that these compensatory antioxidant mechanisms could contribute to the reduced MDA content also observed in these cells. Overall, based on the limited data reviewed so far, some studies (mainly in plasma) showed increased oxidative stress and damage in HD compared with PD, whereas others reported no changes [12–14,31–33,38,44]. By contrast, some authors proposed that PD is associated with a higher accumulation of oxidant compounds and antioxidant depletion [38,57]. Furthermore, a preserved RRF has been related to reduced oxidative stress and lipid damage in PD patients [12,42,43]. However, although PD patients showed higher RRF than HD patients, we did not observe significant correlations between RRF and oxidative-inflammatory parameters analyzed (data not shown) in these patients. Based on our results, we deduce that PD procedure may contribute to increased oxidative stress on CKD patients, and especially in MNs leukocytes.

An optimal redox state is essential for the effectiveness of the immune functions, which can also be influenced by the changes in the leukocyte populations in terms of the cell types composition, quality and proportions [9,10]. Much evidence indicates that lymphocytes impairment is also involved in immune dysregulation in ESRD patients [46,63,64]. We found remarkable changes with respect to MNs populations (mainly lymphocytes), but not in PMNs (>98% were neutrophils in all groups; data not shown). In several studies, lymphocyte subgroups were lower in dialysis patients than in healthy controls [63,64]. In our study, T- and B-cells were also influenced by dialysis modality. Thus, HD and PD patients showed a lower total number of lymphocytes and CD4+/CD8+ ratio than NDD-CKD patients. Interestingly, HD patients also had a marked reduction of CD4+ T-lymphocytes and NK cells than NDD-CKD patients, whereas PD patients had lower B-cell counts. Our results are in agreement with other studies in which dialysis patients exhibited a high incidence of lymphopenia affecting T- and B-cell populations [46,63,64]. The decreased CD4+ T-cells proportion could promote infection and mortality in HD patients [63]. Moreover, the differences in both uremic toxin accumulation and inflammatory status between HD and PD patients could play a key role in the distribution of T- and B-lymphocytes on these subjects [63,65]. Indeed, HD is associated with more sustained inflammation and lymphocyte activation/exhaustion [65], explaining the results observed in HD patients. Because of the cross-sectional nature of our study, we have to confirm these changes in a very large cohort as well as investigate modifications in immunological parameters after and during HD and PD procedures.

Finally, our study has several limitations: (1) the small patient´s sample size, especially in the PD group, which may cause not to find significant differences in several of the parameters analyzed. (2) Our study did not interfere with or evaluate the lifestyle factors of the CKD patients (nutritional status, diet, etc.) and other non-conventional risks (e.g., anemia, etc.), which also contribute to oxidative stress and inflammation. (3) The cross-sectional design and analysis of data, and the small number of subjects, did not allow us to integrate and analyze the influences of all potential covariables (age, gender, type of dialysates, the different dialysis membranes, vascular access, comorbidities, etc.) into the multivariate models. Therefore, additional prospective and comparative clinical studies in a larger sample will be required to validate our findings and confirm the use of these redox state parameters as biomarkers of CKD progression. Moreover, future large studies addressing potential nutritional interventions to decrease oxidative-inflammatory stress in leukocytes from CKD patients are also necessary.

5. Conclusions

The major strength of this study is that we analyzed the changes of numerous redox state markers in both plasma and isolated peripheral blood leukocytes from advanced CKD patients and also analyzed the effect of RRT on redox status. Taken together, our study demonstrated that NDD-CKD, HD and PD patients presented significantly higher oxidative stress and damage in their PMNs and MNs leukocytes than healthy subjects, as well as in plasma. Interestingly, the oxidative stress and damage were more exacerbated in HD and PD patients than in NDD-CKD patients. This enhanced oxidative stress could be mediated by a higher inflammation state, and the changes in lymphocytes populations were also observed in these patients. Interestingly, we also found notable differences in the redox state depending on the localization (plasma/leukocytes) and the immune cell type (PMNs/MNs) analyzed. In fact, several redox state alterations were greatly marked in plasma or PMNs, whereas others were only observed in MNs. Thus, dialysis treatments had a positive effect preserving the GSSG/GSH balance in PMNs, but not in MNs. This study also demonstrated remarkable differences between HD and PD procedures, the PD patients showing greater oxidative stress and damage than the HD patients, especially in MNs. Our results could explain one of the underlying mechanisms of immune dysregulation in CKD patients. Furthermore, since PMNs and MNs are very easy to obtain and due to the simplicity of the assays performed, the assessment of several redox state parameters in these leukocytes, together with the plasma, could be used as potential biomarkers in the CKD progression, as well as in monitoring HD and PD procedures in a clinical setting.

Author Contributions: C.V., E.M. and J.C. contributed in the conception and the design of the study; C.V., C.O., N.C., G.V., C.Y., P.J.C. and I.G.d.P. performed the research experiment and the acquisition of data. C.V. and C.O. performed the statistical analysis; C.Y., P.J.C. and I.G.d.P. provided the samples for the experimental study; C.Y., P.J.C. and E.M. made the clinical diagnosis of NDD-CKD, HD and PD patients and control's selection; E.M. and J.C. acquired funding support; C.V. wrote the manuscript; C.O., N.C., G.V., C.Y., P.J.C., I.G.d.P., E.M. and J.C. revised and completed the final draft of the article. All authors have read and agreed to the published version of the manuscript.

Funding: This study was funded by Instituto de Salud Carlos III through the project "PI17/01029" and "PI20/01321" (Co-funded by European Regional Development Fund "A way to make Europe"). G.V. was funded by a fellowship program "Contratos FiBio-HCR asociados a Proyectos de Investigación" (PEJ-2020-AI/BMD-18141; Comunidad de Madrid y Fondo Social Europeo), Universidad de Alcalá and IRYCIS, Madrid, Spain.

Institutional Review Board Statement: The study was conducted according to the guidelines of the Declaration of Helsinki and approved by the Ethics Committee of Hospital Universitario 12 de Octubre (protocol code: CEIC 17/407; date of approval: 28/11/2017).

Informed Consent Statement: Informed consent was obtained from all subjects involved in the study.

Data Availability Statement: Data is contained within the article.

Acknowledgments: We are grateful for the help of the volunteers who took part in this study. We would also like to thank the medical staff Nephrology Department, Hospital Universitario 12 de Octubre.

Conflicts of Interest: The authors declare no conflict of interest.

References

1. Hill, N.R.; Fatoba, S.T.; Oke, J.L.; Hirst, J.A.; O'Callaghan, C.A.; Lasserson, D.S.; Hobbs, F.R. Global Prevalence of Chronic Kidney Disease-A Systematic Review and Meta-Analysis. *PLoS ONE* **2016**, *11*, e0158765. [CrossRef]
2. Ling, X.C.; Kuo, K. Oxidative stress in chronic kidney disease. *Ren. Replace. Ther.* **2018**, *4*, 53. [CrossRef]
3. Bikbov, B.; Purcell, C.A.; Levey, A.S.; Smith, M.; Abdoli, A.; Abebe, M.; Adebayo, O.M.; Afarideh, M.; Agarwal, S.K.; Agudelo-Botero, M.; et al. Global, regional, and national burden of chronic kidney disease, 1990–2017: A systematic analysis for the Global Burden of Disease Study 2017. *Lancet* **2020**, *395*, 709–733. [CrossRef]
4. Stenvinkel, P.; Larsson, T.E. Chronic kidney disease: A clinical model of premature aging. *Am. J. Kidney Dis.* **2013**, *62*, 339–351. [CrossRef]
5. Carracedo, J.; Alique, M.; Vida, C.; Bodega, G.; Ceprián, N.; Morales, E.; Praga, M.; de Sequera, P.; Ramírez, R. Mechanisms of Cardiovascular Disorders in Patients with Chronic Kidney Disease: A Process Related to Accelerated Senescence. *Front. Cell Dev. Biol.* **2020**, *8*, 185. [CrossRef]
6. Rapa, S.F.; Di Iorio, B.R.; Campiglia, P.; Heidland, A.; Marzocco, S. Inflammation and Oxidative Stress in Chronic Kidney Disease-Potential Therapeutic Role of Minerals, Vitamins and Plant-Derived Metabolites. *Int. J. Mol. Sci.* **2020**, *21*, 263. [CrossRef]
7. Zijlstra, L.E.; Trompet, S.; Mooijaart, S.P.; Van Buren, M.; Sattar, N.; Stott, D.J.; Jukema, W. The association of kidney function and cognitive decline in older patients at risk of cardiovascular disease: A longitudinal data analysis. *BMC Nephrol.* **2020**, *21*, 81–91. [CrossRef]
8. Rysz, J.; Franczyk, B.; Lawinski, J.; Gluba-Brzózka, A. Oxidative Stress in ESRD Patients on Dialysis and the Risk of Cardiovascular Diseases. *Antioxidants* **2020**, *9*, 1079. [CrossRef]
9. De la Fuente, M.; Miquel, J. An update of the Oxidation-Inflammation theory of aging: The involvement of the immune system in oxi-inflamm-aging. *Curr. Pharm. Des.* **2009**, *15*, 3003–3026. [CrossRef] [PubMed]
10. Vida, C.; de Toda, I.M.; Cruces, J.; Garrido, A.; González-Sánchez, M.; De la Fuente, M. Role of macrophages in age-related oxidative stress and lipofuscin accumulation in mice. *Redox Biol.* **2017**, *12*, 423–437. [CrossRef]

11. Vida, C.; de Toda, I.M.; Garrido, A.; Carro, E.; Molina, J.A.; De la Fuente, M. Impairment of Several Immune Functions and Redox State in Blood Cells of Alzheimer's Disease Patients. Relevant Role of Neutrophils in Oxidative Stress. *Front. Immunol.* **2018**, *8*, 1974. [CrossRef]
12. Liakopoulos, V.; Roumeliotis, S.; Gorny, X.; Eleftheriadis, T.; Mertens, P.R. Oxidative stress in patients undergoing peritoneal dialysis: A current review of the literature. *Oxid. Med. Cell. Longev.* **2017**, *2017*, 3494867. [CrossRef] [PubMed]
13. Liakopoulos, V.; Roumeliotis, S.; Bozikas, A.; Eleftheriadis, T.; Dounousi, E. Antioxidant supplementation in renal replacement therapy patients: Is there evidence? *Oxid. Med. Cell. Longev.* **2019**, *2019*, 9109473. [CrossRef]
14. Liakopoulos, V.; Roumeliotis, S.; Zarogiannis, S.; Eleftheriadis, T.; Mertens, P.R. Oxidative stress in hemodialysis: Causative mechanisms, clinical implications, and possible therapeutic interventions. *Semin. Dial.* **2019**, *32*, 58–71. [CrossRef]
15. Duni, A.; Liakopoulos, V.; Roumeliotis, S.; Peschos, D.; Dounousi, E. Oxidative Stress in the Pathogenesis and Evolution of Chronic Kidney Disease: Untangling Ariadne's Thread. *Int. J. Mol. Sci.* **2019**, *20*, 3711. [CrossRef] [PubMed]
16. Podkowinska, A.; Formanowicz, D. Chronic Kidney Disease as Oxidative Stress- and Inflammatory-Mediated Cardiovascular Disease. *Antioxidants* **2020**, *9*, 752. [CrossRef]
17. Grune, T.; Sommerburg, O.; Siems, W.G. Oxidative stress in anemia. *Clin. Nephrol.* **2000**, *53* (Suppl. 1), S18–S22.
18. Nuhul, F.; Bhandari, S. Oxidative Stress and Cardiovascular Complications in Chronic Kidney Disease, the Impact of Anaemia. *Pharmaceuticals* **2018**, *11*, 103.
19. Maraj, M.; Kusnierz-Cabala, B.; Dumnicka, P.; Gawlik, K.; Pawlica-Gosiewska, D.; Gala-Badzinska, A.; Zabek-Adamska, A. Redox balance correlates with nutritional status among patients with End-Stage Renal Disease treated with maintenance hemodialysis. *Oxid. Med. Cell. Longev.* **2019**, *2019*, 6309465. [CrossRef]
20. Garibotto, G.; Picciotto, D.; Saio, M.; Esposito, P.; Verzola, D. Muscle protein turnover and low-protein diets in patients with chronic kidney disease. *Nephrol. Dial. Transplant.* **2020**, *35*, 741–751. [CrossRef] [PubMed]
21. Johnson-Davis, K.L.; Fernelius, C.; Eliason, N.B.; Wilson, A.; Beddhu, S.; Roberts, W.L. Blood enzymes and oxidative stress in chronic kidney disease: A cross sectional study. *Ann. Clin. Lab. Sci.* **2011**, *41*, 331–339.
22. Liakopoulos, V.; Jeron, A.; Shah, A.; Bruder, D.; Mertens, P.R.; Gorny, X. Hemodialysis-related changes in phenotypical features of monocytes. *Sci. Rep.* **2018**, *8*, 13964. [CrossRef] [PubMed]
23. Mihai, S.; Codrici, E.; Popescu, I.D.; Enciu, A.M.; Albulescu, L.; Necula, L.G.; Mambet, C.; Anton, G.; Tanase, C. Inflammation-related mechanisms in chronic kidney disease prediction, progression, and outcome. *J. Immunol. Res.* **2018**, *2018*, 2180373. [CrossRef] [PubMed]
24. Fortuño, A.; Beloqui, O.; San José, G.; Moreno, M.U.; Zalba, G.; Díez, J. Increased phagocytic nicotinamide adenine dinucleotide phosphate oxidase dependent superoxide production in patients with early chronic kidney disease. *Kidney Int. Suppl.* **2005**, *22*, S71–S75. [CrossRef]
25. Puchades-Montesa, M.J.; González-Ricoa, M.A.; Solís-Salgueroa, M.A.; Maicasa-Torregrosa, I.; Juan-García, I.; Miquel-Carrascoa, A.; Tormos-Muñoz, M.C.; Sáez-Tormoc, G. Study of oxidative stress in advanced kidney disease. *Nefrología* **2009**, *29*, 502.
26. Popolo, A.; Autore, G.; Pinto, A.; Marzocco, S. Oxidative stress in patients with cardiovascular disease and chronic renal failure. *Free Radic. Res.* **2013**, *47*, 346–356. [CrossRef] [PubMed]
27. Gondouin, B.; Jourde-Chiche, N.; Sallee, M.; Dou, L.; Cerini, C.; Loundou, A.; Morange, S.; Berland, Y.; Burtey, S.; Brunet, P.; et al. Plasma xanthine oxidase activity is predictive of cardiovascular disease in patients with chronic kidney disease, independently of uric acid levels. *Nephron* **2015**, *131*, 167–174. [CrossRef]
28. Duni, A.; Liakopoulos, V.; Rapsomanikis, K.P.; Dounousi, E. Chronic Kidney Disease and Disproportionally Increased Cardiovascular Damage: Does Oxidative Stress Explain the Burden? *Oxid. Med. Cell. Longev.* **2017**, *2017*, 9036450. [CrossRef] [PubMed]
29. Zhou, Q.; Wu, S.; Jiang, J.; Tian, J.; Chen, J.; Yu, X.; Chen, P.; Mei, C.; Xiong, F.; Shi, W.; et al. Accumulation of circulating advanced oxidation protein products is an independent risk factor for ischaemic heart disease in maintenance haemodialysis patients. *Nephrology* **2012**, *17*, 642–649. [CrossRef]
30. Daenen, K.; Andries, A.; Mekahli, D.; Van Schepdael, A.; Jouret, F.; Bammens, B. Oxidative stress in chronic kidney disease. *Pediatr. Nephrol.* **2019**, *34*, 975–991. [CrossRef]
31. Liakopoulos, V.; Roumeliotis, S.; Gorny, X.; Dounousi, E.; Mertens, P.R. Oxidative Stress in Hemodialysis Patients: A Review of the Literature. *Oxid. Med. Cell. Longev.* **2017**, *2017*, 3081856. [CrossRef]
32. Roumeliotis, S.; Eleftheriadis, T.; Liakopoulos, V. Is oxidative stress an issue in peritoneal dialysis? *Semin. Dial.* **2019**, *32*, 463–466. [CrossRef] [PubMed]
33. Sangeetha-Lakshmi, B.; Harini-Devi, N.; Suchitra, M.M.; Srinivasa-Rao, P.; Siva-Kumar, V. Changes in the inflammatory and oxidative stress markers during a single hemodialysis session in patients with chronic kidney disease. *Ren. Fail.* **2018**, *40*, 534–540. [CrossRef] [PubMed]
34. Pérez-García, R.; Ramírez, R.; de Sequera, P.; Albalate, M.; Puerta, M.; Ortega, M.; Ruiz, M.C.; Alcazar-Arroyo, R. Citrate dialysate does not induce oxidative stress or inflammation in vitro as compared to acetate dialysate. *Nefrología* **2017**, *37*, 630–637. [CrossRef] [PubMed]
35. Vida, C.; Carracedo, J.; Sequera, P.; Bodega, G.; Pérez, R.; Alique, M.; Ramírez, R. Increasing the Magnesium Concentration in Various Dialysate Solutions Differentially Modulates Oxidative Stress in a Human Monocyte Cell Line. *Antioxidants* **2020**, *9*, 319. [CrossRef] [PubMed]

36. Wu, X.; Dai, H.; Liu, L.; Xu, C.; Yin, Y.; Yi, J.; Bielec, M.D.; Han, Y.; Lia, S. Citrate reduced oxidative damage in stem cells by regulating cellular redox signaling pathways and represent a potential treatment for oxidative stress-induced diseases. *Redox Biol.* **2019**, *21*, 101057. [CrossRef]
37. Ogunro, P.S.; Olujombo, F.A.; Ajala, M.O.; Oshodi, T.T. The effect of a membrane dialyzer during hemodialysis on the antioxidant status and lipid peroxidation of patients with end-stage renal disease. *Saudi J. Kidney Dis. Transpl.* **2014**, *25*, 1186–1193. [CrossRef] [PubMed]
38. Poulianiti, K.P.; Kaltsatou, A.; Mitrou, G.I.; Jamurtas, A.Z.; Koutedakis, Y.; Maridaki, M.; Stefanidis, I.; Sakkas, G.K.; Karatzaferi, C. Systemic redox imbalance in chronic kidney disease. *Oxid. Med. Cell. Longev.* **2016**, *2016*, 8598253. [CrossRef] [PubMed]
39. Sepe, V.; Gregorini, M.; Rampino, T.; Esposito, P.; Coppo, R.; Galli, F.; Libetta, C. Vitamin e-loaded membrane dialyzers reduce hemodialysis inflammaging. *BMC Nephrol.* **2019**, *20*, 412. [CrossRef] [PubMed]
40. Yeter, H.H.; Korucu, B.; Akcay, O.F.; Derici, K.; Derici, U.; Arinsoy, T. Effects of medium cut-off dialysis membranes on inflammation and oxidative stress in patients on maintenance hemodialysis. *Int. Urol. Nephrol.* **2020**, *52*, 1779–1789. [CrossRef] [PubMed]
41. Terawaki, H.; Matsuyama, Y.; Era, S.; Matsuo, N.; Ikeda, M.; Ogura, M.; Yokoyama, K.; Yamamoto, H.; Hosoya, T.; Nakayama, M. Elevated oxidative stress measured as albumin redox state in continuous ambulatory peritoneal dialysis patients correlates with small uraemic solutes. *Nephrol. Dial. Transplant.* **2007**, *22*, 968. [CrossRef] [PubMed]
42. Ignace, S.; Fouque, D.; Arkouche, W.; Steghens, J.P.; Guebre-Egziabher, F. Preserved residual renal function is associated with lower oxidative stress in peritoneal dialysis patients. *Nephrol. Dial. Transplant.* **2009**, *24*, 1685–1689. [CrossRef] [PubMed]
43. Furuya, R.; Kumagai, H.; Odamaki, M.; Takahashi, M.; Miyaki, A.; Hishida, A. Impact of residual renal function on plasma levels of advanced oxidation protein products and pentosidine in peritoneal dialysis patients. *Nephron Clin. Pract.* **2009**, *112*, c255–c261. [CrossRef]
44. Krata, N.; Zagozdzon, R.; Foroncewicz, B.; Mucha, K. Oxidative Stress in Kidney Diseases: The Cause or the Consequence? *Arch. Immunol. Ther. Exp. (Warsz.)* **2018**, *66*, 211–220. [CrossRef]
45. Dixon, J.R. The international conference on harmonization good clinical practice guideline. *Qual. Assur.* **1998**, *6*, 65–74. [CrossRef] [PubMed]
46. Molina, M.; Allende, L.M.; Ramos, L.E.; Gutiérrez, E.; Pleguezuelo, D.E.; Hernández, E.R.; Ríos, F.; Fernández, C.; Praga, M.; Morales, E. CD19+ B-Cells, a new biomarker of mortality in hemodialysis patients. *Front. Immunol.* **2018**, *9*, 1221. [CrossRef] [PubMed]
47. Hissin, P.J.; Hilf, R. A fluorometric method for determination of oxidized and reduced glutathione in tissues. *Anal. Biochem.* **1976**, *74*, 214–226. [CrossRef]
48. Russa, D.; Pellegrino, D.; Montesanto, A.; Gigliotti, P.; Perri, A.; Russa, A.; Bonofiglio, R. Oxidative Balance and Inflammation in Hemodialysis Patients: Biomarkers of Cardiovascular Risk? *Oxid. Med. Cell. Longev.* **2019**, *2019*, 8567275. [CrossRef]
49. de Toda, I.M.; Miguélez, L.; Vida, C.; Carro, E.; De la Fuente, M. Altered Redox State in Whole Blood Cells from Patients with Mild Cognitive Impairment and Alzheimer's Disease. *J. Alzheimers Dis.* **2019**, *71*, 153–163. [CrossRef]
50. Tbahriti, H.F.; Kaddous, A.; Bouchenak, M.; Mekki, K. Effect of Different Stages of Chronic Kidney Disease and Renal Replacement Therapies on Oxidant-Antioxidant Balance in Uremic Patients. *Biochem. Res. Int.* **2013**, *2013*, 358985. [CrossRef]
51. Harrison, R. Structure and function of xanthine oxidoreductase: Where are we know? *Free Radic. Biol. Med.* **2002**, *6*, 774–796. [CrossRef]
52. Vida, C.; Rodríguez-Terés, S.; Heras, V.; Corpas, I.; De la Fuente, M.; González, E. The aged-related increase in xanthine oxidase expression and activity in several tissues from mice is not shown in long-lived animals. *Biogerontology* **2011**, *12*, 551–564. [CrossRef] [PubMed]
53. Miric, D.; Kisic, B.; Stolic, R.; Miric, B.; Mitic, R.; Janicijevic-Hudomal, S. The role of xanthine oxidase in hemodialysis-induced oxidative injury: Relationship with nutritional status. *Oxid. Med. Cell. Longev.* **2013**, *2013*, 245253. [CrossRef] [PubMed]
54. Boaz, M.; Matas, Z.; Biro, A.; Katzir, Z.E.; Green, M.; Fainaru, M.; Smetana, S. Serum malondialdehyde and prevalent cardiovascular disease in hemodialysis. *Kidney Int.* **1999**, *56*, 1078–1083. [CrossRef] [PubMed]
55. Terawaki, H.; Hayashi, T.; Murase, T.; Iijima, R.; Waki, K.; Tani, Y.; Nakamura, T.; Yoshimura, K.; Uchida, S.; James, J. Relationship between Xanthine Oxidoreductase Redox and Oxidative Stress among Chronic Kidney Disease Patients. *Oxid. Med. Cell. Longev.* **2018**, *2018*, 9714710. [CrossRef] [PubMed]
56. Sproston, N.R.; Ashworth, J.J. Role of C-reactive protein at sites of inflammation and infection. *Front. Immunol.* **2018**, *9*, 754. [CrossRef] [PubMed]
57. Choi, J.Y.; Yoon, Y.J.; Choi, H.J.; Park, S.H.; Kim, C.D.; Kim, I.S.; Kwon, T.H.; Do, J.Y.; Kim, S.H.; Ryu, D.H.; et al. Dialysis modality-dependent changes in serum metabolites: Accumulation of inosine and hypoxanthine in patients on haemodialysis. *Nephrol. Dial. Transplant.* **2011**, *26*, 1304–1313. [CrossRef]
58. McNally, J.S.; Davis, M.E.; Giddens, D.P.; Saha, A.; Hwang, J.; Dikalov, S.; Jo, H.; Harrison, D.G. Vascular signaling by free radicals role of xanthine oxidoreductase and NAD(P)H oxidase in endothelial superoxide production in response to oscillatory shear stresss. *Am. J. Physiol. Heart Circ. Physiol.* **2003**, *285*, H2290–H2297. [CrossRef] [PubMed]
59. Akiyama, S.; Inagaki, M.; Tsuji, M.; Gotoh, H.; Gotoh, T.; Gotoh, Y.; Oguchi, K. mRNA study on Cu/Zn superoxide dismutase induction by hemodialysis treatment. *Nephron Clin. Pract.* **2005**, *99*, 107–114. [CrossRef]
60. Drögue, W.; Breitkreutz, R. Glutathione and immune function. *Proc. Nutr. Soc.* **2000**, *59*, 595–600. [CrossRef]

61. González-Rico, M.; Puchades-Montesa, J.; García-Ramón, R.; Sáez, G.; Tormos, M.C.; Miguel, A. Effect of hemodialysis therapy on oxidative stress in patients with chronic renal failure. *Nefrología* **2006**, *26*, 218–225. [PubMed]
62. Tarng, D.C.; Chen, T.W.; Huang, T.P.; Chen, C.L.; Liu, T.Y.; Wei, Y.H. Increased oxidative damage to peripheral blood leukocyte DNA in chronic peritoneal dialysis patients. *J. Am. Soc. Nephrol.* **2002**, *13*, 1321–1330. [CrossRef] [PubMed]
63. Xiaoyan, J.; Rongyi, C.; Xuesen, C.; Jianzhou, Z.; Jun, J.; Xiaoquiang, D.; Xiaofang, Y. The difference of T cell phenotypes in end stage renal disease patients under different dialysis modality. *BMC Nephrol.* **2019**, *20*, 301. [CrossRef]
64. Usta, M.; Ersoy, A.; Ayar, Y.; Budak, F. The relationship between lymphocyte subsets, nutritional status and tuberculin reactivity in continuous ambulatory peritoneal dialysis and hemodialysis patients. *Int. Urol. Nephrol.* **2020**, *52*, 1167–1172. [CrossRef] [PubMed]
65. Ducloux, D.; Legendre, M.; Bamoulid, J.; Rebibou, J.M.; Saas, P.; Courivaud, C.; Crepin, T. ESRD-associated immune phenotype depends on dialysis modality and iron status: Clinical implications. *Immun. Ageing* **2018**, *15*, 16. [CrossRef]

MDPI

Article

Glycerol Improves Intracerebral Hemorrhagic Brain Injury and Associated Kidney Dysfunction in Rats

Cheng-Yi Chang [1,2], Ping-Ho Pan [2,3], Jian-Ri Li [4,5], Yen-Chuan Ou [6], Su-Lan Liao [7], Wen-Ying Chen [2], Yu-Hsiang Kuan [8] and Chun-Jung Chen [7,9,*]

1 Department of Surgery, Feng Yuan Hospital, Taichung City 420, Taiwan; c.y.chang.ns@gmail.com
2 Department of Veterinary Medicine, National Chung Hsing University, Taichung City 402, Taiwan; pph.pgi@gmail.com (P.-H.P.); wychen@dragon.nchu.edu.tw (W.-Y.C.)
3 Department of Pediatrics, Tungs' Taichung Metro Harbor Hospital, Taichung City 435, Taiwan
4 Division of Urology, Taichung Veterans General Hospital, Taichung City 407, Taiwan; fisherfishli@yahoo.com.tw
5 Department of Nursing, HungKuang University, Taichung City 433, Taiwan
6 Department of Urology, Tungs' Taichung Metro Harbor Hospital, Taichung City 435, Taiwan; ycou228@gmail.com
7 Department of Medical Research, Taichung Veterans General Hospital, Taichung City 407, Taiwan; slliao@vghtc.gov.tw
8 Department of Pharmacology, Chung Shan Medical University, Taichung City 402, Taiwan; kuanyh@csmu.edu.tw
9 Department of Medical Laboratory Science and Biotechnology, China Medical University, Taichung City 404, Taiwan
* Correspondence: cjchen@vghtc.gov.tw; Tel.: +886-4-23592525 (ext. 4022)

Citation: Chang, C.-Y.; Pan, P.-H.; Li, J.-R.; Ou, Y.-C.; Liao, S.-L.; Chen, W.-Y.; Kuan, Y.-H.; Chen, C.-J. Glycerol Improves Intracerebral Hemorrhagic Brain Injury and Associated Kidney Dysfunction in Rats. *Antioxidants* **2021**, *10*, 623. https://doi.org/10.3390/antiox10040623

Academic Editors: Chiara Nediani and Monica Dinu

Received: 25 March 2021
Accepted: 17 April 2021
Published: 19 April 2021

Abstract: In stroke patients, the development of acute kidney injury (AKI) is closely linked with worse outcomes and increased mortality. In this study, the interplay between post-stroke and AKI and treatment options was investigated in a rodent model of hemorrhagic stroke. Intrastriatal collagenase injection for 24 h caused neurological deficits, hematoma formation, brain edema, apoptosis, blood–brain barrier disruption, oxidative stress, and neuroinflammation in Sprague Dawley rats. Elevation of serum blood urea nitrogen, serum creatinine, urine cytokine-induced neutrophil chemoattractant-1, and urine Malondialdehyde, as well as moderate histological abnormality in the kidney near the glomerulus, indicated evidence of kidney dysfunction. The accumulation of podocalyxin DNA in urine further suggested a detachment of podocytes and structural deterioration of the glomerulus. Circulating levels of stress hormones, such as epinephrine, norepinephrine, corticosterone, and angiotensin II were elevated in rats with intracerebral hemorrhage. Osmotic agent glycerol held promising effects in alleviating post-stroke brain injury and kidney dysfunction. Although the detailed protective mechanisms of glycerol have yet to be determined, the intrastriatal collagenase injection hemorrhagic stroke model in rats allowed us to demonstrate the functional and structural integrity of glomerulus are targets that are vulnerable to post-stroke injury and stress hormones could be surrogates of remote communications.

Keywords: brain edema; hemorrhagic stroke; intracranial pressure; kidney dysfunction; stress hormones

1. Introduction

Stroke is a leading cause of neurological disability and mortality worldwide, resulting in huge public health and socioeconomic burden. There are two main types of stroke, ischemic and hemorrhagic stroke, and approximately 87% of stroke patients are ischemic [1]. Chronic kidney disease (CKD) has long been recognized as a risk factor for stroke. Conversely, stroke may lead to kidney dysfunction, and acute kidney injury (AKI) adversely impacts patient outcomes [2–5]. Thus, a better understanding of post-stroke AKI and identifying intervention strategies may help provide options for ameliorating post-stroke disease progression.

AKI is characterized by kidney dysfunction resulting in disturbance of electrolytes, acid–base, and homeostasis of fluids with a clinical spectrum ranging from mild, asymptomatic injury to severe injury. Serum and urine biomarkers, as well as estimated glomerular filtration rate (eGFR), have been validated for the early detection of AKI. Serum cystatin C level, but not creatinine, blood urea nitrogen (BUN), β2-microglobulin, and eGFR in the intensive care unit have been demonstrated to be important biomarker for predicting AKI and mortality in stroke patients [3]. Therefore, the identification of alternative biomarkers for the early detection of stroke-associated AKI is of interest for therapeutic treatments.

Oxidative stress, neuroinflammation, blood–brain barrier (BBB) disruption, and apoptosis have substantial roles in secondary brain injury after an ischemic and hemorrhagic stroke. The latter includes intracerebral hemorrhage (ICH) and subarachnoid hemorrhage (SAH) [6–9]. Additionally, alteration of the sympathetic nervous system, hypothalamic pituitary axis, and renin–angiotensin–aldosterone system may occur and impact stroke disease outcome involving oxidative stress and neuroinflammation [10–12]. The generation and expansion of hematoma are closely linked to hemorrhagic stroke neurological deficits. Hematoma and derived products act as neurotoxins and mainly contribute to edema formation and brain injury [13]. Clinically, brain edema represents a deteriorative complication after hemorrhagic stroke and may lead to higher intracranial pressure and a worse outcome [14]. Osmotherapeutic agents are useful adjuncts to reduce neuroinflammation following hemorrhagic stroke [15]. Besides its effects on the brain, over-activation of the sympathetic nerve system, hypothalamic–pituitary axis, and renin–angiotensin–aldosterone system impairs kidney function, thereby contributing to developing AKI [12,16,17]. This highlights the possibility that osmotic agents may have an ameliorating effect on stroke-associated kidney dysfunction.

Osmotic agents have been used to treat brain edema in stroke patients. Although its effects on the noninfarcted hemisphere and large hemispheric infarction remain controversial, glycerol possesses the ability to decrease brain edema and intracranial pressure [18–20]. Mannitol and hypertonic saline further reduce ICH brain injury and neuroinflammation [15]. To investigate whether osmotic agents are potentially capable of alleviating post-stroke brain injury and associated kidney dysfunction, the effects of glycerol were evaluated in a rodent model of stroke. Patients with hemorrhagic stroke suffer from a worse outcome than those with ischemic stroke, and hemorrhagic transformation is a common complication of ischemic stroke, which is exacerbated by thrombolytic therapy [21]. Therefore, an ICH animal model was established using Sprague-Dawley to investigate the aforementioned phenomena.

2. Materials and Methods

2.1. Experimental Allocation and ICH Induction

Adult male Sprague-Dawley rats (200–230 g) were purchased from BioLASCO (Taipei, Taiwan) and were kept in conventional cages with free access to food and water. Rats were allocated to four groups: sham with saline (n = 32); ICH with saline (n = 32); sham with glycerol (n = 32); ICH with glycerol (n = 32). Under anesthesia with isoflurane (2–4%), the rat's head was fixed in a stereotactic apparatus, and stereotactic surgery was performed following the relevant study [22]. ICH was induced by injection of type IV collagenase (0.3 U/2 μL saline) through a Hamilton syringe at the coordinates corresponding to the striatum (3 mm lateral to the midline, 0.2 mm posterior to bregma, depth 6 mm below the surface of the skull). After injection, the syringe was remained for an additional 3 min to minimize the leakage of collagenase. The burr holes were then sealed with bone wax. Sham-operated rats received the same stereotactic surgical processes and saline injection. The experimental protocols adhered to the Institute's guidelines and were approved by the Institutional Animal Care and Use Committee of Taichung Veterans General Hospital (IACUC approval code: La-100859, IACUC approval date: 11 November 2011). Glycerol at a dose of 6 mL/kg (10% glycerol) or an equal volume of saline was intraperitoneally injected into rats 30 min after surgery. The dose of glycerol was performed according to a previously described study with modifications [23]. All rats were euthanized for analyses 24 h after surgery.

2.2. Morphological Examination

After euthanasia, the brains (n = 8 per group) were quickly removed, chilled in cold phosphate-buffered saline (PBS), and 2 mm coronal slices were cut using a tissue splicer. Seven sections were photographed, and all visible areas with hematoma were delineated. The percentage of the marked area was calculated.

2.3. Neurological Evaluation

To evaluate sensorimotor performance (n = 8 per group), a modified six-point neurological deficit severity scoring criteria was applied according to our previously reported study [24]. The scoring criteria in the neurological evaluation were as follows: 0, no neurological deficit; 1, difficulty in fully extending the left forepaw; 2, unable to extend the left forepaw; 3, mild circling to the left; 4, severe cycling to the left; and 5, falling to the left.

2.4. Brain Edema Evaluation

After euthanasia, the brains (n = 8 per group) were quickly removed and separated into contralateral and ipsilateral hemispheres to isolate the striatum. The dissected contralateral and ipsilateral striatal tissues were dried in an oven at 110 °C for 24 h. The water content was calculated by the wet/dry weight method [24]. Data are expressed as the subtraction of ipsilateral content with contralateral content in the same rat.

2.5. Caspase 3 Activity Assay

After euthanasia, the brains (n = 8 per group) were quickly removed and separated into contralateral and ipsilateral hemispheres to isolate the striatum. The dissected ipsilateral striatal tissues were subjected to the measurement of caspase 3 activity using a caspase-3 colorimetric assay kit (BioVision, Mountain View, CA, USA).

2.6. Evans Blue Extravasation Assay

Three hours prior to the end of the study, Evans blue (4%, 1 mL/kg) was injected into the rats via the tail vein. After euthanasia, rats (n = 8 per group) were perfused with heparinized saline. The dissected ipsilateral striatal tissues were weighed, homogenized in PBS (500 µL), and centrifuged. The obtained supernatants were mixed with trichloroacetic acid (500 µL, 100%) and stand overnight at 4 °C. After centrifugation at 12,000 rpm at 4 °C for 10 min, the supernatants were collected and subjected to the measurement of Evans blue content using a spectrophotometer (absorbance at 620 nm). The contents of Evans blue were calculated using a standard solution.

2.7. Urine, Blood, and Tissue Sample Collection

Prior to sacrifice for analyses, rats were housed in metabolic cages for 12 h for the collection of urine samples. At the end of the study, rats were euthanized, and the blood samples were withdrawn from the left femoral artery. Tissues of the kidney and brain were rapidly dissected and stored in liquid nitrogen or were soaked in formalin until analyses could be performed.

2.8. Histological Examination

After euthanasia, the left kidney (n = 8 per group) was quickly removed, fixed in 4% buffered formaldehyde, and embedded in paraffin. Sections (4 µm) were then deparaffinized, rehydrated, and stained with Hematoxylin/Eosin (H&E) according to standard procedures. Digitalized images were captured at 200X and 400X magnification using a light microscope equipped with a digital camera (Nikon, ECLIPSE, 50i, Tokyo, Japan).

2.9. Biochemical Analyses

The serum levels of BUN and creatinine were measured using automated, standardized procedures (Roche Hitachi 917/747, Mannheim, Germany). The levels of kidney injury molecule-1 (KIM-1; R&D Systems, Minneapolis, MN, USA), cytokine-induced neutrophil chemoattractant-1 (CINC-1; R&D Systems, Minneapolis, MN, USA), neutrophil gelatinase-associated lipocalin (NGAL, R&D Systems, Minneapolis, MN, USA), corticosterone (BioVendor, Germany), norepinephrine (LDN, Nordhorn, Germany), epinephrine (LDN, Nordhorn, Germany), and angiotensin II (Ang II, R&D Systems, Minneapolis, MN, USA) were determined through enzyme-linked immunosorbent assay (ELISA) kits according to the manufacturer's instructions.

2.10. Lipid Peroxidation Product Measurement

Levels of lipid peroxidation product (n = 8 per group) were measured using a TBARS assay kit (Cayman Chemical, Ann Arbor, MI, USA) according to the manufacturer's instructions. Data are expressed as Malondialdehyde (MDA) equivalents.

2.11. Measurement of Glutathione (GSH)

Levels of GSH (n = 8 per group) in dissected ipsilateral striatal tissues and kidney cortical tissues were measured using a Glutathione Assay Kit (Cayman Chemical, Ann Arbor, MI, USA) according to the manufacturer's instructions.

2.12. Measurement of Tumor Necrosis Factor-α (TNF-α) and Interleukin-1β (IL-1β)

After euthanasia, the brains (n = 8 per group) were quickly removed and separated into contralateral and ipsilateral hemispheres to isolate the striatum. The dissected ipsilateral striatal tissues were subjected to the measurement of TNF-α and IL-1β content using ELISA (R&D Systems, Minneapolis, MN, USA).

2.13. Western Blot

After euthanasia, the brains (n = 8 per group) were quickly removed and separated into contralateral and ipsilateral hemispheres to isolate the striatum. The dissected ipsilateral striatal tissues were subjected to the extraction of proteins (tissue protein extraction reagents, Pierce Biotechnology, Rockford, IL, USA) and conduction of a standardized SDS–PAGE and PVDF membrane transfer. After incubation with 5% skim milk for 30 min, the specific proteins on the membranes were recognized with the corresponding antibodies, including Matrix Metalloproteinase-9 (MMP-9, mouse monoclonal antibody, 1:1000), MMP-2 (mouse monoclonal antibody, 1:1000), Zonula Occludens-1 (ZO-1, rat monoclonal antibody, 1:1000), cluster of differentiation 68 (CD68, mouse monoclonal antibody, 1:1000), myeloperoxidase (MPO, mouse monoclonal antibody, 1:500), p65 (rabbit polyclonal antibody, 1:1000), phospho-p65 (mouse monoclonal antibody, 1:500), and glyceraldehyde 3-phosphate dehydrogenase (GAPDH, mouse monoclonal antibody, 1:3000) (Santa Cruz Biotechnology, Santa Cruz, CA, USA). Proteins on the membranes were visualized by the sequential incubation with horseradish peroxidase-conjugated IgG and enhanced chemiluminescence Western blotting reagents. The chemiluminescent blots were scanned using the G:BOX mini multi fluorescence and chemiluminescence imaging system (Syngene, Frederick, MD, USA). The intensity of immunoreactive signals was quantitated by ImageJ software (National Institute of Health, Bethesda, MD, USA).

2.14. Zymography Assay

After euthanasia, the brains (n = 8 per group) were quickly removed and separated into contralateral and ipsilateral hemispheres to isolate the striatum. The dissected ipsilateral striatal tissues were subjected to the extraction of proteins (tissue protein extraction reagents, Pierce Biotechnology, Rockford, IL, USA) and conduction of a standardized SDS–PAGE (8%). After separation, the gels were washed twice for 30 min with 2.5% Triton X-100 and then incubated in buffer (25 mM Tris, 150 mM NaCl, 10 mM $CaCl_2$, 0.2% Brij-35, pH 7.5) overnight at 37 °C. Afterward, the gels were stained with Coomassie brilliant blue R-250 (0.2%). The intensities of visualized bands were quantitated by ImageJ software (National Institute of Health, Bethesda, MD, USA).

2.15. Electrophoretic Mobility Shift Assay (EMSA)

After euthanasia, the brains (n = 8 per group) were quickly removed and separated into contralateral and ipsilateral hemispheres to isolate the striatum. The dissected ipsilateral striatal tissues were subjected to the extraction of nuclear proteins (NE-PER nuclear and cytoplasmic extraction kit, Thermo Fisher Scientific, Waltham, MA, USA) and conduction of EMSA (LightShift^{TM} chemiluminescent EMSA Kit, Thermo Fisher Scientific, Waltham, MA, USA) according to the manufacturer's instructions. The sequences of oligonucleotide were: NF-κB, 5′-AGTTGAGGGGACTTTCCCAGGC. The intensity of the bands was analyzed by ImageJ software (National Institute of Health, Bethesda, MD, USA).

2.16. Semi-Quantitative Polymerase Chain Reaction (PCR)

The pooled urine samples (n = 8 per group) were centrifuged at 2000 rpm for 5 min at 4 °C. Total DNA was extracted from the cell pellets using a DNA isolation kit (Abcam, Cambridge, UK). The PCR consisted of 30 cycles of reaction (denaturation at 95 °C for 45 s, annealing at 60 °C for 30 s, and extension at 72 °C for 1 min) and a final extension at 72 °C for 10 min. The amplified DNA products were separated by 1.5% agarose gel electrophoresis and stained with ethidium bromide. The intensity of the bands was analyzed by ImageJ software (National Institute of Health, Bethesda, MD, USA). The oligonucleotides for PCR were: 5′-GCAGGGCTTTGAACCTCTTG and 5′-GCTCTGTGACACTCGGATTT for podocalyxin; 5′-AGATCCACAACGGATACATT and 5′-TCCCTCAAGATTGTCAGCAA for GAPDH.

2.17. Statistical Analysis

Experimental data were analyzed by SPSS software and expressed as mean values ± standard deviation. All data were first analyzed by one-way or two-way analysis of variance (ANOVA), and the statistical differences were determined by Dunnett post hoc analysis. The level of significance was set at $p < 0.05$.

3. Results

3.1. Glycerol Alleviated Hemorrhagic Stroke Brain Injury

Intrastriatal collagenase injection for 24 h caused neurological deficits (Figure 1A), brain hematoma formation (Figure 1B,C), brain edema (Figure 1D), and caspase 3 activations (Figure 1E) in rats. A dose of intraperitoneal glycerol injection alleviated brain injury in hemorrhagic stroke rats (Figure 1). The findings suggest a beneficial effect of glycerol against hemorrhagic stroke brain injury.

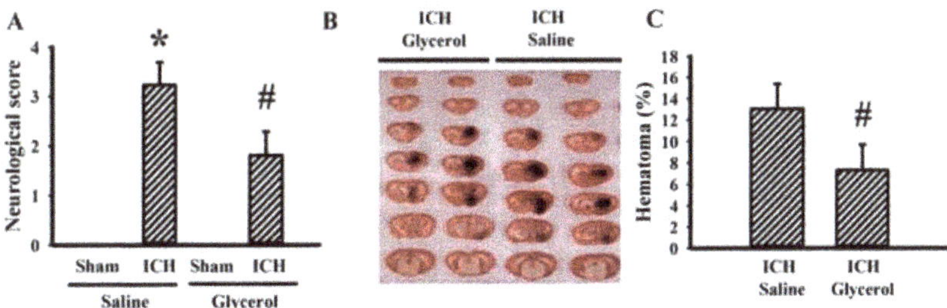

Figure 1. *Cont.*

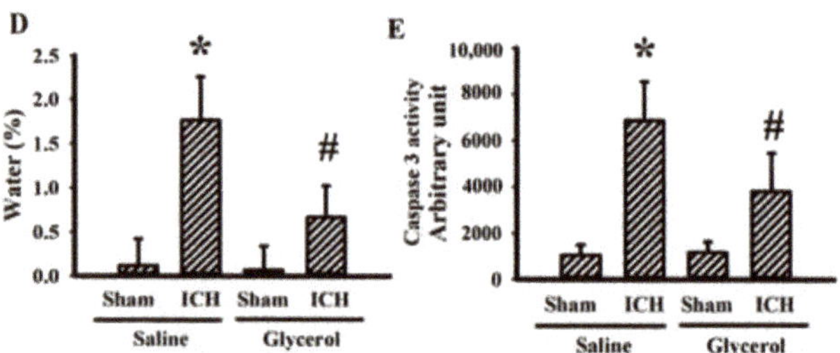

Figure 1. Glycerol alleviated hemorrhagic stroke brain injury. ICH and sham rats were intraperitoneally injected with saline or glycerol and housed for an additional 24 h. (**A**) Neurological deficits were evaluated by neurological score. (**B**) Representative photographs show histological examination of hematoma. (**C**) The percentage of hematoma in the ipsilateral striatum is depicted. (**D**) The water content differences between ipsilateral and contralateral striatum were measured. (**E**) Proteins were extracted from the ipsilateral striatal tissues and subjected to the measurement of caspase 3 activity. * $p < 0.05$ vs. sham/saline and # $p < 0.05$ vs. ICH/saline, n = 8.

3.2. Glycerol Alleviated Hemorrhagic Stroke BBB Disruption

ICH is associated with the disruption of BBB [7,8,25]. Parameters of BBB integrity were examined. An apparent Evans blue extravasation (Figure 2A) was observed in the ipsilateral striatal tissues of hemorrhagic stroke rats, and the elevation was paralleled with enhanced MMP-9 activity (Figure 2B), increased MMP-9 protein expression (Figure 2C), and reduced tight junction ZO-1 protein expression (Figure 2C). However, the change of MMP-2 was not of significance (Figure 2C). Glycerol injection alleviated all the changes in hemorrhagic stroke rats (Figure 2). In other words, ICH rats display increased cerebrovascular permeability, and the disruption of the BBB could be alleviated by glycerol.

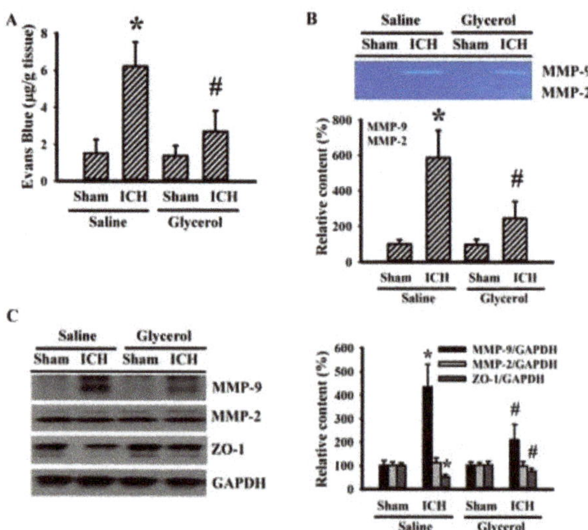

Figure 2. Glycerol alleviated hemorrhagic stroke BBB disruption. ICH and sham rats were intraperitoneally injected with saline or glycerol and housed for an additional 24 h. (**A**) The contents of Evans blue in the ipsilateral striatal tissues were measured. (**B**) Proteins were extracted from the ipsilateral striatal tissues and subjected to zymography assay. Representative gels and the quantitative data are shown. (**C**) Proteins were extracted from the ipsilateral striatal tissues and subjected to Western blot assay with indicated antibodies. Representative blots and the quantitative data are shown. Specific protein content was normalized with GAPDH. * $p < 0.05$ vs. sham/saline and # $p < 0.05$ vs. ICH/saline, n = 8.

3.3. Glycerol Alleviated Hemorrhagic Stroke Oxidative Stress and Neuroinflammation

Oxidative stress and neuroinflammation are pivotal to the pathogenesis of hemorrhagic stroke. Agents show antioxidant and/or anti-inflammatory potential displaying neuroprotective effects against hemorrhagic stroke [8,26,27]. ICH caused a decreased content of GSH (Figure 3A) and increased content of lipid peroxidation product MDA (Figure 3B), TNF-α and IL-1β inflammatory cytokine (Figure 3C), macrophage/microglia-related CD68 protein, neutrophil-related MPO protein, NF-κB p65 protein phosphorylation (Figure 3D), and NF-κB DNA-binding activity (Figure 3E) in ipsilateral striatal tissues. ICH-induced alterations could be reversed by glycerol (Figure 3). The findings indicate a resolution of oxidative stress and neuroinflammation in hemorrhagic stroke rats by glycerol.

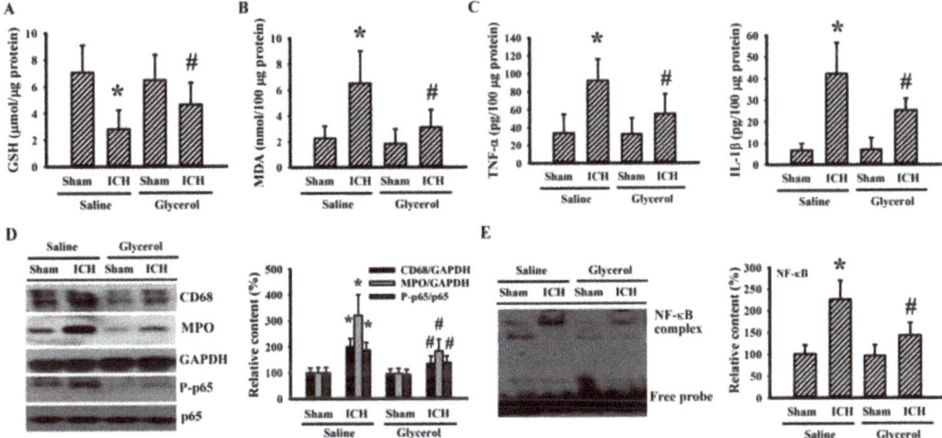

Figure 3. Glycerol alleviated hemorrhagic stroke oxidative stress and neuroinflammation. ICH and sham rats were intraperitoneally injected with saline or glycerol and housed for an additional 24 h. The contents of GSH (**A**) and MDA (**B**) in the ipsilateral striatal tissues were measured. (**C**) Proteins were extracted from the ipsilateral striatal tissues and subjected to ELISA for the measurement of TNF-α and IL-1β. (**D**) Proteins were extracted from the ipsilateral striatal tissues and subjected to Western blot assay with indicated antibodies. (**E**) Nuclear proteins were extracted from the ipsilateral striatal tissues and subjected to EMSA for the measurement of NF-κB DNA-binding activity. Representative blots and the quantitative data are shown. Specific protein content was normalized with the corresponding total protein or GAPDH. * $p < 0.05$ vs. sham/saline and # $p < 0.05$ vs. ICH/saline, n = 8.

3.4. Glycerol Alleviated Hemorrhagic Stroke-Associated Kidney Dysfunction

To identify any kidney dysfunction in hemorrhagic stroke rats, several biochemical and histological examinations centered on kidney structural and functional integrity were conducted. Hemorrhagic stroke rats had an increased level of serum BUN (Figure 4A) and creatinine (Figure 4B), as well as a level of urinary GAPDH (Figure 4C) and podocalyxin (Figure 4D) DNA content. Moreover, histological examination with H&E further revealed a mild dilation of the Bowman's capsule, tubular dilation, and cast formation (Figure 4E). The biochemical and histological changes in the hemorrhagic stroke rats could be alleviated by glycerol injection (Figure 4). Hemorrhagic stroke-associated kidney dysfunction was further validated by serum and urinary KIM-1, NGAL, CINC-1, and MDA, early biomarkers of AKI [3]. There was no remarkable change in the measurements except urinary CINC-1 and urinary MDA. Hemorrhagic stroke rats had increased urine levels of CINC-1 and MDA, and the increments could be alleviated by glycerol injection (Figure 5). Moreover, an increase in MDA content (Figure 6A) and a reduction in GSH content (Figure 6B) were found in kidney cortical tissues of hemorrhagic stroke rats. The changes could be alleviated by glycerol injection (Figure 6). Current findings indicate a concurrent kidney dysfunction in hemorrhagic stroke rats and an alleviating effect of glycerol.

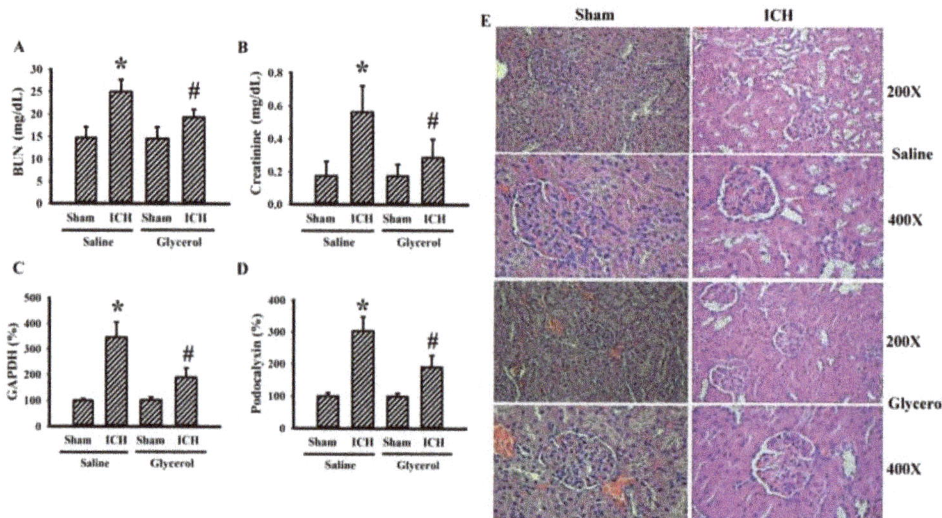

Figure 4. Glycerol alleviated hemorrhagic stroke-associated kidney dysfunction. ICH and sham rats were intraperitoneally injected with saline or glycerol and housed for an additional 24 h. Serum samples were subjected to the measurement of BUN (**A**) and creatinine (**B**). Urine samples were subjected to DNA isolation and PCR for the measurement of GAPDH (**C**) and podocalyxin (**D**) DNA content. Paraffin sections of kidney tissues were subjected to histological staining with H&E (**E**). Representative photomicrographs showed one of eight independent rats. * $p < 0.05$ vs. sham/saline and # $p < 0.05$ vs. ICH/saline, n = 8.

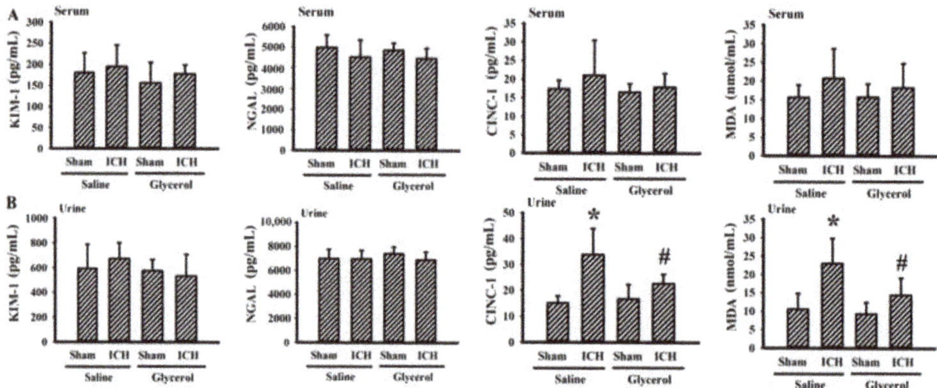

Figure 5. Glycerol alleviated urine biomarkers in hemorrhagic stroke rats. ICH and sham rats were intraperitoneally injected with saline or glycerol and housed for an additional 24 h. Serum (**A**) and urine (**B**) samples were subjected to ELISA for the measurement of KIM-1, NGAL, and CINC-1 as well as to TBARS assay for the measurement of MDA. * $p < 0.05$ vs. sham/saline and # $p < 0.05$ vs. ICH/saline, n = 8.

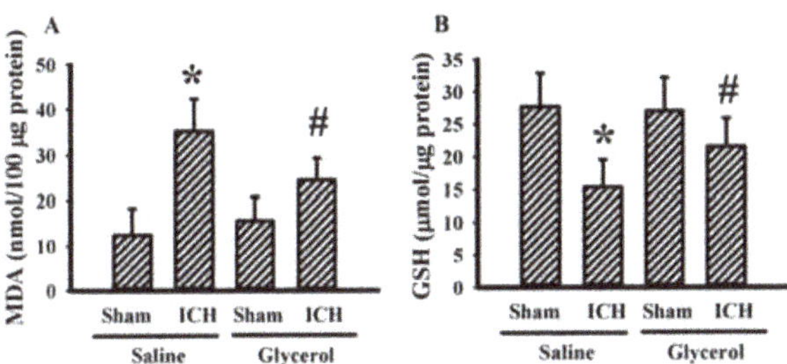

Figure 6. Glycerol alleviated kidney oxidative stress in hemorrhagic stroke rats. ICH and sham rats were intraperitoneally injected with saline or glycerol and housed for an additional 24 h. Kidney cortical tissues were isolated and subjected to the measurement of MDA (**A**) and GSH (**B**) * $p < 0.05$ vs. sham/saline and # $p < 0.05$ vs. ICH/saline, n = 8.

3.5. Glycerol Alleviated Stress Hormones in Hemorrhagic Stroke Rats

Stress hormones are adaptive molecules functioning in tissue homeostasis, while they also cause tissue/organ injury [10–12,16,17]. Hemorrhagic stroke rats exhibited increased levels of serum epinephrine, norepinephrine, corticosterone, and Ang II. Glycerol injection alleviated elevated stress hormones in hemorrhagic stroke rats (Figure 7). The findings suggest that ICH could increase stress hormone production, and the increments could be alleviated by glycerol.

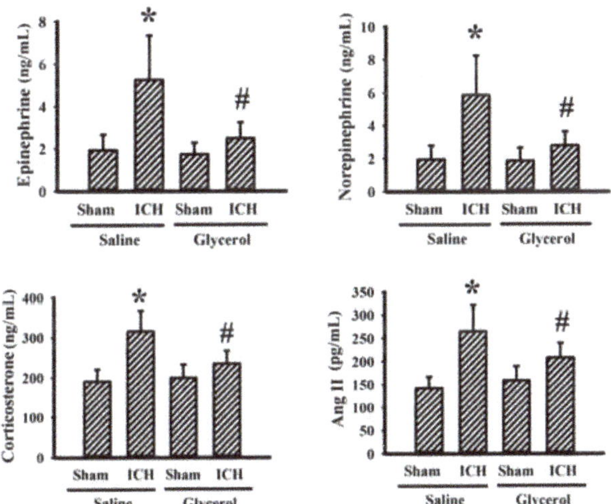

Figure 7. Glycerol alleviated stress hormones in hemorrhagic stroke rats. ICH and sham rats were intraperitoneally injected with saline or glycerol and housed for an additional 24 h. Serum samples were subjected to ELISA for the measurement of epinephrine, norepinephrine, corticosterone, and Ang II. * $p < 0.05$ vs. sham/saline and # $p < 0.05$ vs. ICH/saline, n = 8.

4. Discussion

CKD is recognized as a risk factor for stroke with deteriorative effects by itself and is commonly accompanied by hypertension, hypercholesterolemia, and diabetes mellitus [2]. Having an episode of AKI after stroke is associated with worse outcomes and increased in-hospital mortality. Clinically, the reported rate of AKI after stroke varies widely with a range of 0.82% to 30.18% [2,3,5,28–30]. Khatri et al. [29] report that stroke patients with ICH in intensive care units and wards are more likely than patients with ischemia to develop AKI. In an intrastriatal collagenase injection hemorrhagic stroke model, we demonstrated that hemorrhagic stroke rats developed kidney dysfunction, as evidenced by elevated serum levels of BUN and creatinine. The elevations reached a peak at 24 h after collagenase injection and returned to normal ranges 72 h later (data not shown). Although the worst effects of AKI on stroke brain injury have not been investigated, the current data clearly demonstrate an occurrence of kidney dysfunction after a hemorrhagic stroke. However, the effects of an AKI in a rat model of ischemic stroke were not evaluated in this study.

To date, several serum and urine biomarkers have been validated to detect kidney injury at various stages. KIM-1, NGAL, CINC-1, lipid peroxidation product, cystatin C, BUN, creatinine, IL-18, β2-microglobulin, and eGFR are common biomarkers [3,31–33]. Among stroke patients in intensive care units, serum cystatin C is an important biomarker for predicting AKI and in-hospital mortality [3]. Elevated levels of serum BUN, serum creatinine, urinary CINC-1, and urinary MDA were found in intrastriatal collagenase injection hemorrhagic stroke rats. KIM-1, NGAL, serum CINC-1, and serum MDA appeared to be less strongly correlated. CINC-1 is a counterpart of human IL-8 family members, mediating inflammatory reactions by attracting neutrophils. Disease progression of AKI is associated well with renal neutrophil infiltration, cytokine expression, and oxidative stress [34]. The current study suggests that urinary CINC-1 and MDA are sensitive biomarkers than their serum counterparts in hemorrhagic stroke rats.

Besides serum and urinary biomarkers, the hemorrhagic stroke affected kidney structural integrity, particularly near the glomerulus. The glomerulus is a globular structure, and the filtering unit of the kidney is surrounded by the Bowman's capsule. Podocytes are specialized epithelial cells located outside the glomerular basement membrane, wrapping glomerular capillaries to form the filtration barrier. Podocyte detachment and its urine accumulation represent alternative biomarkers of kidney injury [35]. The presence of GAPDH DNA in urine samples of intrastriatal collagenase injection hemorrhagic stroke rats reflected a leakage of cells from the kidneys. Our data further indicated the detached cells could be podocytes due to the successful detection of podocalyxin DNA, a sialoglycoprotein of podocytes [35]. The alteration of hemodynamics or kidney oxidative stress is considered a possible cause of podocyte detachment [31,35]. The elevation of renal and urinary MDA and reduction of renal GSH found in this study may highlight a potential involvement of oxidative stress in kidney structural changes.

Following a stroke, various changes and deterioration in function, structure, and remote communications occur. Among the remote communications, the sympathetic nervous system, hypothalamic pituitary axis, and renin–angiotensin–aldosterone system are pivotal in regulating renal blood flow, free radical generation, inflammatory response, and glomerular filtration [12,16,17,31]. Intrastriatal collagenase injection hemorrhagic stroke rats showed an elevated circulating level of epinephrine, norepinephrine, corticosterone, and Ang II. Therefore, prolonged elevation of stress hormones after stroke may lead to kidney dysfunction observed in this study. The roles and importance of epinephrine, norepinephrine, corticosterone and Ang II in hemorrhagic stroke-associated kidney dysfunction can be investigated by introducing pharmacological antagonists. However, their involvements were not addressed in the current study. Collagenase has been implicated in the deterioration of blood vessels and kidney glomerulus [36]. Despite the small dose and intrastriatal injection, the remote effects of collagenase on kidney dysfunction could not be totally excluded.

Usually, hemorrhagic stroke causes primary and secondary brain injury as well as the above-mentioned remote organ dysfunction. The mass effect of hematoma is the main cause of primary brain injury. In contrast, the extravasated blood components induce a wave of inflammatory and oxidative change leading to BBB disruption, edema, neuronal cell dysfunction/destruction, and other events of secondary brain injury. Agents or strategies intervening in hematoma expansion, oxidative stress, inflammation, BBB disruption, edema, apoptosis, or associated hemorrhagic stroke changes have promising effects in the prevention and treatment of hemorrhagic stroke brain injury and complications [6–9,13,15,25–27,37]. Brain edema and increased intracranial pressure severely threaten the disease progression and outcome of stroke patients [14]. In rodent models, short hypothermia was found to be effective at decreasing intracranial pressure after ischemic stroke [38]. Clinically, osmotic agents, such as mannitol and glycerol, are commonly used to treat increased intracranial pressure. Reduction of brain water content, enhancement of cerebrospinal fluid absorption, and increased cerebral blood flow are proposed beneficial mechanisms of osmotic agents [2,18,20]. Mannitol and hypertonic saline further reveal beneficial outcomes against hemorrhagic stroke brain injury and neuroinflammation [15]. The neurological deficit, hematoma formation, brain edema, BBB disruption, apoptosis, oxidative stress, and inflammation were demonstrated in the ipsilateral striatal tissues of ICH rats. A dose of intraperitoneal glycerol injection after stroke not only protected the brain from hemorrhagic injury but also improved accompanying kidney dysfunction. An NF-κB-dominant axis has been implicated in the expression and activation of macrophages/microglia, neutrophils, an inflammatory cytokine, and MMP after hemorrhagic stroke [7,8,25,27]. Consistent with the findings, data of NF-κB DNA-binding activity, p65 protein phosphorylation, CD68, MPO, and MMP-9 protein expression, and TNF-α and IL-1β production demonstrated the occurrence of NF-κB axis in ICH rats and could be intervened by glycerol. We further identified a concurrent change in BBB disruption, MMP-9 activation, and ZO-1 degradation in ICH rats, implying a contribution of the MMP-9/ZO-1 pathway in hemorrhagic stroke cerebrovascular permeability change. Although using glycerol in treating brain edema remains controversial, the current rodent study provides evidence demonstrating its beneficial effects again post-stroke brain injury and kidney dysfunction. Despite the encouraging findings, there are some limitations in interpreting experimental data. Clinically, stroke patients have been prescribed glycerol as an infusion dose of 250 mL (10%) at intervals of 4–6 h [18,20]. Rats show no significant kidney injury upon treatment with intraperitoneal hypertonic glycerol solution injection (10 mL/kg, 50%) followed by 24 h water deprivation. However, there are signs of oxidative stress observed in plasma and the kidney [23]. In this study, glycerol at a dose of 6 mL/kg (10%) was administrated intraperitoneally 30 min after surgery. Therefore, further investigation is needed to determine the optimal therapeutic regimen and to monitor the criteria for the application of glycerol.

5. Conclusions

CKD has long been considered a risk factor for stroke, while AKI after stroke worsens patient outcomes. Although the deteriorative correlation has yet to be determined, the intrastriatal collagenase injection hemorrhagic stroke model in rats presented herein demonstrated the occurrence of kidney dysfunction after stroke onset. Data from histological, biochemical, and molecular studies identified the functional and structural integrity of the glomerulus are targets vulnerable to post-stroke injury, and stress hormones could be surrogates of remote communications. Although the osmotic agent glycerol held promising effects in alleviating post-stroke brain injury and kidney dysfunction demonstrated by this study, its specific targets of action and protective mechanisms had not been fully investigated. Therefore, the clinical translation of current rodent model studies warrants further investigation.

Author Contributions: Conceptualization, C.-Y.C. and C.-J.C.; investigation, P.-H.P., J.-R.L., Y.-C.O., S.-L.L., W.-Y.C. and Y.-H.K.; writing—original draft preparation, C.-Y.C.; writing—review and editing, C.-J.C.; supervision, C.-J.C.; funding acquisition, C.-Y.C. All authors have read and agreed to the published version of the manuscript.

Funding: This work was supported by grants from Feng Yuan Hospital and the Ministry of Health and Welfare, Taiwan (101).

Institutional Review Board Statement: The experimental protocols adhered to the Institute's guidelines and were approved by the Institutional Animal Care and Use Committee of Taichung Veterans General Hospital (IACUC approval code: La-100859, IACUC approval date: 11 November 2011).

Informed Consent Statement: Not applicable.

Data Availability Statement: No new data were created or analyzed in this study. Data sharing is not applicable to this article.

Conflicts of Interest: The authors declare no conflict of interest. The funders had no role in the design of the study; in the collection, analyses, or interpretation of data; in the writing of the manuscript, or in the decision to publish the results.

References

1. Barthels, D.; Das, H. Current advances in ischemic stroke research and therapies. *Biochim. Biophys. Acta Mol. Basis Dis.* **2020**, *1866*, 165260. [CrossRef] [PubMed]
2. Arnold, J.; Ng, K.P.; Sims, D.; Gill, P.; Cockwell, P.; Ferro, C. Incidence and impact on outcomes of acute kidney injury after a stroke: A systematic review and meta-analysis. *BMC Nephrol.* **2018**, *19*, 283. [CrossRef] [PubMed]
3. Jiang, F.; Su, L.; Xiang, H.; Zhang, X.; Xu, D.; Zhang, Z.; Peng, P. Incidence, risk factors, and biomarkers predicting ischemic or hemorrhagic stroke associated acute kidney injury and outcome: A retrospective study in a general intensive care unit. *Blood Purif.* **2019**, *47*, 317–326. [CrossRef] [PubMed]
4. Zhao, Q.; Yan, T.; Chopp, M.; Venkat, P.; Chen, J. Brain-kidney interaction: Renal dysfunction following ischemic stroke. *J. Cereb. Blood Flow Metab.* **2020**, *40*, 246–262. [CrossRef]
5. Zorrilla-Vaca, A.; Ziai, W.; Connolly, E.S.; Geocadin, R.; Thompson, R.; Rivera-Lara, L. Acute kidney injury following acute ischemic stroke and intracerebral hemorrhage: A meta-analysis of prevalence rate and mortality risk. *Cerebrovasc. Dis.* **2018**, *45*, 1–9. [CrossRef]
6. Liu, X.C.; Wu, C.Z.; Hu, X.F.; Wang, T.L.; Jin, X.P.; Ke, S.F.; Wang, E.; Wu, G. Gastrodin attenuates neuronal apoptosis and neurological deficits after experimental intracerebral hemorrhage. *J. Stroke Cerebrovasc. Dis.* **2020**, *29*, 104483. [CrossRef]
7. Xi, T.; Jin, F.; Zhu, Y.; Wang, J.; Tang, L.; Wang, Y.; Liebeskind, D.S.; Scalzo, F.; He, Z. miR-27a-3p protects against blood-brain barrier disruption and brain injury after intracerebral hemorrhage by targeting endothelial aquaporin-11. *J. Biol. Chem.* **2018**, *293*, 20041–20050. [CrossRef]
8. Zeng, Z.; Gong, X.; Hu, Z. L-3-n-butylphthalide attenuates inflammation response and brain edema in rat intracerebral hemorrhage model. *Aging (Albany NY)* **2020**, *12*, 11768–11780. [CrossRef]
9. Zhang, Z.; Wu, Y.; Yuan, S.; Zhang, P.; Zhang, J.; Li, H.; Li, X.; Shen, H.; Wang, Z.; Chen, G. Glutathione peroxidase 4 participates in secondary brain injury through mediating ferroptosis in a rat model of intracerebral hemorrhage. *Brain Res.* **2018**, *1701*, 112–125. [CrossRef]
10. Abdel-Fattah, M.M.; Messiha, B.A.S.; Mansour, A.M. Modulation of brain ACE and ACE2 may be a promising protective strategy against cerebral ischemia/reperfusion injury: An experimental trial in rats. *Naunyn. Schmiedebergs Arch. Pharmacol.* **2018**, *391*, 1003–1020. [CrossRef]
11. Chen, W.Y.; Mao, F.C.; Liu, C.H.; Kuan, Y.H.; Lai, N.W.; Wu, C.C.; Chen, C.J. Chromium supplementation improved post-stroke brain infarction and hyperglycemia. *Metab. Brain Dis.* **2016**, *31*, 289–297. [CrossRef]
12. Zhao, Y.; Zeng, H.; Liu, B.; He, X.; Chen, J.X. Endothelial prolyl hydroxylase 2 is necessary for angiotensin II-mediated renal fibrosis and injury. *Am. J. Physiol. Renal. Physiol.* **2020**, *319*, F345–F357. [CrossRef]
13. Wang, J.; Wang, G.; Yi, J.; Xu, Y.; Duan, S.; Li, T.; Sun, X.G.; Dong, L. The effect of monascin on hematoma clearance and edema after intracerebral hemorrhage in rats. *Brain Res. Bull.* **2017**, *134*, 24–29. [CrossRef]
14. Lietke, S.; Zausinger, S.; Patzig, M.; Holtmanspötter, M.; Kunz, M. CT-based classification of acute cerebral edema: Association with intracranial pressure and outcome. *J. Neuroimaging* **2020**, *30*, 640–647. [CrossRef]
15. Schreibman, D.L.; Hong, C.M.; Keledjian, K.; Ivanova, S.; Tsymbalyuk, S.; Gerzanich, V.; Simard, J.M. Mannitol and hypertonic saline reduce swelling and modulate inflammatory markers in a rat model of intracerebral hemorrhage. *Neurocrit. Care* **2018**, *29*, 253–263. [CrossRef]
16. Li, J.R.; Ou, Y.C.; Wu, C.C.; Wang, J.D.; Lin, S.Y.; Wang, Y.Y.; Chen, W.Y.; Chen, C.J. Ischemic preconditioning improved renal ischemia/reperfusion injury and hyperglycemia. *IUBMB Life* **2019**, *71*, 321–329. [CrossRef]
17. Tsutsui, H.; Shimokawa, T.; Miura, T.; Takama, M.; Nishinaka, T.; Terada, T.; Yamagata, M.; Yukimura, T. Effect of monoamine oxidase inhibitors on ischaemia/reperfusion-induced acute kidney injury in rats. *Eur. J. Pharmacol.* **2018**, *818*, 38–42. [CrossRef]
18. Berger, C.; Sakowitz, O.W.; Kiening, K.L.; Schwab, S. Neurochemical monitoring of glycerol therapy in patients with ischemic brain edema. *Stroke* **2005**, *36*, e4–e6. [CrossRef]
19. Ishikawa, M.; Sekizuka, E.; Sato, S.; Yamaguchi, N.; Inamasu, J.; Kawase, T. Glycerol attenuates the adherence of leukocytes in rat pial venules after transient middle cerebral artery occlusion. *Neurol. Res.* **1999**, *21*, 785–790. [CrossRef]
20. Sakamaki, M.; Igarashi, H.; Nishiyama, Y.; Hagiwara, H.; Ando, J.; Chishiki, T.; Curran, B.C.; Katayama, Y. Effect of glycerol on ischemic cerebral edema assessed by magnetic resonance imaging. *J. Neurol. Sci.* **2003**, *209*, 69–74. [CrossRef]
21. Cole, L.; Dewey, D.; Letourneau, N.; Kaplan, B.J.; Chaput, K.; Gallagher, C.; Hodge, J.; Floer, A.; Kirton, A. Clinical characteristics, risk factors, and outcomes associated with neonatal hemorrhagic stroke: A population-based case-control study. *JAMA Pediatr.* **2017**, *171*, 230–238. [CrossRef] [PubMed]
22. Marques, M.S.; Cordeiro, M.F.; Marinho, M.A.G.; Vian, C.O.; Vaz, G.R.; Alves, B.S.; Jardim, R.D.; Hort, M.A.; Dora, C.L.; Horn, A.P. Curcumin-loaded nanoemulsion improves haemorrhagic stroke recovery in wistar rats. *Brain Res.* **2020**, *1746*, 147007. [CrossRef] [PubMed]

23. Rieger, E.; Rech, V.C.; Feksa, L.R.; Wannmacher, C.M. Intraperitoneal glycerol induces oxidative stress in rat kidney. *Clin. Exp. Pharmacol. Physiol.* **2008**, *35*, 928–933. [CrossRef] [PubMed]
24. Lin, S.Y.; Wang, Y.Y.; Chang, C.Y.; Wu, C.C.; Chen, W.Y.; Kuan, Y.H.; LiaoL, S.L.; Chen, C.J. Effects of b-adrenergic blockade on metabolic and inflammatory responses in a rat model of ischemic stroke. *Cells* **2020**, *9*, 1373. [CrossRef]
25. Song, Y.; Yang, Y.; Cui, Y.; Gao, J.; Wang, K.; Cui, J. Lipoxin A4 methyl ester reduces early brain injury by inhibition of the nuclear factor kappa B (NF-kappaB)-dependent matrix metallopeptidase 9 (MMP-9) pathway in a rat model of intracerebral hemorrhage. *Med. Sci. Monit.* **2019**, *25*, 1838–1847. [CrossRef]
26. Fan, X.; Mu, L. The role of heme oxygenase-1 (HO-1) in the regulation of inflammatory reaction, neuronal cell proliferation and apoptosis in rats after intracerebral hemorrhage (ICH). *Neuropsychiatr. Dis. Treat.* **2016**, *13*, 77–85. [CrossRef]
27. Qu, X.; Wang, N.; Cheng, W.; Xue, Y.; Chen, W.; Qi, M. MicroRNA-146a protects against intracerebral hemorrhage by inhibiting inflammation and oxidative stress. *Exp. Ther. Med.* **2019**, *18*, 3920–3928. [CrossRef]
28. El Husseini, N.; Fonarow, G.C.; Smith, E.E.; Ju, C.; Schwamm, L.H.; Hernandez, A.F.; Schulte, P.J.; Xian, Y.; Goldstein, L.B. Renal dysfunction is associated with poststroke discharge disposition and in-hospital mortality: Findings from get with the guidelines-stroke. *Stroke* **2017**, *48*, 327–334. [CrossRef]
29. Khatri, M.; Himmelfarb, J.; Adams, D.; Becker, K.; Longstreth, W.T.; Tirschwell, D.L. Acute kidney injury is associated with increased hospital mortality after stroke. *J. Stroke Cerebrovasc. Dis.* **2014**, *23*, 25–30. [CrossRef]
30. Tsagalis, G.; Akrivos, T.; Alevizaki, M.; Manios, E.; Theodorakis, M.; Laggouranis, A.; Vemos, K.N. Long-term prognosis of acute kidney injury after first acute stroke. *Clin. J. Am. Soc. Nephrol.* **2009**, *4*, 616–622. [CrossRef]
31. Kang, J.S.; Lee, S.J.; Lee, J.H.; Kim, J.H.; Son, S.S.; Cha, S.K.; Lee, E.S.; Chung, C.H.; Lee, E.Y. Angiotensin II-mediated MYH9 downregulation causes structural and functional podocyte injury in diabetic kidney disease. *Sci. Rep.* **2019**, *9*, 7679. [CrossRef]
32. Polidori, N.; Giannini, C.; Salvatore, R.; Pelliccia, P.; Parisi, A.; Chiarelli, F.; Mohn, A. Role of urinary NGAL and KIM-1 as biomarkers of early kidney injury in obese prepubertal children. *J. Pediatr. Endocrinol. Metab.* **2020**, *33*, 1183–1189. [CrossRef]
33. Sato, H.; Ueki, M.; Asaga, T.; Chujo, K.; Maekawa, M. D-ribose attenuates ischemia/reperfusion-induced renal injury by reducing neutrophil activation in rats. *Tohoku. J. Exp. Med.* **2009**, *218*, 35–40. [CrossRef]
34. Ahmad, A.; Olah, G.; Szczesny, B.; Wood, M.E.; Whiteman, M.; Szabo, C. AP39, a mitochondrially targeted hydrogen sulfide donor, exerts protective effects in renal epithelial cells subjected to oxidative stress in vitro and in acute renal injury in vivo. *Shock* **2016**, *45*, 88–97. [CrossRef]
35. Hara, M.; Yamagata, K.; Tomino, Y.; Saito, A.; Hirayama, Y.; Ogasawara, S.; Kurosawa, H.; Sekine, S.; Yan, K. Urinary podocalyxin is an early marker for podocyte injury in patients with diabetes: Establishment of a highly sensitive ELISA to detect urinary podocalyxin. *Diabetologia* **2012**, *55*, 2913–2919. [CrossRef]
36. Sakamaki, Y.; Sasamura, H.; Hayashi, K.; Ishiguro, K.; Takaishi, H.; Okada, Y.; D'Armiento, J.M.; Saruta, T.; Itoh, H. Absence of gelatinase (MMP-9) or collagenase (MMP-13) attenuates adriamycin-induced albuminuria and glomerulosclerosis. *Nephron. Exp. Nephrol.* **2010**, *115*, e22–e32. [CrossRef]
37. Wang, Y.; Song, Y.; Pang, Y.; Yu, Z.; Hua, W.; Gu, Y.; Qi, J.; Wu, H. miR-183-5p alleviates early injury after intracerebral hemorrhage by inhibiting heme oxygenase-1 expression. *Aging* **2020**, *12*, 12869–12895. [CrossRef]
38. Murtha, L.A.; Beard, D.J.; Bourke, J.T.; Pepperall, D.; McLeod, D.D.; Spratt, N.J. Intracranial pressure elevation 24 h after ischemic stroke in aged rats is prevented by early, short hypothermia treatment. *Front. Aging Neurosci.* **2016**, *8*, 124. [CrossRef]

 *antioxidants* 

Article

Activated Histone Acetyltransferase p300/CBP-Related Signalling Pathways Mediate Up-Regulation of NADPH Oxidase, Inflammation, and Fibrosis in Diabetic Kidney

Alexandra-Gela Lazar, Mihaela-Loredana Vlad, Adrian Manea, Maya Simionescu and Simona-Adriana Manea *

Institute of Cellular Biology and Pathology, "Nicolae Simionescu" of the Romanian Academy, 050568 Bucharest, Romania; alexandra.lazar@icbp.ro (A.-G.L.); loredana.antonescu@icbp.ro (M.-L.V.); adrian.manea@icbp.ro (A.M.); maya.simionescu@icbp.ro (M.S.)
* Correspondence: simona.manea@icbp.ro

Abstract: Accumulating evidence implicates the histone acetylation-based epigenetic mechanisms in the pathoetiology of diabetes-associated micro-/macrovascular complications. Diabetic kidney disease (DKD) is a progressive chronic inflammatory microvascular disorder ultimately leading to glomerulosclerosis and kidney failure. We hypothesized that histone acetyltransferase p300/CBP may be involved in mediating diabetes-accelerated renal damage. In this study, we aimed at investigating the potential role of p300/CBP in the up-regulation of renal NADPH oxidase (Nox), reactive oxygen species (ROS) production, inflammation, and fibrosis in diabetic mice. Diabetic C57BL/6J mice were randomized to receive 10 mg/kg C646, a selective p300/CBP inhibitor, or its vehicle for 4 weeks. We found that in the kidney of C646-treated diabetic mice, the level of H3K27ac, an epigenetic mark of active gene expression, was significantly reduced. Pharmacological inhibition of p300/CBP significantly down-regulated the diabetes-induced enhanced expression of Nox subtypes, pro-inflammatory, and pro-fibrotic molecules in the kidney of mice, and the glomerular ROS overproduction. Our study provides evidence that the activation of p300/CBP enhances ROS production, potentially generated by up-regulated Nox, inflammation, and the production of extracellular matrix proteins in the diabetic kidney. The data suggest that p300/CBP-pharmacological inhibitors may be attractive tools to modulate diabetes-associated pathological processes to efficiently reduce the burden of DKD.

Keywords: diabetes; nephropathy; epigenetics; histone acetylation; p300/CBP; NADPH oxidase; oxidative stress

Citation: Lazar, A.-G.; Vlad, M.-L.; Manea, A.; Simionescu, M.; Manea, S.-A. Activated Histone Acetyltransferase p300/CBP-Related Signalling Pathways Mediate Up-Regulation of NADPH Oxidase, Inflammation, and Fibrosis in Diabetic Kidney. *Antioxidants* **2021**, *10*, 1356. https://doi.org/10.3390/antiox10091356

Academic Editors: Chiara Nediani and Monica Dinu

Received: 21 July 2021
Accepted: 21 August 2021
Published: 26 August 2021

1. Introduction

Diabetic kidney disease (DKD), the major microvascular complication of both type I and type II diabetes, is a complex multifactorial renal disorder having a detrimental impact on the patient's quality of life and life-span expectation [1–3]. Dysfunction of glomerular endothelial cells and podocytes, thickening of the glomerular basement membrane, mesangial cells hypertrophy and proliferation, progressive accumulation of mesangial extracellular matrix (ECM) components, podocyte damage, and disruption of glomerular endothelium fenestrations are the main structural alterations ultimately leading to glomerulosclerosis; a pathological condition that is further responsible for increased intraglomerular capillary pressure, hyperfiltration, and eventually kidney failure [4–6].

Regardless of major achievements in DKD therapeutics, the current pharmacological strategies that include blood glucose, lipids, and blood pressure control can only delay the progression of renal damage. Consequently, dialysis, and ultimately kidney transplantation, remain the main therapeutic options for kidney failure in end-stage DKD [7,8]. Thus, the development of additional or supportive, glomerular cell-oriented pharmacological interventions acting in conjunction with standard anti-hyperglycaemic and anti-hypertensive drugs may contribute to formulating novel treatment algorithms to efficiently reduce the burden of DKD than is currently possible.

Hyperglycaemia, the main metabolic abnormality in diabetes, induces glomerular structural-functional dysfunction via multiple mechanisms that broadly include the formation of advanced glycation end products (AGEs), overproduction of detrimental reactive oxygen species (ROS), hemodynamic alterations, metabolic dysfunction, inflammation, phenotypic changes of the glomerular cells, excess synthesis, and accumulation of ECM components [7,9–11].

Studies on human diabetic kidney biopsies and experimental models of diabetes provide reliable evidence that ROS overproduction, typically driven by the up-regulated NADPH oxidase (Nox) family, are the major triggers of the proinflammatory and profibrotic signalling pathways [12–18]. Noteworthy, hyperglycaemia-induced metabolic memory via Nox-enhanced ROS production provides a possible explanation of the limitations of the current glucose-lowering therapeutic strategies in diabetic patients [19]. Consequently, Nox subtypes, and their upstream regulators, may become important candidates as pharmacological targets in DKD.

Accumulating evidence demonstrates that the dysregulation of epigenetic mechanisms, namely, DNA methylation, post-translational modifications of histones, and non-coding RNA, plays a major role in the pathology of cardiovascular disorders (CVD), and potentially in DKD [20–23]. Among other epigenetic modifiers, histone acetylation-based mechanisms have emerged as attractive therapeutic targets in experimental CVD. Lysine-specific acetylation of nucleosomal histones is tightly regulated by the coordinated activities of two major enzymatic systems, namely, histone acetyltransferases (HAT) and histone deacetylases (HDAC). Canonically, HAT-mediated histone acetylation induces chromatin relaxation due to electric charge neutralization of the histone tails, a condition that enables DNA-transcription factors interactions and activation of gene expression. Other than histones, several HAT subtypes regulate the function of a range of transcription factors to induce or repress the gene expression [24].

Abnormal expression levels of both HAT and HDAC subtypes and lysine-specific histone acetylation patterns have been mechanistically implicated in an increasing number of human and experimental models of CVD. Histone acetyltransferase p300/CBP is a ubiquitously expressed transcriptional co-activator and regulates the expression of genes by several interconnected mechanisms comprising histone acetylation-induced chromatin relaxation owing to the intrinsic HAT activity of p300, recruitment of RNA polymerase II, and also serves as a scaffold protein to form active transcriptional complexes within gene promoter and enhancer regions. Moreover, p300 directly regulates the function of several transcription factors (e.g., NF-kB) to enhance the expression of the target genes [25–27].

Histone acetyltransferase p300/CBP is an important transcriptional co-activator regulating the expression of oxidative stress- and inflammation-related genes. Previously, we have demonstrated a mechanistic connection among activated histone-acetylation pathways and Nox-derived ROS overproduction in experimental models of diabetes and atherosclerosis [28–30]. Yet, the role of p300/CBP in DKD remains elusive. Based on the important functions of the dysregulated histone acetylation-related mechanisms in vascular pathology, we hypothesized that in DKD, p300/CBP may act as an upstream regulator of Nox expression, ROS production, inflammation, and fibrosis. We provide here evidence that activation of p300/CBP mediates the up-regulation of Nox subtypes, ROS overproduction, and increases the expression of proinflammatory molecules and extracellular matrix proteins in the mouse diabetic kidney.

Our data propose p300/CBP-pharmacological inhibitors as attractive tools to modulate molecular and cellular processes mechanistically linked to the pathology of DKD.

2. Materials and Methods

2.1. Materials

Unless specifically mentioned, standard chemicals and reagents were obtained from Sigma-Aldrich (Darmstadt, Germany). C646 {4-[4-[[5-(4,5-Dimethyl-2-nitrophenyl)-2-furanyl]methylene]-4,5-dihydro-3-methyl-5-oxo-1H-pyrazol-1-yl] benzoic acid}, a cell-permeable, highly selective, reversible inhibitor of histone acetyltransferase p300/CBP (purity $\geq$ 98%, high-performance liquid chromatography analysis) was obtained from Sigma-Aldrich/Calbiochem (Darmstadt, Germany). Primary and secondary antibodies were from Santa Cruz Biotechnology (Heidelberg, Germany), Thermo Fisher Scientific (Vienna, Austria), Abcam (Cambridge, UK), and Diagenode (Seraing, Belgium). The pNF-κB-Luc (#219078), pC/EBP-Luc (#240112), pGAS-Luc (#219091), and pISRE-Luc (#219089) *cis*-reporter plasmids were purchased from Stratagene (La Jolla, San Diego, CA, USA). The American Type Culture Collection (ATCC) (Manassas, VA, USA) derived human embryonic kidney 293 (HEK293) cell line was used.

2.2. Procedure for Induction of Experimental Diabetes and Treatment Strategy in Mice

Male C57BL/6J (Stock No: 000664) mice obtained from The Jackson Laboratory were used. The mice reproduced in our SPF animal facility, were exposed to 12-h of light/dark cycles and had access to a standard rodent chow diet and water ad libitum. At 8 weeks of age, the mice were intraperitoneally (i.p.) injected with 55 mg/kg streptozotocin (STZ), for 5 successive days to induce experimental diabetes [9,31,32]. Age-matched C57BL/6J mice injected with citrate buffer (pH 4.5) were employed as non-diabetic controls. The installation of hyperglycaemia in the diabetic animal group was confirmed by measuring the glucose level in the blood collected by tail puncture, one week after the last STZ injection. The animals were further randomized into three experimental groups to receive (i.p.) 10 mg/kg C646 or its vehicle (5% DMSO + 95% PBS, pH 7.4), every other day for 4 weeks: (i) non-diabetic mice + vehicle (n = 10/group), (ii) diabetic mice + vehicle (n = 10/group), and (iii) diabetic mice + C646 (n = 10/group). The chosen dose and the procedure of C646 administration to mice were in good agreement with previous reports [33,34]. At the end of the treatment procedure, the animals were sacrificed and the blood was collected by cardiac puncture. After perfusion of mice with calcium-containing PBS (pH 7.4), the kidneys were surgically harvested. The animal studies were conducted in accordance with the guidelines of EU Directive 2010/63/EU and the experimental protocols were approved by the ethical committee of the Institute of Cellular Biology and Pathology "Nicolae Simionescu" (#11/29.06.2016, #384/09.02.2018, #6/07.04.2021).

2.3. Quantitative Real-Time Polymerase Chain Reaction (Real Time-PCR) Assay

The kidneys were suspended in RNase inhibitor-containing lysis buffer (Sigma) and subjected to glass bead-based (1.0 mm diameter) tissue homogenization (BioSpec, Bartlesville, OK, USA). Total RNA was extracted from the tissue homogenates using a column-based purification kit (Sigma). The synthesis of complementary DNA (cDNA) was performed by reverse transcribing 0.5 μg of total RNA using the MMLV reverse transcriptase enzyme (Thermo Fisher). The relative mRNA transcript levels corresponding to specific Nox subtypes, inflammatory, and fibrotic markers were determined by amplification of the cDNA employing a real-time PCR machine (LightCycler ^{TM} 480 II, Roche, Basel, Switzerland) and quantified by the C_T comparative method [35]. The β-actin mRNA expression level was used for internal normalization. The oligonucleotide primer sequences are shown in Table S1 in the Supplementary Materials file.

2.4. Western Blot Assay

Kidneys derived from each animal were suspended in RIPA lysis buffer containing a protease inhibitor cocktail (Sigma) and subjected to homogenization as indicated above. Protein denaturation was performed in Laemmli electrophoresis sample buffer (Serva) at 95 °C for 20 min. Protein samples (30 μg/lane) were separated by SDS-PAGE and transferred onto nitrocellulose membranes (Bio-Rad). The membranes were incubated for 12 h (4 °C) with primary antibodies against H3K27ac (rabbit polyclonal, C15410174, dilution 1:1000), H3pan (mouse monoclonal, C15200011, dilution 1:1000), Nox1 (rabbit polyclonal, ab131088, concentration 0.5 µg/mL), Nox2 (rabbit polyclonal, PA5-79118, concentration 0.5 µg/mL), Nox4 (rabbit polyclonal, sc-30141, 1:200), or β-actin (mouse monoclonal, sc-47778, 1:500) followed by one-hour exposure (room temperature) to anti-rabbit IgG-HRP (sc-2370, 1:2000) or anti-mouse IgG-HRP (sc-2031, dilution 1:2000) secondary antibodies. Protein bands were detected by chemiluminescence imaging (ImageQuant LAS 4000 system, Fujifilm, Tokyo, Japan). TotalLab ™ (Newcastle upon Tyne, UK)-based densitometric analysis of the protein levels was carried out using the expression level of β-actin protein as internal normalization.

2.5. Histochemistry and Immunofluorescence Microscopy

After harvesting, the kidneys were fixed overnight at 4 °C in paraformaldehyde 4% solution in phosphate buffer 0.1 M, pH 7.4, cryoprotected in 5%, 10%, 20%, and 50% glycerol solutions in phosphate buffer 0.1 M, pH 7.4, and embedded in optimal cutting temperature compound. Kidney cryosections (5 µm-thick) were mounted onto SuperFrost Plus ™ microscope slides (Thermo Scientific) and stained with hematoxylin-eosin solution or subjected to fluorescence immunolabeling for Nox1 (rabbit polyclonal, ab131088, dilution 1:250), Nox2 (rabbit polyclonal, PA5-79118, dilution 1:250), Nox4 (rabbit polyclonal, sc-30141, dilution 1:50), collagen IV (rabbit polyclonal, ab6586, dilution 1:80), fibronectin (rabbit polyclonal, ab2413, dilution 1:250) or laminin (rabbit polyclonal, PA5-16287, dilution 1:250). As a secondary antibody, Alexa Fluor ™ 594 goat anti-rabbit IgG (A11037, dilution 1:500) was used. In these experiments, 4′,6-diamidino-2-phenylindole (DAPI) stain was used to detect the cell nuclei in the specimens. Sections were examined and photographed with an inverted fluorescence microscope (Zeiss Axio Observer). Quantitative analysis of the staining was carried out using ImageJ software (NIH Image, Bethesda, MD, USA).

2.6. Detection of ROS In Situ

The redox-sensitive probe, dihydroethidium (DHE)/hydroethidine, was used for in situ detection of ROS on fresh-frozen 5 µm-thick kidney cryosections as previously described [17]. Briefly, the sections were rinsed for 30 min in PBS (pH 7.4) at room temperature (to remove the OCT) and then incubated with 5 µM DHE in PBS (pH 7.4) at 37 °C for 30 min in a dark humidified chamber slide. After washing in PBS (pH 7.4) to remove the residual DHE, the fluorescence images were taken with an inverted fluorescence microscope (Zeiss Axio Observer). Quantitative analysis of the DHE staining was performed employing ImageJ software (NIH Image, USA).

2.7. Luciferase Reporter Assay

HEK293 cells were employed as the luciferase expression system due to their high efficiency of transfection. Twenty-four hours prior to transfection, HEK293 cells were seeded at 1×10^5 cells/well into 12-well tissue culture plates. The Superfect ™ reagent was used for transient transfection of the cells in accordance with the manufacturer's protocol (Qiagen). The enhancer element configuration and sequence of the *cis*-reporter plasmids were as follows: pNF-κB-Luc [(NF-κB (5×), $(TGGGGACTTTCCGC)_5$], pC/EBP-Luc [C/EBP (3×), $(ATTGCGCAAT)_3$], pGAS-Luc [GAS (4×), $(AGTTTCATATTACTCTAAATC)_4$], and pISRE-Luc [ISRE (5×), $(TAGTTTCACTTTCCC)_5$]. The optimized plasmid concentrations were: 0.9 µg/mL of *cis*-reporter plasmid and 0.1 µg/mL pSV-β-galactosidase expression vector (Promega, Walldorf, Germany). The transcription factor/luciferase activity was calculated from the ratio of firefly luciferase (luciferase assay system/Promega) to β-galactosidase level (o-nitrophenyl-β-D-galactopyranoside-based β-galactosidase enzyme assay/Promega).

2.8. Statistical Analysis

Data obtained from at least three independent experiments were expressed as the mean ± standard deviation. Statistical analysis was performed by two-tailed *t*-test and one-way analysis of variance (ANOVA) followed by Tukey's post hoc test; $p < 0.05$ was considered statistically significant.

3. Results

3.1. In Diabetic Mice Long-Term Pharmacological Inhibition of p300/CBP Has No Effect on Plasma Level of Glucose and Body Weight

STZ-induced diabetic C57BL/6J mice were employed as an experimental model to find out whether p300/CBP has a role in the up-regulation of oxidative stress-, inflammation-, and fibrosis-related gene or protein expression levels in the kidney. As schematically depicted in Figure 1A, the animals (n = 30) were distributed into experimental groups to receive either C646 pharmacological inhibitor or its vehicle for 4 weeks. We found that after 4 weeks of diabetes, the plasma glucose level was significantly increased (≈4-fold) and the bodyweight reduced (≈20%) in vehicle-treated diabetic mice as compared with vehicle-treated non-diabetic control counterparts. As compared with vehicle-treated diabetic animals, no significant changes in blood glucose levels and body weights were detected following long-term pharmacological blockade of p300/CBP in diabetic mice (Figure 1B,C).

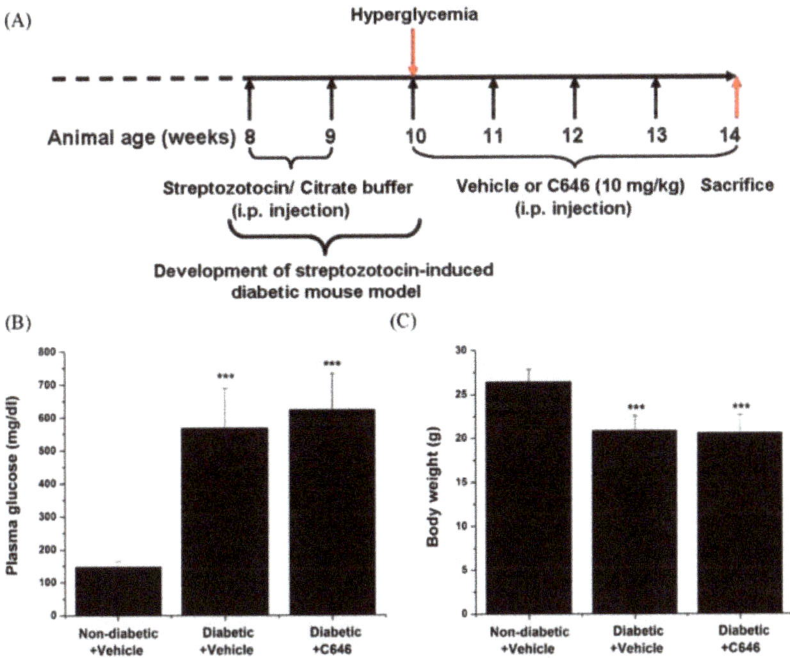

Figure 1. (**A**) Schematic depiction of the experimental set-up to induce diabetes in C57BL/6J mice and the duration of C646/vehicle administration to mice. (**B**,**C**) Plasma glucose levels and body weights were assessed for each animal group at the end of the treatment procedure. n = 5–7, *** $p < 0.001$. p-values were taken in relation to non-diabetic + vehicle condition.

3.2. Histone Acetyltransferase p300/CBP Mediates the Up-Regulation of H3K27ac Level in the Diabetic Kidney

To investigate the impact of diabetic conditions on p300/CBP-induced histone acetylation in the kidney, we examined the relative level of H3K27ac, an important substrate of p300 biochemical activity and an epigenetic marker of active gene expression [29]. Western blot analysis of the whole kidney protein extracts revealed a significant increase (≈1.5-fold) in H3K27ac levels in the diabetic mice group as compared with non-diabetic animals. Importantly, the diabetes-induced increase in the H3K27ac level was suppressed by 4-week administration of C646 pharmacological inhibitor to diabetic mice. The relative expression of the total H3 protein level in the kidney, analyzed in parallel experiments, remained steady in all experimental groups (Figure 2).

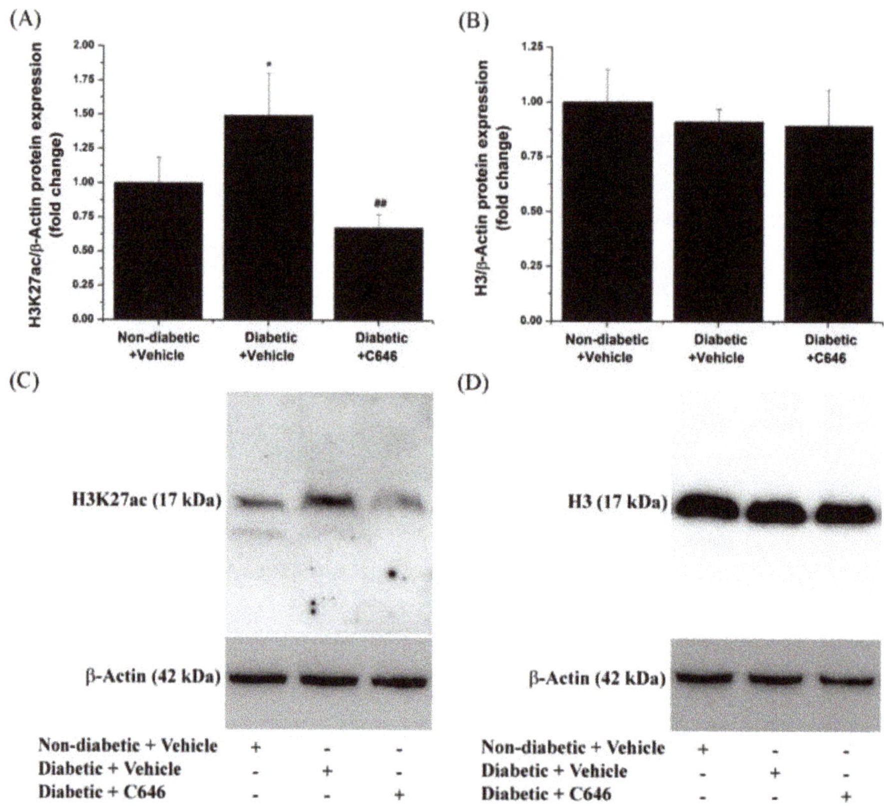

Figure 2. Pharmacological inhibition of histone acetyltransferase p300/CBP suppresses the up-regulation of H3K27ac level in the kidney of diabetic mice. n = 3–4, * $p < 0.05$, p-value taken in relation to non-diabetic + vehicle condition; ## $p < 0.01$, p-value taken in relation to diabetic + vehicle condition. (**A**,**B**) Densitometric analysis of the Western blot data showing the modulation of overall renal H3K27ac and H3 protein expression levels in each animal group. (**C**,**D**) Representative immunoblots depicting the relative expression levels of H3K27ac and H3 in the whole kidney protein extracts.

3.3. Diabetes-Activated p300/CBP Signalling Pathways Mediate the Up-Regulation of Renal Nox Expression

ROS overproduction driven by up-regulated Nox enzymes is acknowledged as the main trigger of oxidative stress-induced renal injury in DKD [12]. To determine whether p300/CBP is implicated in the regulation of the overall Nox expression, the gene and protein expression levels of the Nox1, Nox2, and Nox4 subtypes, the catalytic subunits of the Nox complex, were assessed in the whole kidney extracts. A significant increase in the mRNA and protein levels of Nox1 (mRNA ≈ 3-fold; protein ≈ 2-fold), Nox2 (mRNA ≈ 2-fold; protein ≈ 2.7-fold), and Nox4 (mRNA ≈ 4-fold; protein ≈ 1.85-fold) isoforms were determined in the kidney of diabetic mice after 4 weeks of hyperglycaemia.

Long-term blockade of p300/CBP by C646 suppressed the up-regulation of Nox1, Nox2, and Nox4 transcript levels in the diabetic kidney. Furthermore, the protein expression levels of Nox1 and Nox4 subtypes, except the Nox2 isoform, were significantly diminished in response to C646 administration to diabetic mice as compared to non-treated diabetic animals (Figure 3).

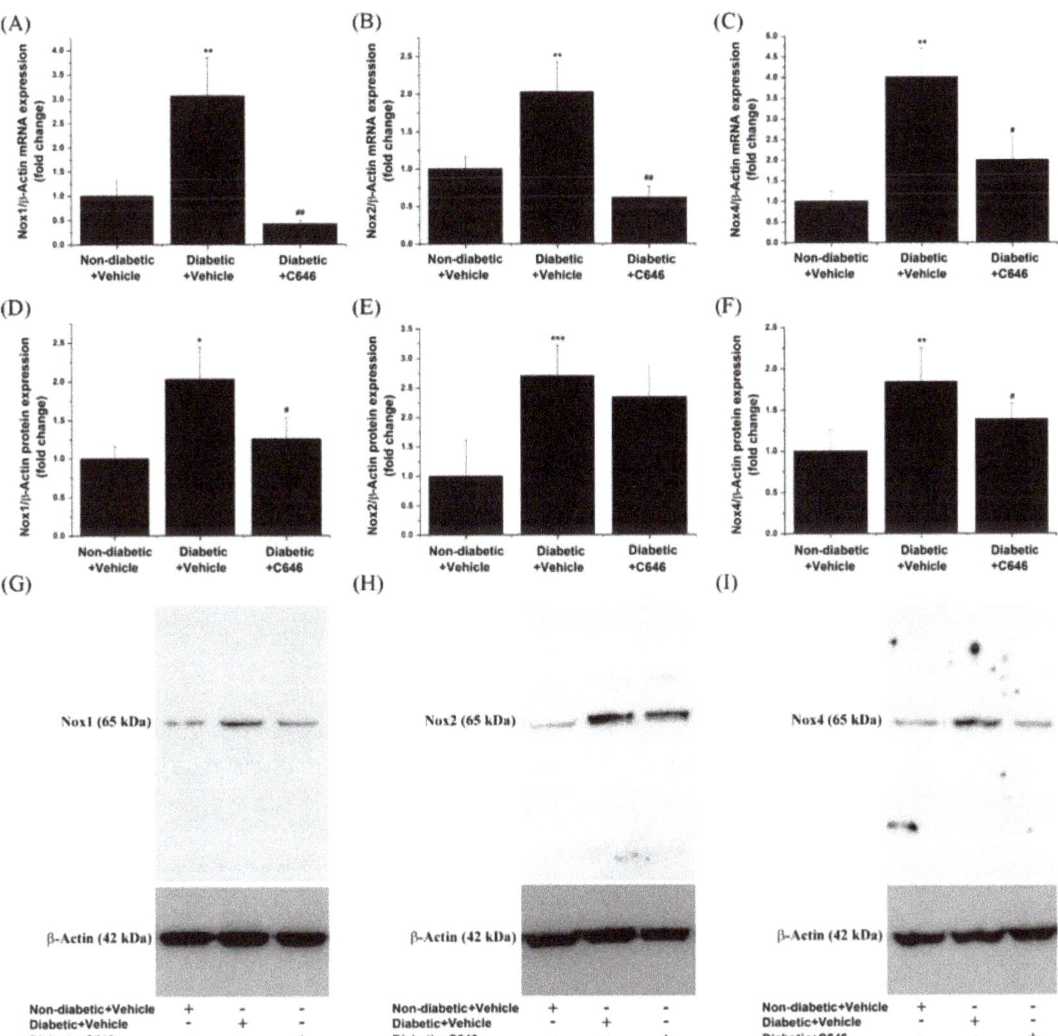

Figure 3. Histone acetyltransferase p300/CBP mediates the up-regulation of Nox expression in the kidney of diabetic mice. (**A–C**) Quantitative real-time PCR analysis indicating the suppressive effects of p300/CBP blockade on diabetes-induced enhanced mRNA levels of Nox1, Nox2, and Nox4 subtypes. (**D–F**) Densitometric analysis of the Western blot data showing the relative expression levels of Nox1, Nox2, and Nox4 proteins in each experimental condition. (**G–I**) Representative immunoblots depicting the modulation of Nox proteins expression. $n = 3–4$, * $p < 0.05$, ** $p < 0.01$, *** $p < 0.001$, p-values taken in relation to non-diabetic + vehicle condition; # $p < 0.05$, ## $p < 0.01$, p-values taken in relation to diabetic + vehicle condition.

Previous comprehensive studies provided compelling evidence that the up-regulation of Nox enzymes is a key pathogenic mechanism associated with glomerulosclerosis [14,15]. Hitherto, the role of p300/CBP in the regulation of glomerular Nox expression in diabetic conditions remains elusive. Thus, we further examined the relative expression of Nox catalytic subunits in the glomeruli by immunofluorescence (IF) microscopy. Compared to controls (non-diabetic animals), enhanced glomerular immunolabeling of Nox1 (≈1.65-fold), Nox2 (≈1.9-fold), and Nox4 (≈1.95-fold) proteins was detected in diabetic mice. Blockade of p300/CBP resulted in a significant decrease in Nox1, Nox2, and Nox4 protein levels in the renal glomeruli of diabetic mice. Noteworthy, there was an apparent increase in Nox4 immunodetection, suggesting that the Nox4 subtype, rather than Nox1 and Nox2, is abundant in the glomeruli. These data are consistent and extend a previous study demonstrating that Nox4-containing Nox, rather than Nox1 or Nox2, is a major source of ROS in the glomeruli and contributes to oxidative stress-induced renal damage in experimental diabetes [15]. However, the superior antibody-recognition performance of Nox4 protein as well as the accessibility of the antibody to its specific epitopes in IF microscopy assays should be considered (Figure 4).

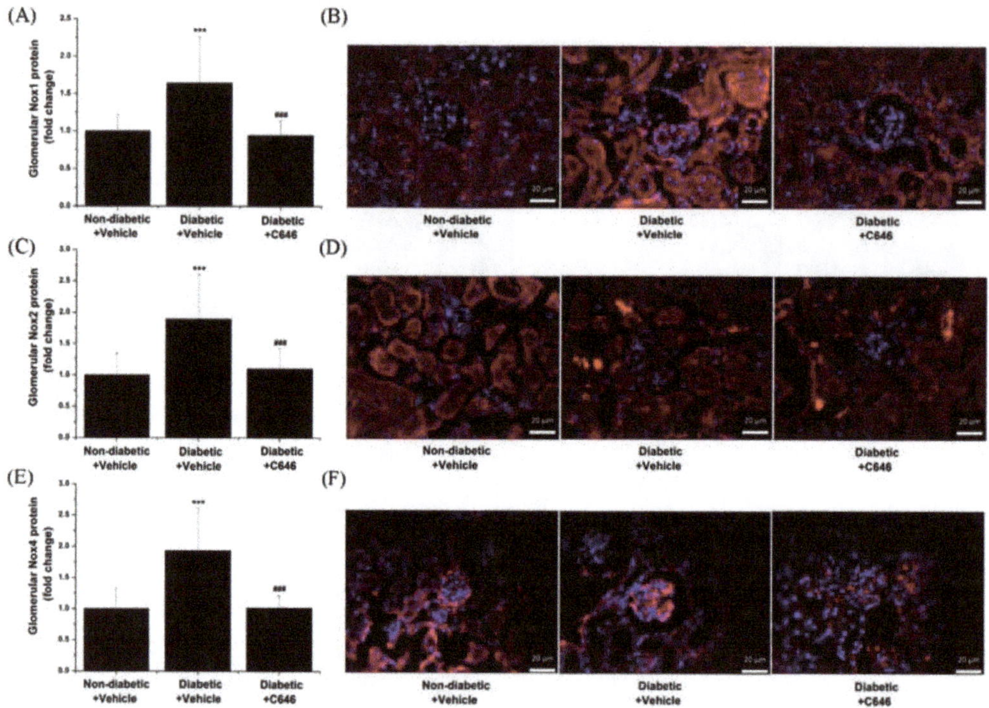

Figure 4. Pharmacological inhibition of histone acetyltransferase p300/CBP by C646 mitigates the glomerular expression of Nox subtypes in diabetic mice. (**A**,**C**,**E**) Quantification of fluorescence immunolabeling for glomerular Nox1, Nox2, and Nox4 subtypes. (**B**,**D**,**F**) Representative IF microscopy images taken at 40× magnification depicting the immunofluorescence staining (red) for Nox subtypes. Sections were counterstained with DAPI (blue) stain to detect the cell nuclei in the specimens. Note that the up-regulation of glomerular Nox1, Nox2, and Nox4 protein levels in diabetic mice is blunted following C646 treatment. n = 14–19/condition quantified glomeruli, *** $p < 0.001$, p-values taken in relation to non-diabetic + vehicle condition; ### $p < 0.001$, p-values taken in relation to diabetic + vehicle condition.

To further examine the possible association of Nox upregulation with ROS overproduction, the glomerular formation of ROS was assessed by in situ labelling employing the DHE redox-sensitive probe. The results showed that pharmacological blockade of p300/CBP significantly reduced the glomerular ROS production in diabetic mice as compared with vehicle-treated diabetic mice (Figure 5).

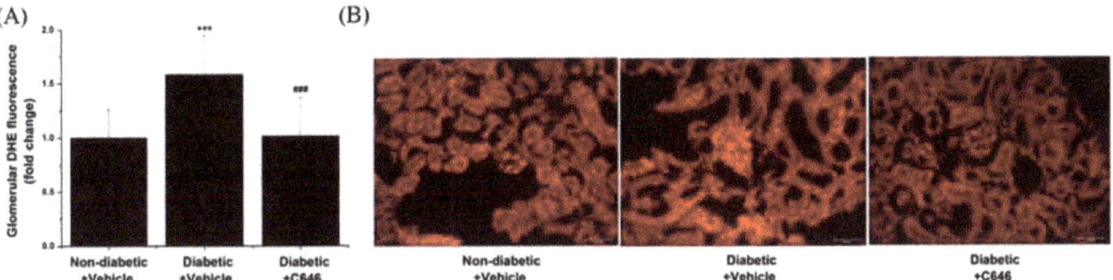

Figure 5. Histone acetyltransferase p300/CBP-dependent induction of in situ glomerular production of ROS in diabetic mice. (**A**) Quantification of glomerular intensity of ROS-induced DHE specific fluorescence signal. (**B**) Representative fluorescence microscopy images (40× magnification) depicting the down-regulatory effects of p300/CBP pharmacological blockade on glomerular ROS formation. n = 17–21/condition quantified glomeruli, *** $p < 0.001$, p-values taken in relation to non-diabetic + vehicle condition; ### $p < 0.001$, p-values taken in relation to diabetic + vehicle condition.

3.4. Pharmacological Inhibition of p300/CBP Suppresses the Diabetes-Induced Enhanced Expression of Pro-Inflammatory and Pro-Fibrotic Molecules in the Kidney

Multiple pathogenic mechanisms, including Nox-activated redox-sensitive signalling pathways, leading to augmented expression of pro-inflammatory and pro-fibrotic mediators, have been shown to promote and accelerate renal injury in DKD [16–18]. Thus, our next query was to uncover the function of histone acetyltransferase p300/CBP as a co-transcriptional regulator of both inflammation- and fibrosis-related genes in the diabetic kidney. To this purpose, we examined the mRNA levels of selective molecules mediating glomerular monocyte/macrophage infiltration and inflammation, as well as ECM components whose enhanced accumulation contributes to glomerulosclerosis in diabetes.

Compared to controls (non-diabetic animals), whole kidney gene expression analysis demonstrated a significant increase in mRNA levels of pro-inflammatory mediators. These are: monocyte chemotactic protein 1/MCP-1 (≈45-fold), tumour necrosis factor α/TNFα (≈12-fold), nitric oxide synthase 2/NOS2 (≈8-fold), intercellular adhesion molecule 1/ICAM-1 (≈4-fold), vascular cell adhesion molecule 1/VCAM-1 (≈8-fold), E-selectin (≈12-fold). In addition, significant augmented mRNA levels of pro-fibrotic mediators [collagen IV (≈4 fold), fibronectin (≈6-fold), and laminin (≈8-fold)] was detected. Interestingly, the latter diabetes-induced up-regulatory gene expression effects were significantly diminished in C646-treated diabetic mice (Figure 6).

3.5. Pharmacological Inhibition of p300/CBP Attenuates the Activity of NF-kB and STAT Transcription Factors

Reportedly, p300 plays a key role in regulating the expression of pro-inflammatory molecules by modulating the function of the NF-kB transcription factor [36,37]. Hence, we questioned the involvement of other transcription factors known to regulate both Nox subtypes and selective diabetes-relevant pro-inflammatory and pro-fibrotic molecules [38–43]. To this purpose, we analyzed the role of p300/CBP in mediating C/EBP and STAT transcription factor activities employing transient transfection of HEK293 reporter cells, with pNF-κB-Luc, pC/EBP-Luc, pGAS-Luc, and pISRE-Luc *cis*-reporter plasmids containing highly conserved NF-kB, C/EBP, GAS, or ISRE *cis*-acting elements. Twenty-four hours after transfection, the cells were exposed for an additional 24 h to 5 μM C646, vehicle (DMSO), or culture medium alone and then subjected to luciferase level detection as described above. The pharmacological blockade of p300/CBP resulted in a significant down-regulation of luciferase level directed by NF-kB, GAS or ISRE elements, except C/EBP. Interestingly, the blockade of p300/CBP virtually abolished the luciferase level directed by GAS elements, the typically bind STAT1/STAT3 heterodimer as compared with ISRE elements that characteristically mediate the transcriptional response of STAT1/STAT2 complex (Figure 7).

3.6. Activation of p300/CBP-Related Signalling Pathways Mediates Diabetes-Induced Glomerular Hypertrophy and Accumulation of Extracellular Matrix Proteins

Glomerular hypertrophy typically induced by mesangial cell hypertrophy and proliferation, and accumulation of the ECM proteins is the main structural feature associated with the pathology of DKD. Morphometric analysis of the kidney specimens demonstrated that blockade of p300/CBP significantly reduces diabetes-induced (≈1.7-fold) glomerular hypertrophy (Figure 8).

Considering the aforementioned findings highlighting that pharmacological blockade of p300/CBP prevents the up-regulation of collagen IV, fibronectin, and laminin transcript levels in the kidney of diabetic mice, we next examined the effect of p300/CBP blockade on the glomerular protein levels. As revealed by IF microscopy, in the diabetic kidney, a significant increase in the glomerular collagen IV/COL4A (≈2-fold), fibronectin/FN (≈1.5-fold), and laminin/LM (≈2.5-fold) was detected. The level of ECM proteins, namely, collagen IV, fibronectin, and laminin was significantly lower in the glomeruli of C646-treated diabetic mice as compared with vehicle-injected diabetic animals (Figure 9). Altogether, these data point to p300/CBP as an important component of a complex transcriptional hub that integrates and transduces signals of diabetic factors to induce renal structural and functional alterations in DKD.

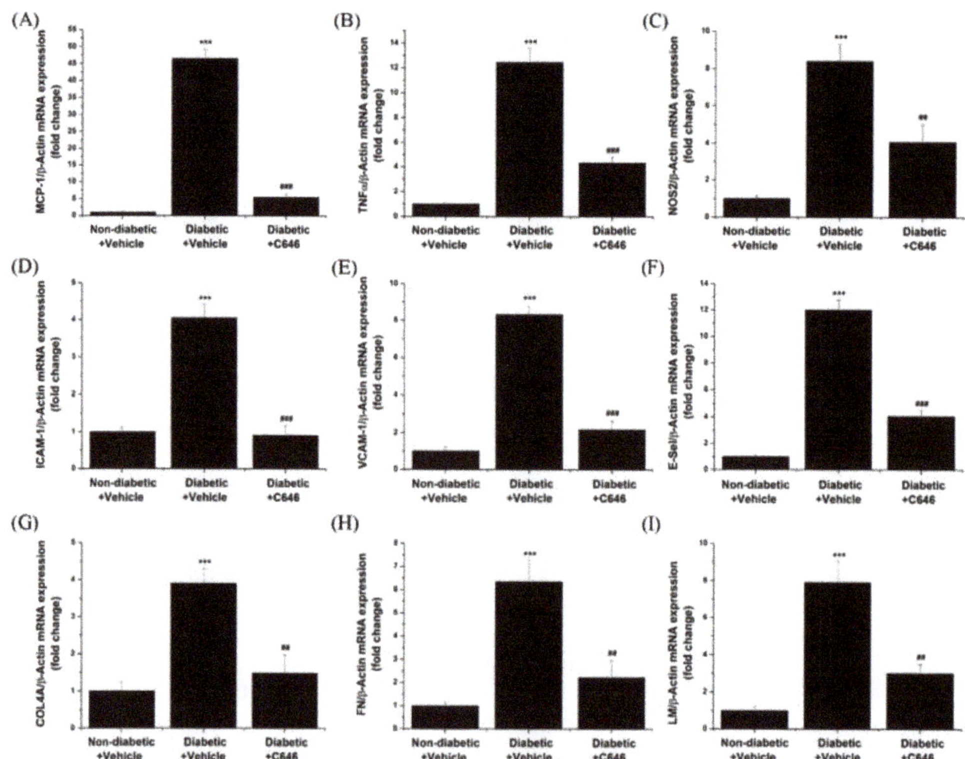

Figure 6. Quantitative real-time PCR analysis indicating the down-regulatory effects of C646-induced p300/CBP inhibition on diabetes-induced augmented mRNA levels of (**A**) MCP-1, (**B**) TNFα, (**C**) NOS2, (**D**) ICAM-1, (**E**) VCAM-1, (**F**) E-selectin, (**G**) collagen IV/COL4A, (**H**) fibronectin/FN, and (**I**) laminin/LM. n = 3–4, *** $p < 0.001$, p-values taken in relation to vehicle-treated non-diabetic mice condition; ## $p < 0.01$, ### $p < 0.001$, p-values taken in relation to vehicle-treated diabetic mice condition.

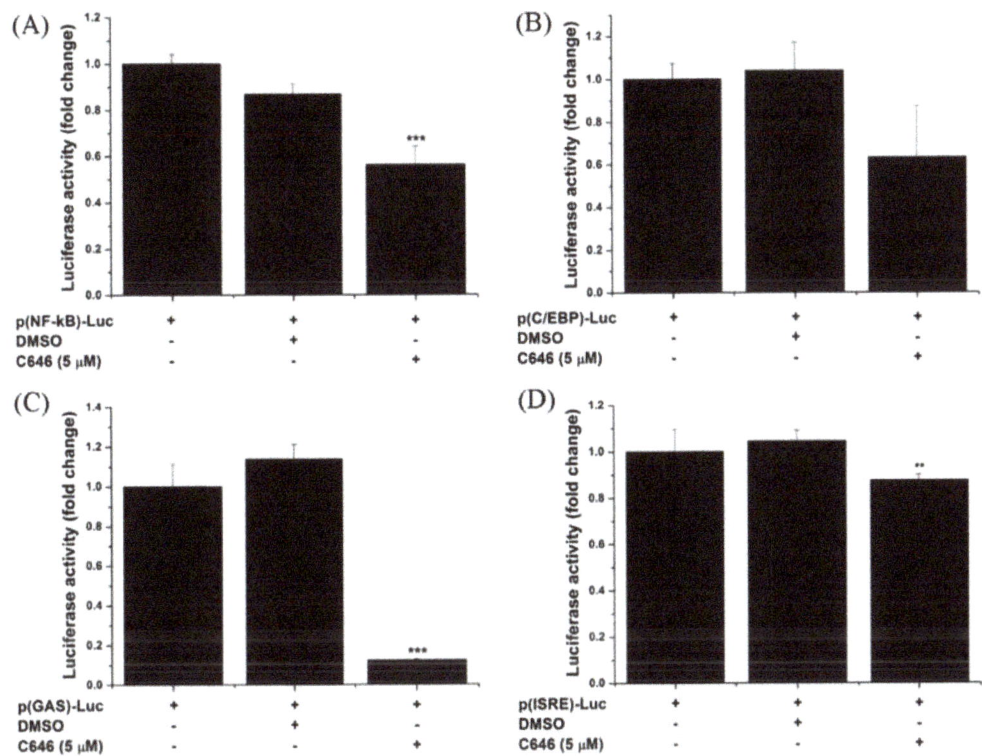

Figure 7. P300/CBP-dependent regulation of the luciferase level directed by the activation of (**A**) NF-kB, (**B**) C/EBP, (**C**) GAS, and (**D**) ISRE enhancer elements in HEK293 cells transfected with the corresponding transcription factor *cis*-reporter plasmids. $n = 3–6$, ** $p < 0.01$, *** $p < 0.001$, p-values taken in relation to vehicle (DMSO)-treated cells condition.

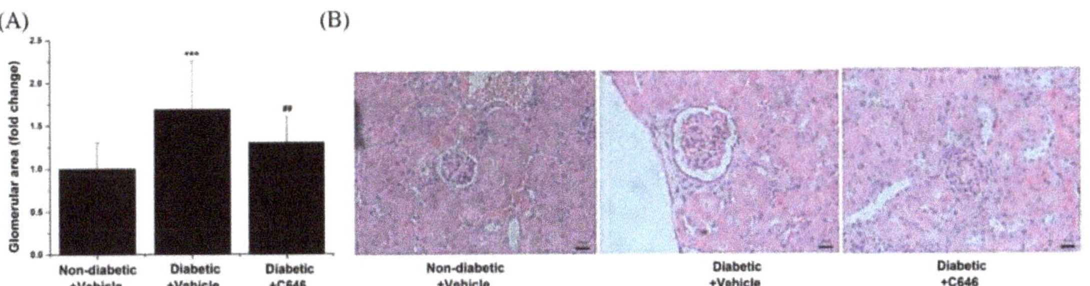

Figure 8. Long-term administration of C646 pharmacological inhibitor of p300/CBP prevents glomerular hypertrophy in diabetic mice. (**A**) Quantification of the relative glomerular area in each experimental animal group. (**B**) Representative hematoxylin-eosin phase-contrast microscopy images taken at 40× magnification depicting the staining of glomeruli. Note that C646 treatment of diabetic mice reduced significantly the relative glomerular hypertrophy to the level of control non-diabetic mice. $n = 23$/condition quantified glomeruli, *** $p < 0.001$, p-values taken in relation to non-diabetic + vehicle condition; ## $p < 0.01$, p-values taken in relation to diabetic + vehicle condition.

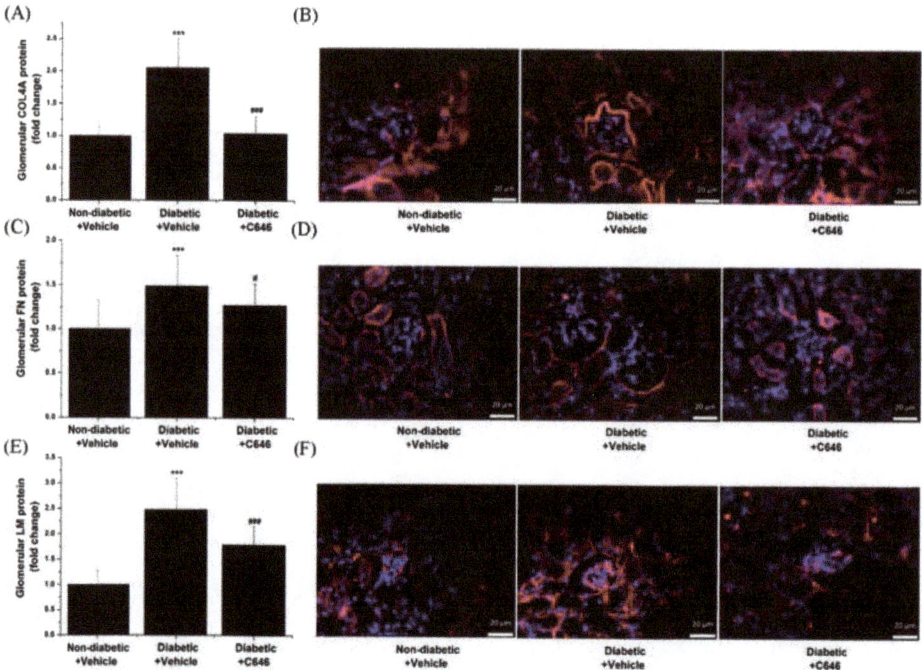

Figure 9. Blockade of histone acetyltransferase p300/CBP reduces the glomerular expression of extracellular matrix proteins collagen IV, fibronectin, and laminin in diabetic mice. (**A**,**C**,**E**) Quantification of fluorescence immunolabeling for glomerular collagen IV/COL4A, fibronectin/FN, and laminin/LM. (**B**,**D**,**F**) Representative immunofluorescence (IF) microscopy images taken at 40× magnification depicting the IF staining (red) for COL4A, FN, and LM. Sections were counterstained with the DAPI stain (blue) to detect the cell nuclei in the specimens. n = 18–23/condition quantified glomeruli, *** $p < 0.001$, p-values taken in relation to non-diabetic + vehicle condition; # $p < 0.05$, ### $p < 0.001$, p-values taken in relation to diabetic + vehicle condition.

4. Discussion

The importance of the complex functional networking among oxidative stress, inflammation, and fibrosis in diabetes-induced renal disorders cannot be overemphasized. Over the past two decades, multiple molecular triggers and signalling pathways of glomerular capillary dysfunction contributing to kidney failure have been determined [1–3]. Yet, despite the significant advances in clinical and experimental research in diabetes and its complications, the diagnosis and effective treatment of DKD remains challenging [4–8]. Noteworthy, due to the metabolic memory in diabetes (an emergent and rapidly evolving epigenetic-related mechanistic concept) and in spite of the therapeutic control of hyperglycaemia, diabetes pathological long-lasting effects persist and continue to promote systemic cellular detrimental effects [44–47]. Consequently, in order to efficiently counteract/attenuate the diabetes-accelerated renal failure, other intrinsic glomerular cell-specific signalling pathways should be considered for drug targeting.

Emerging evidence demonstrates that the diabetes-associated histone acetylation-based epigenetic alterations induce specific transcriptomic instructions in the aorta of mice that regulate important pathological aspects of diabetes-related vascular disorders, including inflammation and oxidative stress [30]. To further uncover potential culprit epigenetic-based pathways underlying renal damage in DKD, we design experiments to find out whether the transcriptional co-activator p300/CBP is mechanistically implicated in the process of oxidative stress, inflammation, and fibrosis in the kidney of diabetic mice.

The novel findings generated by this study on the diabetic kidney disease are: (i) pharmacological inhibition of p300/CBP reduces the level of H3K27ac, an epigenetic mark of active gene expression; (ii) histone acetyltransferase p300/CBP mediates the diabetes-induced mRNA and protein level of Nox subtypes; (iii) blockade of p300/CBP suppresses diabetes-induced glomerular ROS production in situ; (iv) pharmacological inhibition of p300/CBP reduces the transcript levels of pro-inflammatory (MCP-1, TNFα, NOS2, ICAM-1, VCAM-1, E-selectin) and pro-fibrotic molecules (collagen IV, fibronectin, laminin); (v) blockade of p300/CBP reduces the activity of NF-kB and STAT pro-inflammatory transcription factors in vitro; (vi) diabetes-activated p300/CBP-related signalling pathways induce glomerular hypertrophy and an enhanced level of ECM proteins, key pathological features of mesangial expansion leading to kidney failure in diabetes.

Previous comprehensive studies demonstrated that renal Nox subtypes are up-regulated in the human and STZ-induced diabetic mice, a clinically relevant experimental model of DKD and that pharmacological targeting or genetic manipulation of the expression of different Nox subtypes reduces oxidative stress-mediated renal injury in diabetes [15,16]. Yet, other than direct targeting of individual Nox subtype function, reducing the up-regulated Nox expression and the ensuing oxidative stress as well as down-regulating the expression of proinflammatory and profibrotic molecules by selective pharmacological targeting of common up-stream regulators may represent attractive therapeutic strategies in reducing the burden of DKD.

Pharmacological blockade of p300/CBP co-transcriptional activator emerged as a potential way to intervene in several human malignancies [34,48]; however, its possible role in DKD is yet to be determined. Evidence indicates that in addition to histone acetylation-induced chromatin relaxation, p300 activates the function of the NF-kB proinflammatory transcription factor, and consequently may induce the expression of a large number of genes potentially contributing to multiple pathological aspects of CVD such as inflammation and Nox-derived excessive production of ROS [29,37]. Thus, complex co-operative mechanisms involving epigenetic changes in chromatin conformation and lysine-acetylation of transcription factors are likely to be involved in p300-mediated gene transcription.

Our experiments were intended to uncover new mechanisms responsible for the diabetes-induced renal disease such as the role of p300/CBP in mediating renal oxidative stress, inflammation, and fibrosis. To this purpose, we employed C646, a reversible, cell-permeable pyrazolone-based inhibitor that competes with acetyl-CoA for the histone acetyltransferase p300/CBP-containing Lys-CoA binding pocket. In addition, in accordance with the manufacturer's specifications (Calbiochem), C646 exhibits high selectivity for p300/CBP activity and less inhibitory effects against other HAT subtypes including aralkylamine *N*-acetyltransferase, p300/CBP-associated factor (PCAF), GCN5, Rtt109, Sas, and MOZ histone acetyltransferases.

Apart from beneficial effects demonstrated in different experimental models of cancer [49], it has been increasingly shown that pharmacological targeting of the up-regulated p300/CBP activity with C646 inhibitor represents a promising way to intervene in the progression of metabolic and cardiovascular disorders. Reportedly, C646 reduced skeletal muscle atrophy in *db/db* mice, an experimental model of type 2 diabetes [50]. In addition, C646-mediated blockade of p300 improved coronary flow in an experimental model of heart failure [33] and reduced cardiac fibrosis in hypertensive mice [51].

Considering the fact that the level of H3K27ac was found significantly elevated in diabetic kidneys, indicative of p300/CBP overactivity, the diabetic mice were subjected to C646 treatment in order to attenuate the p300/CBP-induced renal epigenomic alterations and transcription factor activation potentially contributing to up-regulation of pro-oxidant, pro-inflammatory, and profibrotic genes in response to diabetic insults. Noteworthy, whole-kidney analysis revealed that C646-induced pharmacological inhibition of p300/CBP suppresses the up-regulation of global histone acetylation level, namely, H3K27ac, an epigenetic mark of active gene expression, indicating the implication of p300/CBP in mediating these effects.

Previously, we reported that under pro-inflammatory conditions, p300/CBP mediates the up-regulation of Nox1, Nox2, Nox4, and Nox5 expression in cultured human macrophages [29]. In addition, we demonstrated that the p300 protein physically interacts with Nox1-5 gene promoters at the sites of active transcription and that a high concentration of glucose induces H3K27ac enrichment at the level of Nox1, Nox4, and Nox5 gene promoters correlated with the up-regulation of both mRNA and protein levels in cultured human aortic smooth muscle cells [30]. Considering the fact that p300/CBP mediates the induction of Nox expression in response to pro-inflammatory or diabetic factors, we hypothesized that p300/CBP could become an additional therapeutic target in DKD to attenuate oxidative stress-induced kidney damage.

Whole renal tissue gene and protein expression analysis revealed that p300/CBP-activated signalling pathways contribute to the overall up-regulation of Nox1, Nox2, and Nox4 mRNA expression and protein levels in the diabetic kidney. Consistent with these findings, pharmacological blockade of p300/CBP suppressed the diabetes-induced glomerular ROS production in situ. Regarding the functional significance of Nox subtypes up-regulation, in a previous comprehensive study, it was demonstrated that the induction of Nox4 rather than Nox1 or Nox2 is implicated in ROS overproduction and oxidative stress-induced kidney failure [15]. Our data are in good agreement and extend this study, and further demonstrate that Nox4, rather than Nox1 and Nox2 subtypes is abundantly expressed in the mouse glomeruli. Yet, as revealed by several in vitro and in vivo studies, the expression of all Nox subtypes and the ensuing ROS overproduction are induced in various cell types in the kidney (i.e., endothelial cells, mesangial cells, podocytes) in response to diabetic conditions by molecular mechanisms that broadly implicate the activation of pro-inflammatory signalling pathways [52–56]. Consistent with this evidence, it was recently demonstrated that ROS overproduction driven by both Nox1 and Nox2 contribute to oxidative stress and kidney injury in diabetic mice [18]. Collectively, the data of our study suggest that the activation of p300/CBP in response to diabetic conditions induces oxidative stress in the kidney via multiple, Nox subtype-dependent, excessive production of ROS. Moreover, pharmacological blockade of p300/CBP significantly reduced the diabetes-associated inflammatory response in the kidney, as demonstrated by the knock-down effects on MCP-1, TNFα, NOS2, ICAM-1, VCAM-1, and E-selectin transcript levels. We also found that in addition to NF-kB, the function of STAT pro-inflammatory transcription factors is reduced in response to p300/CBP inhibition. Noteworthy, NF-kB and STAT are important transcriptional regulators of Nox expression, a condition that may partially explain the down-regulatory effects of p300/CBP pharmacological blockade on Nox1, Nox2, and Nox4 gene expression and protein levels in the diabetic mice kidney [42,43]. This evidence further supports and expands the potential role of p300/CBP as a key modulator of both oxidative stress- and inflammation-related gene expression levels in DKD.

Glomerular hypertrophy generally caused by mesangial cell proliferation and accumulation of ECM proteins is a reliable index of mesangial expansion and subsequent glomerulosclerosis, the main structural abnormality inducing kidney failure in diabetes [1–3]. Evidence exists that ROS overproduction, typically generated by activated Nox, and excess production of inflammatory mediators are involved in the mesangial cell phenotypic alterations and increased synthesis of matrix proteins [14]. We provide evidence that in the diabetic kidney, glomeruli enlargement correlates with increased expression of collagen IV, fibronectin, and laminin at both mRNA and protein levels. Importantly, pharmacological inhibition of p300/CBP significantly reduced the diabetes-associated glomerular enlargement and the expression of profibrotic markers of mesangial expansion. Our data are in good agreement and extend a previous study demonstrating that p300/CBP mediates the function of TGFβ1, an important profibrotic factor in mesangial cells [57].

There is evidence that HDAC-dependent epigenetic mechanisms are functionally involved in multiple pathological aspects linked to DKD. It was demonstrated that pharmacological inhibition of HDAC attenuates renal oxidative stress, inflammation, and fibrosis, and improves kidney function in experimental diabetes [58–63]. In line with these studies, we have previously reported that HDAC-related signalling pathways mediate the up-regulation of vascular Nox expression and markers of oxidative stress both in vitro and in vivo experimental models of diabetes [30]. Interestingly, the pan-HDAC inhibitor trichostatin A activates the ubiquitin-proteasome pathway to induce p300 protein degradation, thus preventing the formation of transcriptional activation complexes [64]. In sum, these findings could explain the occurrence of similar pharmacological effects driven by HAT or HDAC inhibitors; considering the fact that these enzymatic systems have opposite biochemical functions.

5. Conclusions

In conclusion, we provide here evidence that, in the diabetic kidney, the activation of histone acetyltransferase p300/CBP enhances ROS production (most probable generated by the up-regulated Nox), inflammation, and the production of extracellular matrix proteins. Based on the fact that p300/CBP is an important co-transcriptional activator regulating the expression of a wide range of pro-oxidant, pro-inflammatory, and pro-fibrotic genes, we postulate that pharmacological targeting of p300/CBP could become a promising supportive therapeutic option in DKD. Further preclinical and clinical studies should address the relevance of p300/CBP in the pathophysiological context of human DKD in order to translate these findings to human pathology.

Supplementary Materials: The following supporting material is available online at https://www.mdpi.com/article/10.3390/antiox10091356/s1, Table S1: Sequences and GenBank® accession number of oligonucleotide primers used in real-time PCR assays.

Author Contributions: Conceptualization, A.M. and S.-A.M.; methodology, A.-G.L., M.-L.V., A.M. and S.-A.M.; data curation and analysis, A.-G.L., M.-L.V., A.M. and S.-A.M.; manuscript writing and revision, A.-G.L., A.M., M.S. and S.-A.M.; funding acquisition, A.M., M.S. and S.-A.M.; supervision, S.-A.M. All authors have read and agreed to the published version of the manuscript.

Funding: Work supported by grants from the Romanian National Authority for Scientific Research and Innovation, UEFISCDI [PN-III-P2-2.1-PED-2019-2512 (Contract number 265PED/2020), PN-III-P2-2.1-PED-2019-2497 (Contract number 342PED/2020), PN-III-P4-ID-PCE-2016-0665 (Contract number PCE69/2017), PN-III-P4-ID-PCE-2020-1898 (Contract number PCE81/2021)] and the Romanian Academy.

Institutional Review Board Statement: The animal studies were carried out in accordance with the guidelines of EU Directive 2010/63/EU and the animal experimental protocols were approved by the Institutional Animal Ethic Committee of the Institute of Cellular Biology and Pathology "Nicolae Simionescu" (#11/29.06.2016, #384/09.02.2018, #6/07.04.2021).

Informed Consent Statement: Not applicable.

Data Availability Statement: Data is contained within the article or Supplementary Material.

Acknowledgments: The authors are indebted to Nae Sanda and Constanta Stan for the helpful technical assistance.

Conflicts of Interest: The authors declare they have no conflict of interest.

References

1. Thomas, M.C.; Brownlee, M.; Susztak, K.; Sharma, K.; Jandeleit-Dahm, K.A.M.; Zoungas, S.; Rossing, P.; Groop, P.H.; Cooper, M.E. Diabetic kidney disease. *Nat. Rev. Dis. Primers* **2015**, *1*, 15018. [CrossRef]
2. Umanath, K.; Lewis, J.B. Update on diabetic nephropathy: Core curriculum 2018. *Am. J. Kidney Dis.* **2018**, *71*, 884–895. [CrossRef]
3. Hong, G.L.; Kim, K.H.; Lee, C.H.; Kim, T.W.; Jung, J.Y. NQO1 deficiency aggravates renal injury by dysregulating Vps34/ATG14L complex during autophagy initiation in diabetic nephropathy. *Antioxidants* **2021**, *10*, 333. [CrossRef]

4. Zoja, C.; Xinaris, C.; Macconi, D. Diabetic nephropathy: Novel molecular mechanisms and therapeutic targets. *Front. Pharmacol.* **2020**, *11*, 586892. [CrossRef]
5. Yamazaki, T.; Mimura, I.; Tanaka, T.; Nangaku, M. Treatment of diabetic kidney disease: Current and future. *Diabetes Metab. J.* **2021**, *45*, 11–26. [CrossRef] [PubMed]
6. Gil, C.L.; Hooker, E.; Larrivée, B. Diabetic kidney disease, endothelial damage, and podocyte-endothelial crosstalk. *Kidney Med.* **2020**, *3*, 105–115. [CrossRef] [PubMed]
7. Fu, H.; Liu, S.; Bastacky, S.I.; Wang, X.; Tian, X.J.; Zhou, D. Diabetic kidney diseases revisited: A new perspective for a new era. *Mol. Metab.* **2019**, *30*, 250–263. [CrossRef] [PubMed]
8. Sharma, D.; Bhattacharya, P.; Kalia, K.; Tiwari, V. Diabetic nephropathy: New insights into established therapeutic paradigms and novel molecular targets. *Diabetes Res. Clin. Pract.* **2017**, *128*, 91–108. [CrossRef] [PubMed]
9. Manea, S.A.; Fenyo, I.M.; Manea, A. c-Src tyrosine kinase mediates high glucose-induced endothelin-1 expression. *Int. J. Biochem. Cell Biol.* **2016**, *75*, 123–130. [CrossRef]
10. Manea, S.A.; Todirita, A.; Manea, A. High glucose-induced increased expression of endothelin-1 in human endothelial cells is mediated by activated CCAAT/enhancer-binding proteins. *PLoS ONE* **2013**, *8*, e84170. [CrossRef]
11. Manea, S.A.; Manea, A.; Heltianu, C. Inhibition of JAK/STAT signaling pathway prevents high-glucose-induced increase in endothelin-1 synthesis in human endothelial cells. *Cell Tissue Res.* **2010**, *340*, 71–79. [CrossRef]
12. Lee, S.R.; An, E.J.; Kim, J.; Bae, Y.S. Function of NADPH oxidases in diabetic nephropathy and development of Nox inhibitors. *Biomol. Ther.* **2020**, *28*, 25–33. [CrossRef] [PubMed]
13. Laddha, A.P.; Kulkarni, Y.A. NADPH oxidase: A membrane-bound enzyme and its inhibitors in diabetic complications. *Eur. J. Pharmacol.* **2020**, *881*, 173206. [CrossRef] [PubMed]
14. Lee, E.S.; Kim, H.M.; Lee, S.H.; Ha, K.B.; Bae, Y.S.; Lee, S.J.; Moon, S.H.; Lee, E.Y.; Lee, J.H.; Chung, C.H. APX-115, a pan-NADPH oxidase inhibitor, protects development of diabetic nephropathy in podocyte specific NOX5 transgenic mice. *Free Radic. Biol. Med.* **2020**, *161*, 92–101. [CrossRef] [PubMed]
15. Gray, S.P.; Jha, J.C.; Kennedy, K.; van Bommel, E.; Chew, P.; Szyndralewiez, C.; Touyz, R.M.; Schmidt, H.H.H.W.; Cooper, M.E.; Jandeleit-Dahm, K.A.M. Combined NOX1/4 inhibition with GKT137831 in mice provides dose-dependent reno- and atheroprotection even in established micro- and macrovascular disease. *Diabetologia* **2017**, *60*, 927–937. [CrossRef]
16. Li, Y.; Li, Y.; Zheng, S. Inhibition of NADPH oxidase 5 (NOX5) suppresses high glucose-induced oxidative stress, inflammation and extracellular matrix accumulation in human glomerular mesangial cells. *Med. Sci. Monit.* **2020**, *26*, e919399. [CrossRef] [PubMed]
17. Jha, J.C.; Banal, C.; Okabe, J.; Gray, S.P.; Hettige, T.; Chow, B.S.M.; Thallas-Bonke, V.; De Vos, L.; Holterman, C.E.; Coughlan, M.T.; et al. NADPH oxidase Nox5 accelerates renal injury in diabetic nephropathy. *Diabetes* **2017**, *66*, 2691–2703. [CrossRef]
18. Muñoz, M.; López-Oliva, M.E.; Rodríguez, C.; Martínez, M.P.; Sáenz-Medina, J.; Sánchez, A.; Climent, B.; Benedito, S.; García-Sacristán, A.; Rivera, L.; et al. Differential contribution of Nox1, Nox2 and Nox4 to kidney vascular oxidative stress and endothelial dysfunction in obesity. *Redox Biol.* **2020**, *28*, 101330. [CrossRef]
19. Ihnat, M.A.; Thorpe, J.E.; Kamat, C.D.; Szabó, C.; Green, D.E.; Warnke, L.A.; Lacza, Z.; Cselenyák, A.; Ross, K.; Shakir, S.; et al. Reactive oxygen species mediate a cellular 'memory' of high glucose stress signalling. *Diabetologia* **2007**, *50*, 1523–1531. [CrossRef]
20. Reddy, M.A.; Zhang, E.; Natarajan, R. Epigenetic mechanisms in diabetic complications and metabolic memory. *Diabetologia* **2015**, *58*, 443–455. [CrossRef]
21. Sankrityayan, H.; Kulkarni, Y.A.; Gaikwad, A.B. Diabetic nephropathy: The regulatory interplay between epigenetics and microRNAs. *Pharm. Res.* **2019**, *141*, 574–585. [CrossRef]
22. Keating, S.T.; Diepen, J.A.; Riksen, N.P.; El-Osta, A. Epigenetics in diabetic nephropathy, immunity and metabolism. *Diabetologia* **2018**, *61*, 6–20. [CrossRef] [PubMed]
23. Zhou, H.; Ni, W.J.; Meng, X.M.; Tang, L.Q. MicroRNAs as Regulators of immune and inflammatory responses: Potential therapeutic targets in diabetic nephropathy. *Front. Cell Dev. Biol.* **2021**, *25*, 618536. [CrossRef] [PubMed]
24. Wang, Y.; Miao, X.; Liu, Y.; Li, F.; Liu, Q.; Sun, J.; Cai, L. Dysregulation of histone acetyltransferases and deacetylases in cardiovascular diseases. *Oxid. Med. Cell Longev.* **2014**, *2014*, 641979. [CrossRef] [PubMed]
25. Kee, H.J.; Kook, H. Roles and targets of class I and IIa histone deacetylases in cardiac hypertrophy. *J. Biomed. Biotechnol.* **2011**, *2011*, 928326. [CrossRef]
26. McKinsey, T.A. Targeting inflammation in heart failure with histone deacetylase inhibitors. *Mol. Med.* **2011**, *17*, 434–441. [CrossRef]
27. Chen, F.; Li, X.; Aquadro, E.; Haigh, S.; Zhou, J.; Stepp, D.W.; Weintraub, N.L.; Barman, S.A.; Fulton, D.J.R. Inhibition of histone deacetylase reduces transcription of NADPH oxidases and ROS production and ameliorates pulmonary arterial hypertension. *Free Radic. Biol. Med.* **2016**, *99*, 167–178. [CrossRef]
28. Manea, S.A.; Vlad, M.L.; Fenyo, I.M.; Lazar, A.G.; Raicu, M.; Muresian, H.; Simionescu, M.; Manea, A. Pharmacological inhibition of histone deacetylase reduces NADPH oxidase expression, oxidative stress and the progression of atherosclerotic lesions in hypercholesterolemic apolipoprotein E-deficient mice; potential implications for human atherosclerosis. *Redox Biol.* **2020**, *28*, 101338. [CrossRef]

29. Vlad, M.L.; Manea, S.A.; Lazar, A.G.; Raicu, M.; Muresian, H.; Simionescu, M.; Manea, A. Histone acetyltransferase-dependent pathways mediate upregulation of NADPH oxidase 5 in human macrophages under inflammatory conditions: A potential mechanism of reactive oxygen species overproduction in atherosclerosis. *Oxid. Med. Cell Longev.* **2019**, *2019*, 3201062. [CrossRef]
30. Manea, S.A.; Antonescu, M.L.; Fenyo, I.M.; Raicu, M.; Simionescu, M.; Manea, A. Epigenetic regulation of vascular NADPH oxidase expression and reactive oxygen species production by histone deacetylase-dependent mechanisms in experimental diabetes. *Redox Biol.* **2018**, *16*, 332–343. [CrossRef]
31. Hsueh, W.; Abel, E.D.; Breslow, J.L.; Maeda, N.; Davis, R.C.; Fisher, R.A.; Dansky, H.; McClain, D.A.; McIndoe, R.; Wassef, M.K.; et al. Recipes for creating animal models of diabetic cardiovascular disease. *Circ. Res.* **2007**, *100*, 1415–1427. [CrossRef]
32. López, A.G.; Molina-Van den Bosch, M.; Vergara, A.; García-Carro, C.; Seron, D.; Jacobs-Cachá, C.; Soler, M.J. Revisiting experimental models of diabetic nephropathy. *Int. J. Mol. Sci.* **2020**, *21*, 3587. [CrossRef]
33. Su, H.; Zeng, H.; He, X.; Zhu, S.H.; Chen, J.X. Histone acetyltransferase p300 inhibitor improves coronary flow reserve in SIRT3 (sirtuin 3) knockout mice. *J. Am. Heart Assoc.* **2020**, *9*, e017176. [CrossRef] [PubMed]
34. Wang, Y.M.; Gu, M.L.; Meng, F.S.; Jiao, W.R.; Zhou, X.X.; Yao, H.P.; Ji, F. Histone acetyltransferase p300/CBP inhibitor C646 blocks the survival and invasion pathways of gastric cancer cell lines. *Int. J. Oncol.* **2017**, *51*, 1860–1868. [CrossRef] [PubMed]
35. Pfaffl, M.W. A new mathematical model for relative quantification in real-time RT-PCR. *Nucleic Acids Res.* **2001**, *29*, e45. [CrossRef] [PubMed]
36. Gerritsen, M.E.; Williams, A.J.; Neish, A.S.; Moore, S.; Shi, Y.; Collins, T. CREB-binding protein/p300 are transcriptional coactivators of p65. *Proc. Natl. Acad. Sci. USA* **1997**, *94*, 2927–2932. [CrossRef] [PubMed]
37. Pons, D.; de Vries, F.R.; van den Elsen, P.J.; Heijmans, B.T.; Quax, P.H.A.; Jukema, J.W. Epigenetic histone acetylation modifiers in vascular remodelling: New targets for therapy in cardiovascular disease. *Eur. Heart J.* **2009**, *30*, 266–277. [CrossRef] [PubMed]
38. Manea, S.A.; Constantin, A.; Manda, G.; Sasson, S.; Manea, A. Regulation of Nox enzymes expression in vascular pathophysiology: Focusing on transcription factors and epigenetic mechanisms. *Redox Biol.* **2015**, *5*, 358–366. [CrossRef] [PubMed]
39. Manea, S.A.; Todirita, A.; Raicu, M.; Manea, A. C/EBP transcription factors regulate NADPH oxidase in human aortic smooth muscle cells. *J. Cell Mol. Med.* **2014**, *18*, 1467–1477. [CrossRef]
40. Manea, A.; Manea, S.A.; Florea, I.C.; Luca, C.M.; Raicu, M. Positive regulation of NADPH oxidase 5 by proinflammatory-related mechanisms in human aortic smooth muscle cells. *Free Radic. Biol. Med.* **2012**, *52*, 1497–1507. [CrossRef] [PubMed]
41. Manea, A.; Tanase, L.I.; Raicu, M.; Simionescu, M. Transcriptional regulation of NADPH oxidase isoforms, Nox1 and Nox4, by nuclear factor-kappaB in human aortic smooth muscle cells. *Biochem. Biophys. Res. Commun.* **2010**, *396*, 901–907. [CrossRef] [PubMed]
42. Manea, A.; Tanase, L.I.; Raicu, M.; Simionescu, M. Jak/STAT signaling pathway regulates nox1 and nox4-based NADPH oxidase in human aortic smooth muscle cells. *Arter. Thromb. Vasc. Biol.* **2010**, *30*, 105–112. [CrossRef]
43. Manea, A.; Manea, S.A.; Gafencu, A.V.; Raicu, M. Regulation of NADPH oxidase subunit p22(phox) by NF-kB in human aortic smooth muscle cells. *Arch. Physiol. Biochem.* **2007**, *113*, 163–172. [CrossRef] [PubMed]
44. Kushwaha, K.; Sharma, S.; Gupta, J. Metabolic memory and diabetic nephropathy: Beneficial effects of natural epigenetic modifiers. *Biochimie* **2020**, *170*, 140–151. [CrossRef]
45. Testa, R.; Bonfigli, A.R.; Prattichizzo, F.; La Sala, L.; De Nigris, V.; Ceriello, A. The "metabolic memory" theory and the early treatment of hyperglycemia in prevention of diabetic complications. *Nutrients* **2017**, *9*, 437. [CrossRef]
46. Prattichizzo, F.; Giuliani, A.; De Nigris, V.; Pujadas, G.; Ceka, A.; La Sala, L.; Genovese, S.; Testa, R.; Procopio, A.D.; Olivieri, F.; et al. Extracellular microRNAs and endothelial hyperglycaemic memory: A therapeutic opportunity? *Diabetes Obes. Metab.* **2016**, *18*, 855–867. [CrossRef] [PubMed]
47. Ceriello, A. The emerging challenge in diabetes: The "metabolic memory". *Vasc. Pharmacol.* **2012**, *57*, 133–138. [CrossRef]
48. Zheng, S.; Koh, X.Y.; Goh, H.C.; Rahmat, S.A.B.; Hwang, L.A.; Lane, D.P. Inhibiting p53 acetylation reduces cancer chemotoxicity. *Cancer Res.* **2017**, *77*, 4342–4354. [CrossRef]
49. Ono, H.; Kato, T.; Murase, Y.; Nakamura, Y.; Ishikawa, Y.; Watanabe, S.; Akahoshi, K.; Ogura, T.; Ogawa, K.; Ban, D.; et al. C646 inhibits G2/M cell cycle-related proteins and potentiates anti-tumor effects in pancreatic cancer. *Sci. Rep.* **2021**, *11*, 10078. [CrossRef] [PubMed]
50. Fan, Z.; Wu, J.; Chen, Q.N.; Lyu, A.K.; Chen, J.L.; Sun, Y.; Lyu, Q.; Zhao, Y.X.; Guo, A.; Liao, Z.Y.; et al. Type 2 diabetes-induced overactivation of P300 contributes to skeletal muscle atrophy by inhibiting autophagic flux. *Life Sci.* **2020**, *258*, 118243. [CrossRef] [PubMed]
51. Rai, R.; Sun, T.; Ramirez, V.; Lux, E.; Eren, M.; Vaughan, D.E.; Ghosh, A.K. Acetyltransferase p300 inhibitor reverses hypertension-induced cardiac fibrosis. *J. Cell Mol. Med.* **2019**, *23*, 3026–3031. [CrossRef]
52. Manea, A.; Manea, S.A.; Todirita, A.; Albulescu, I.C.; Raicu, M.; Sasson, S.; Simionescu, M. High-glucose-increased expression and activation of NADPH oxidase in human vascular smooth muscle cells is mediated by 4-hydroxynonenal-activated PPARα and PPARβ/δ. *Cell Tissue Res.* **2015**, *361*, 593–604. [CrossRef]
53. Xie, X.; Chen, Y.; Liu, J.; Zhang, W.; Zhang, X.; Zha, L.; Liu, W.; Ling, Y.; Li, S.; Tang, S. High glucose induced endothelial cell reactive oxygen species via OGG1/PKC/NADPH oxidase pathway. *Life Sci.* **2020**, *256*, 117886. [CrossRef] [PubMed]
54. Xia, L.; Wang, H.; Goldberg, H.J.; Munk, S.; Fantus, I.G.; Whiteside, C.I. Mesangial cell NADPH oxidase upregulation in high glucose is protein kinase C dependent and required for collagen IV expression. *Am. J. Physiol. Ren. Physiol.* **2006**, *290*, F345–F356. [CrossRef]

55. Chen, S.; Meng, X.F.; Zhang, C. Role of NADPH oxidase-mediated reactive oxygen species in podocyte injury. *Biomed. Res. Int.* **2013**, *2013*, 839761. [CrossRef] [PubMed]
56. Sedeek, M.; Nasrallah, R.; Touyz, R.M.; Hébert, R.L. NADPH oxidases, reactive oxygen species, and the kidney: Friend and foe. *J. Am. Soc. Nephrol.* **2013**, *24*, 1512–1518. [CrossRef]
57. Yuan, H.; Reddy, M.A.; Sun, G.; Lanting, L.; Wang, M.; Kato, M.; Natarajan, R. Involvement of p300/CBP and epigenetic histone acetylation in TGF-β1-mediated gene transcription in mesangial cells. *Am. J. Physiol Ren. Physiol.* **2013**, *304*, F601–F613. [CrossRef] [PubMed]
58. Khan, S.; Jena, G.; Tikoo, K. Sodium valproate ameliorates diabetes-induced fibrosis and renal damage by the inhibition of histone deacetylases in diabetic rat. *Exp. Mol. Pathol.* **2015**, *98*, 230–239. [CrossRef]
59. Wang, X.; Liu, J.; Zhen, J.; Zhang, C.; Wan, Q.; Liu, G.; Wei, X.; Zhang, Y.; Wang, Z.; Han, H.; et al. Histone deacetylase 4 selectively contributes to podocyte injury in diabetic nephropathy. *Kidney Int.* **2014**, *86*, 712–725. [CrossRef] [PubMed]
60. Lee, H.B.; Noh, H.; Seo, J.Y.; Yu, M.R.; Ha, H. Histone deacetylase inhibitors: A novel class of therapeutic agents in diabetic nephropathy. *Kidney Int. Suppl.* **2007**, *72*, S61–S66. [CrossRef] [PubMed]
61. Gilbert, R.E.; Huang, Q.; Thai, K.; Advani, S.L.; Lee, K.; Yuen, D.A.; Connelly, K.A.; Advani, A. Histone deacetylase inhibition attenuates diabetes-associated kidney growth: Potential role for epigenetic modification of the epidermal growth factor receptor. *Kidney Int.* **2011**, *79*, 1312–1321. [CrossRef] [PubMed]
62. Khan, S.; Jena, G. Sodium butyrate, a HDAC inhibitor ameliorates eNOS, iNOS and TGF-β1-induced fibrogenesis, apoptosis and DNA damage in the kidney of juvenile diabetic rats. *Food Chem. Toxicol.* **2014**, *73*, 127–139. [CrossRef] [PubMed]
63. Noh, H.; Oh, E.Y.; Seo, J.Y.; Yu, M.R.; Kim, Y.O.; Ha, H.; Lee, H.B. Histone deacetylase-2 is a key regulator of diabetes- and transforming growth factor-beta1-induced renal injury. *Am. J. Physiol. Ren. Physiol.* **2009**, *297*, F729–F739. [CrossRef] [PubMed]
64. Hakami, N.Y.; Dusting, G.J.; Peshavariya, H.M. Trichostatin A, a histone deacetylase inhibitor suppresses NADPH oxidase 4-derived redox signalling and angiogenesis. *J. Cell Mol. Med.* **2016**, *20*, 1932–1944. [CrossRef]

 *antioxidants*

Article

p47phox-Dependent Oxidant Signalling through ASK1, MKK3/6 and MAPKs in Angiotensin II-Induced Cardiac Hypertrophy and Apoptosis

Fangfei Liu [1], Lampson M. Fan [2], Li Geng [1,3] and Jian-Mei Li [1,3,*]

[1] School of Biological Sciences, University of Reading, Reading RG6 6AS, UK; liufangfei97@gmail.com (F.L.); gengli@jiangnan.edu.cn (L.G.)
[2] The Royal Wolverhampton NHS Trust, Wolverhampton WV10 0QP, UK; lampson.fan1@nhs.net
[3] Faculty of Health and Medical Sciences, University of Surrey, Guildford GU2 7XH, UK
* Correspondence: jian-mei.li@reading.ac.uk

Abstract: The p47^{phox} is a key regulatory subunit of Nox2-containing NADPH oxidase (Nox2) that by generating reactive oxygen species (ROS) plays an important role in Angiotensin II (AngII)-induced cardiac hypertrophy and heart failure. However, the signalling pathways of p47^{phox} in the heart remains unclear. In this study, we used wild-type (WT) and p47^{phox} knockout (KO) mice (C57BL/6, male, 7-month-old, $n = 9$) to investigate p47^{phox}-dependent oxidant-signalling in AngII infusion (0.8 mg/kg/day, 14 days)-induced cardiac hypertrophy and cardiomyocyte apoptosis. AngII infusion resulted in remarkable high blood pressure and cardiac hypertrophy in WT mice. However, these AngII-induced pathological changes were significantly reduced in p47^{phox} KO mice. In WT hearts, AngII infusion increased significantly the levels of superoxide production, the expressions of Nox subunits, the expression of PKCα and C-Src and the activation of ASK1 (apoptosis signal-regulating kinase 1), MKK3/6, ERK1/2, p38 MAPK and JNK signalling pathways together with an elevated expression of apoptotic markers, i.e., γH2AX and p53 in the cardiomyocytes. However, in the absence of p47^{phox}, although PKCα expression was increased in the hearts after AngII infusion, there was no significant activation of ASK1, MKK3/6 and MAPKs signalling pathways and no increase in apoptosis biomarker expression in cardiomyocytes. In conclusion, p47^{phox}-dependent redox-signalling through ASK1, MKK3/6 and MAPKs plays a crucial role in AngII-induced cardiac hypertrophy and cardiomyocyte apoptosis.

Keywords: redox-signalling; p47^{phox}; knockout mice; Angiotensin II; cardiac hypertrophy; apoptosis

Citation: Liu, F.; Fan, L.M.; Geng, L.; Li, J.-M. p47^{phox}-Dependent Oxidant Signalling through ASK1, MKK3/6 and MAPKs in Angiotensin II-Induced Cardiac Hypertrophy and Apoptosis. *Antioxidants* **2021**, *10*, 1363. https://doi.org/10.3390/antiox10091363

Academic Editors: Chiara Nediani and Monica Dinu

Received: 14 July 2021
Accepted: 23 August 2021
Published: 26 August 2021

1. Introduction

Nicotinamide adenine dinucleotide phosphate oxidase (NADPH oxidase, or Nox) is a membrane-bound enzyme that by generating reactive oxygen species (ROS) plays important role in the regulation of cellular function. So far, seven isoforms of the catalytic component of Nox have been identified namely Nox1–5, and durox 1–2 [1]. Angiotensin II is a vasoconstricting peptide (Asp-Arg-Val-Tyr-Ile-His-Pro-Phe) of the renin-angiotensin-aldosterone system involved in the regulation of blood pressure and other aspects of organ functions [2]. Oxidative stress and inflammation due to the activation of a Nox2-contaninig NADPH oxidase (Nox2) has been found to play an essential role in mediating AngII-induced cardiac hypertrophy and failure [1–5]. Nox2 is a multi-subunit enzyme consisting of two membrane-bound subunits, p22^{phox} and Nox2 (also named gp91^{phox}), and four cytosolic regulatory subunits, i.e., p40^{phox}, p47^{phox}, p67^{phox} and rac1. The p47^{phox} is a key regulatory subunit of Nox2 enzyme [2,6,7]. The phosphorylation of p47^{phox} initiates the process of coordination and association of regulatory subunits with membrane-bound p22^{phox}/Nox2 complex, and the subsequent $O_2^{\bullet -}$ production [2,6].

In the mammalian heart, the p47^{phox} is expressed in the myocardium, epicardium and coronary vessels [1]. In cardiomyocytes p47^{phox} had been reported to co-localise with

F-actin and cortactin in order to facilitate the translocation of the cytosolic regulatory subunits to the p22^{phox}/Nox2 complex [8,9]. Under pathological conditions, p47^{phox} was suggested to link oxidative stress with the hypertrophic growth of cardiomyocytes through the activation of mitogen-activated protein kinases (MAPKs), i.e., extracellular signal-regulated kinase (ERK), c-Jun N-terminal kinase (JNK) and p38 mitogen-activated protein kinase (p38MAPK) [10,11]. In response to oxidative stress, the redox-sensitive MAPK kinase (MKK) to MAPKs signalling pathways are activated, which in turn promote the activities of pro-apoptotic signalling molecules such as p53, γH2AX and apoptosis signal-regulating kinase 1 (ASK1) leading to cardiac damage [12–14]. Genetic ablation of p47^{phox} attenuated AngII-induced abdominal aortic aneurysm formation in apolipoprotein E-deficient mice [15], and reduced the level of cardiac hypertrophy after experimental myocardial infarction [16].

Despite the importance of p47^{phox} as a key regulator of Nox2-derived ROS production in the heart, its signalling pathways and functional complexity in AngII-induced cardiac hypertrophy and cardiomyocyte damage remained unclear. There was insufficient information about the upstream and downstream signalling pathways of p47^{phox} in the hearts. In the current study, we investigated the role of p47^{phox} and its oxidant-signalling pathways in the hearts using a murine model of AngII-infusion-induced cardiac hypertrophy and cardiomyocyte apoptosis in WT and p47^{phox} KO mice. The complex role of p47^{phox} in the myocardium was investigated by examining AngII-induced cardiac oxidative stress, the expressions of Nox subunits, the expression of PKCα and C-Src (both were involved in p47^{phox} phosphorylation). We also examined the levels of AngII-induced p47^{phox} phosphorylation, the activation of redox-sensitive ASK1, MKK3/6 and MAPKs and the expression of pro-apoptotic markers, i.e., γH2AX and p53 in the cardiomyocytes. Our results suggested that p47^{phox} oxidant-signalling through ASK1, MKK3/6 and MAPKs played a vital role in mediating cardiac hypertrophic response and the expression of apoptotic markers in cardiomyocytes in response to AngII challenge. Knockout of p47^{phox} inhibited the activation of these stress signalling pathways and protected hearts from AngII-induced oxidative damages.

2. Materials and Methods

2.1. Chemicals and Reagents

AngII was purchased from Sigma-Aldrich (Amersham, UK); NADPH was purchased from Fisher Scientific (Loughborough, UK); dihydroethidium (DHE) was purchased from Invitrogen (Loughborough, UK); FITC-labelled wheat germ agglutinin (WGA, Catalogue No. L-4895) was from Sigma-Aldrich. Primary antibodies to p47^{phox}, p22^{phox}, Nox1, Nox2, Nox4, p38-MAPK, ERK1/2, phos-JNK (Thr183/Tyr185) and total JNK, phos-Akt (Ser473) and total Akt were purchased from Santa Cruz Biotechnology (Dallas, TX, USA); antibodies to β-actin, phos-MKK3(Ser189)/6(Ser207) and phos-ASK1 (Thr845), total MKK3/6, γH2AX (Ser139/Tyr142) were purchased from Cell Signalling Technology (London, UK); Antibodies to phos-p47^{phox} (Ser359), phos-p38-MAPK (Thr180/Tyr182) and phos-ERK1/2 (Thr202/Tyr204) were purchased from Sigma-Aldrich. Nox2-ds-tat (Nox2tat, [H]-RKKRRQRRRCSTRVRRQL-[NH2]) were provided by PeptideSynthetics (PPR Ltd., Fareham, UK). Other reagents, chemicals and antibodies, unless specified, were purchased from Sigma-Aldrich.

2.2. Animals

All studies were performed following protocols approved by the Ethics Committees of the Surrey and Reading Universities and the Home Office under the Animals (Scientific Procedures) Act 1986 UK. The p47^{phox} KO mice on a 129sv background were initially obtained from the European mouse mutant archive, and backcrossed to C57BL/6 for ten generations at the animal units in the University of Surrey [17]. Littermates of wild-type and p47^{phox} KO mice at the age of 7-months were randomly grouped into control and AngII groups (n = 9 per group). The dose of AngII (0.8 mg/kg/day) was chosen based on the literature and our own pilot experiments to produce significant cardiac hypertrophy effectively. AngII was delivered to mice through osmotic minipumps (ALZET osmotic pumps, DURECT Corporation, Cupertino, CA, USA) for 14 days. The control group was infused with saline. Systolic and diastolic blood pressure (BP) were measured using a computerised tail-cuff system (CODA, Kent Scientific, Torrington, CT, USA) on conscious mice following one week of training with the instrument [18]. Mice were used at the end of two weeks of AngII infusion. Bodyweight and heart weight were measured, and the tissues were harvested and stored in −80 °C freezer for experimental use.

2.3. Measurement of Cross-Sectional Cardiomyocyte Sizes

Left ventricular cryosections (8 μm) were prepared and fixed in freshly prepared 1% formaldehyde phosphate-buffered solution. Cardiomyocytes in the cardiac sections were outlined by FITC-conjugated wheat germ agglutinin (WGA) that binds to glycoproteins of the cell membrane, and is routinely used for the staining of cardiac sarcolemma to determine cross sectional area or myocyte density [19]. Staining was visualised under the A1R confocal microscope (Nikon, Chiyoda, Japan) (20–40× magnification, 1024 × 1024 pixels)). Cross-sectional cardiomyocyte sizes were measured according to the method published previously [19,20] using software of ImageJ 1.50i (NIH, Bethesda, MD, USA). For statistical analyses, cardiomyocyte sizes were obtained from at least three microscopic areas per section, three cross sections per heart and nine mice per group.

2.4. Measurement of ROS Production

ROS production was measured using the homogenates of left ventricular tissues. The homogenates were used immediately for the ROS measurement as described previously using three independent methods [4]. Lucigenin (5 μM)-chemiluminescence was used for measuring real-time NADPH-dependent $O_2^{\bullet-}$ production in heart homogenates detected using a 96-well microplate luminometer (Molecular Devices, Wokingham, UK). Catalase (300 U/mL)-inhibitable amplex red (6.25 μM) assay was used for measuring the H_2O_2 production in heart homogenates detected using FluoStar OPTIMA (BMG LabTech, Aylesbury, UK). DHE (2 μM)-fluorescence was used to measure in situ ROS production by cardiac sections, and images were captured using Nikon Eclipse Ti2-E inverted microscope and the DHE fluorescence intensities were quantified. The specificity of the lucigenin and DHE assays for the detection of $O_2^{\bullet-}$ was confirmed by using tiron (10 mM), a non-enzymatic $O_2^{\bullet-}$ scavenger, and superoxide dismutase (SOD) (200 U/mL). The enzymatic sources of $O_2^{\bullet-}$ production were investigated using different enzyme inhibitors, i.e., L-NAME (N-nitroarginine methyl ester, 100 μM, nitric oxide synthase inhibitor), rotenone (100 μM, mitochondrial complex-1 enzyme inhibitor), diphenyleneiodonium (DPI) (20 μM, flavoprotein inhibitor), oxypurinol (100 μM, xanthine oxidase inhibitor) and Nox2tat (a specific peptide Nox2 inhibitor, 10 μM) [21]. Individual inhibitor was added into the wells loaded with homogenates and incubated for 10 min at room temperature before the measurement of ROS production.

2.5. Immunoblotting

Immunoblotting was performed exactly as described previously [4,18] using the left ventricular tissue homogenates. β-actin detected in the same sample was used as a loading control. For the quantification of phosphorylation of MAPKs, the total levels of the same protein in the same sample were pre-tested and justified for equal loading and used as loading controls for the quantification of phosphorylated proteins. The results were captured by BioSpectrum AC imaging system (UVP, Upland, CA, USA). The optical density of the bands was quantified and normalised to the relevant loading controls.

2.6. Immunofluorescence Microscopy

These experiments were performed as described previously [18]. The left ventricular tissue cryosections (8 μm) were fixed with 1:1 methanol: acetone solution for 10 min at

−20 degree. All buffers and reagents were freshly prepared and kept on ice before use. Sections were then blocked using 2% bovine albumin serum (BSA) in PBS with 0.1% Triton X-100. BSA (2%) was used in the place of primary antibodies as a negative control. Primary antibodies were used at 1:100 dilutions in 0.2% BSA/PBS. Biotin-conjugated secondary antibodies were used at 1:1000 dilution in 0.2% BSA/PBS and detected using streptavidin-FITC or streptavidin-Cy3. Images were captured by Nikon A1R confocal microscope, and the fluorescent intensities (Fluo-intensity) were quantified. For statistical analysis, at least five random fields per section with three sections per heart were used per animal and nine animals were used per group. The control background fluorescence captured from sections without primary antibody was deducted and the results were expressed as Fluo-intensity.

2.7. Statistics

The Statistical analysis was performed using GraphPad Prism 7.0. Two-way ANOVA plus Tukey's multiple comparison test were used for multiple-group significance testing and for testing repeated measures of blood pressure. One-way ANOVA followed by a Bonferroni post-hoc test was employed for other data analyses where it was appropriate. $p \leq 0.05$ was denoted as statistically significant. Nine mice per group were used for statistical analysis. Results were presented as mean ± SD unless specified in the figure legends.

3. Results

3.1. Knockout $p47^{phox}$ Attenuated AngII Infusion-Induced High Blood Pressure and Cardiac Hypertrophy

The mice used in this study were middle-aged (7-months) which were more susceptible to AngII-induced cardiovascular damages than mice at younger ages. At day 0 (before AngII challenge), there was no significant difference in BP between WT and $p47^{phox}$ KO mice. AngII infusion (14 days) of WT mice markedly increased the systolic BP to an average of 180.3 ± 7.5 mmHg and the diastolic BP to an average of 142.6 ± 10.8 mmHg as compared to saline-infused controls (Figure 1A,B). However, in the absence of $p47^{phox}$, AngII infusion only caused mild but significant increases in the systolic BP to an average of 150 ± 6 mmHg and the diastolic blood pressure to an average of 118.6 ± 9.7 mmHg (Figure 1A,B). The levels of AngII-induced cardiac hypertrophy were expressed as the increases in heart weight (HW) and the HW/body weight (BW) ratios. In WT mice, AngII infusion significantly increased the heart weights (Figure 1C) and the HW/BW ratios (Figure 1D). However, in the $p47^{phox}$ KO mice, AngII induced cardiac hypertrophy was significantly reduced in comparison to WT mice. Although there were increases in HW/BW ratio in AngII-infused $p47^{phox}$ KO mice, these were not statistically significant (Figure 1D). AngII-induced cardiac hypertrophy was further examined by measuring cardiomyocyte cross sectional area in the left ventricular tissue sections. The cardiomyocytes were labelled with FITC-WGA, which binds to glycoproteins of the cardiomyocyte membrane and outlines the cardiomyocytes on cross sections [19]. In comparison to saline-infused controls, there were significant increases in the cross-sectional areas of cardiomyocytes in AngII-infused WT hearts, which were significantly reduced in $p47^{phox}$ KO hearts (Figure 1E).

3.2. Knockout $p47^{phox}$ Inhibited AngII-Induced Cardiac Oxidative Stress

The effect of genetic ablation of $p47^{phox}$ on AngII-induced cardiac oxidative stress were first examined by measuring NADPH-dependent $O_2^{\bullet-}$ production in heart homogenates using lucigenin chemiluminescence. A representative example of real-time measurements of $O_2^{\bullet-}$ production by heart homogenates is shown in the left panel of Figure 2A. Tiron (an $O_2^{\bullet-}$ scavenger) was used to confirm the assay specificity. The statistical analyses were shown in the right panel of Figure 2A. Compared to saline-infused WT controls, AngII infusion resulted in 2.6-folds increases in the levels of $O_2^{\bullet-}$ production in the WT hearts. However, this was significantly inhibited by knockout of $p47^{phox}$. Although there were some increases in the levels of $O_2^{\bullet-}$ production in AngII infused $p47^{phox}$ KO hearts, they were not statistically significant. The enzymatic sources of AngII-induced

$O_2^{\bullet-}$ production found in WT hearts were examined using different enzyme inhibitors including L-NAME (nitric oxide synthase inhibitor), rotenone (mitochondrial respiratory chain inhibitor), oxypurinol (xanthine oxidase inhibitor), apocynin (NADPH oxidase inhibitor), DPI (flavoprotein inhibitor) and Nox2tat (a specific peptide inhibitor of Nox2) (Figure 2B). The $O_2^{\bullet-}$ production detected in AngII-infused WT hearts was not affected by rotenone and oxypuronol, but was significantly inhibited by apocynin, Nox2tat or DPI suggesting Nox2 as a major enzymatic source of AngII-induced $O_2^{\bullet-}$ production. There was some inhibition of AngII-induced $O_2^{\bullet-}$ production by L-NAME, indicating nitric oxide synthase disfunction. SOD (superoxide dismutase) was used to double confirm the detection of $O_2^{\bullet-}$.

$O_2^{\bullet-}$ is not stable and can be quickly converted to H_2O_2 in cells. Therefore, we examined cardiac H_2O_2 production using catalase-inhibitable amplex red assay (Figure 2C). There was no significant difference in the basal (without AngII) levels of H_2O_2 production between WT and p47^{phox} KO hearts. Compared to saline infused controls, the level of H_2O_2 production was significantly increased in AngII-infused WT hearts, which might link to the high level of $O_2^{\bullet-}$ production. However, in the p47^{phox} KO hearts, although the level of $O_2^{\bullet-}$ production showed no change in response to AngII challenge, there was a significant increase in the levels of H_2O_2 production in AngII-infused p47^{phox} KO hearts as compared to saline controls.

The levels of AngII-induced $O_2^{\bullet-}$ production in the hearts were further examined by in situ DHE fluorescence on cardiac cryosections (Figure 2D). There were significant high levels of $O_2^{\bullet-}$ production in AngII-infused WT hearts in comparison to saline-infused WT controls. However, there was no significant difference in DHE fluorescence intensities between AngII-infused and saline-infused p47^{phox} KO hearts.

3.3. *AngII-Induced Upregulation of Nox Subunits, PKCα and C-Src Protein Kinases and p47^{phox} Phosphorylation in Murine Hearts*

The levels of expression of p47^{phox}, p22^{phox}, p67^{phox}, rac1, Nox1, Nox2 and Nox4 in response to AngII infusion were examined in WT and p47^{phox} KO hearts by immunoblotting (Figure 3). The levels of β-actin detected in the same sample were used as loading controls. The p47^{phox} was highly expressed in the WT hearts, but was barely detectable in the p47^{phox} KO hearts. AngII infusion resulted in a great upregulation of the levels of p22^{phox} expression in both WT and p47^{phox} KO hearts without significant difference between the two groups. In comparison to saline infused WT controls, AngII-infusion increased significantly the levels of expression of p47^{phox}, p67^{phox}, rac1 and Nox2 in the WT hearts. However, in the absence of p47^{phox} KO, AngII infusion had no significant effect on the levels of expression of p67^{phox} and Nox2, but increased significantly the levels of p22^{phox}, Nox4 and rac1 expression. Nox1 expression remained the same without significant difference between WT and p47^{phox}KO hearts.

Protein kinase C alpha (PKCα) is highly expressed in the myocardium [22], and phosphorylates p47^{phox} at multiple serine sites in response to AngII stimulation. [23]. C-Src had been proposed to be an upstream tyrosine kinase that phosphorylates p47^{phox} in response to AngII stimulation [24]. Therefore, we examined the levels of expressions of (PKCα) and C-Src together with the levels of p47^{phox} phosphorylation in WT and p47^{phox} KO hearts by immunoblotting (Figure 4A). Compared to saline-infused controls, AngII infusion increased the levels of PKCα expression in both WT and p47^{phox} KO hearts without significant difference between these two groups. However, AngII-induced C-Src expression was only found in WT hearts, but not in p47^{phox} KO hearts suggesting a key role of oxidative stress in cardiac C-Src activation (Figure 4A). Accompanied with increased PKCα expression, there were significant increases in p47^{phox} serine phosphorylation detected using specific antibodies to phos-p47^{phox}.

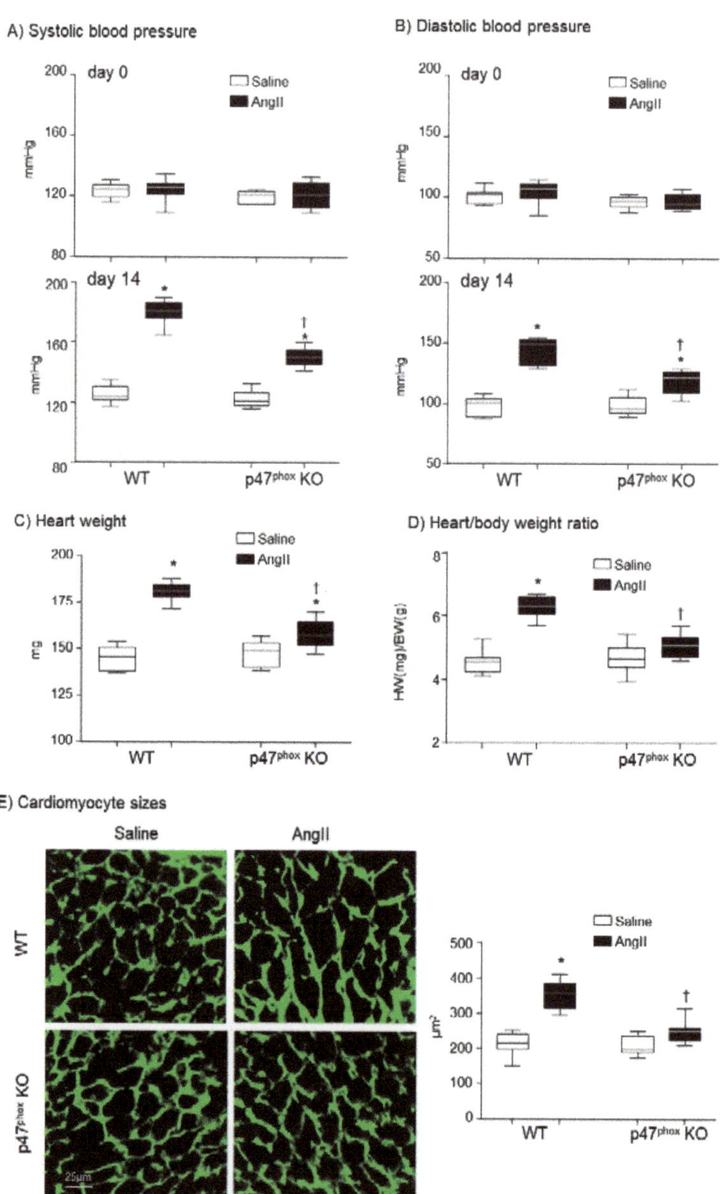

Figure 1. Development of hypertension and cardiac hypertrophy in AngII-infused WT and p47^{phox} KO mice. (**A**) Systolic blood pressure. (**B**) Diastolic blood pressure. Day 0: day of minipump implantation. Day 14: day of minipump removal. (**C**) Heart weights. (**D**) Heart weight (HW, mg)/body weight (BW, g) ratio. (**E**) Left panels: Representative images of cardiomyocyte sizes on the cross-sections of left ventricular tissues. The cardiomyocytes membranes were labelled with WGA-FITC (green). Right panel: Statistical analysis of cardiomyocyte cross-sectional areas (µm^2). $n = 9$ mice per group. Statistical analyses were performed using two-way ANOVA. * $p < 0.05$ for AngII values versus saline values in the same genetic group; † $p < 0.05$, for p47^{phox} KO AngII values versus WT AngII values.

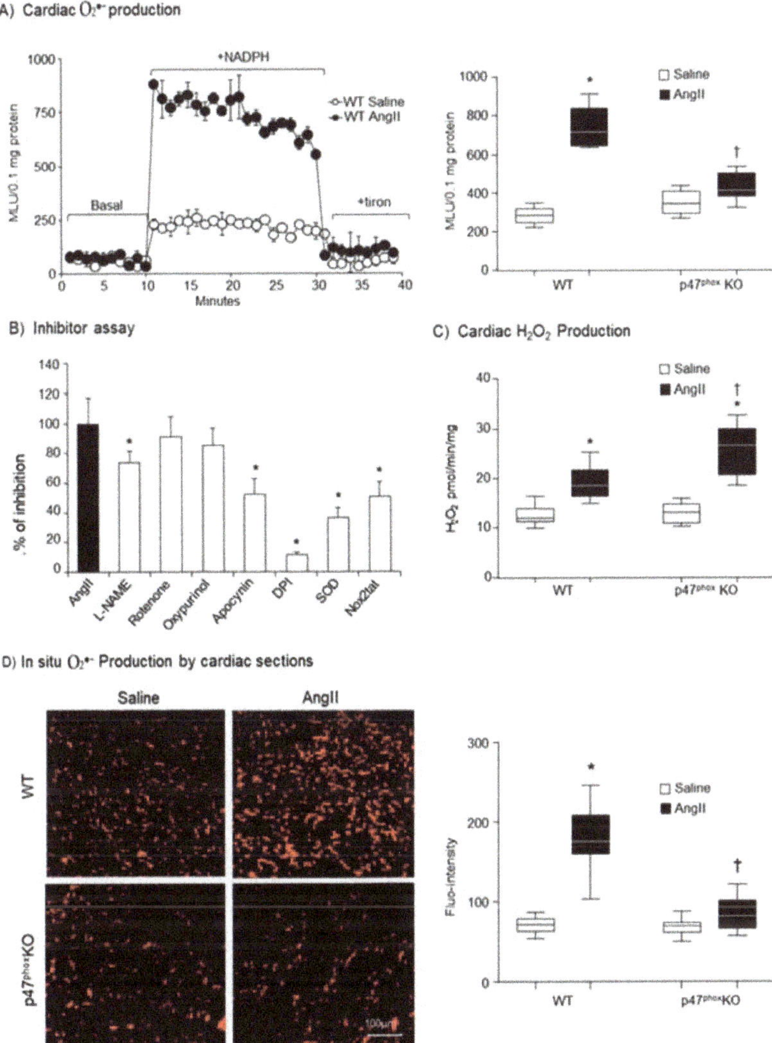

Figure 2. Cardiac ROS production. (**A**) Levels of $O_2^{\bullet-}$ production measured by lucigenin-chemiluminescence. Left panel: Representative examples of kinetic measurements of $O_2^{\bullet-}$ production by WT heart homogenates. NADPH (0.1 mM) was added at 10 min. Tiron (10 mM) was added at 30 min to scavenge $O_2^{\bullet-}$. Right panel: Differences in NADPH-dependent $O_2^{\bullet-}$ production measured between 10–30 min shown in the left panel. (**B**) The effects of different enzyme inhibitors on the levels of $O_2^{\bullet-}$ by AngII-infused WT heart homogenates. L-NAME: (NG-nitroarginine methyl ester), NOS inhibitor; Rotenone: mitochondrial respiratory chain inhibitor; Oxypurinol, xanthine oxidase inhibitor; DPI: (diphenyleneiodonium), flavoprotein inhibitor; SOD: (superoxide dismutase). (**C**) Cardiac H_2O_2 production detected by amplex red assay. (**D**) In situ detection of reactive oxygen species production by DHE fluorescence. Left panel: Representative images of DHE staining on cardiac sections; right panel: Quantification of DHE fluorescence intensity. $n = 9$ per group. Statistical comparisons were done using one-way ANOVA for inhibitor assay, and two-way ANOVA for the rests (panels A, C, D). * $p < 0.05$ for indicated AngII values versus saline values in the same genetic group; † $p < 0.05$ for p47^{phox} KO AngII values versus WT AngII values.

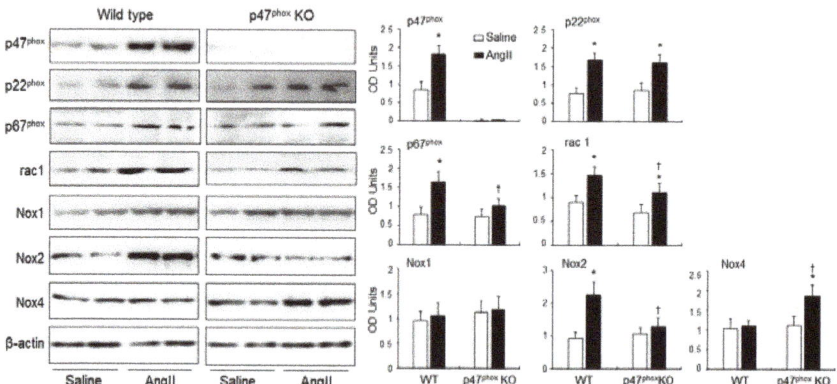

Figure 3. Expressions of isoforms of the catalytic subunit of Nox (i.e., Nox1. Nox2, Nox4) and other subunits of Nox ($p47^{phox}$, $p22^{phox}$, $p67^{phox}$ and rac1) in murine hearts. Left panels: Representative Western blot images. β-actin detected in the same samples were used as loading controls. Right panels: Optical densities (OD) of Western blot bands were quantified and normalised to the levels of β-actin detected in the same samples. $n = 9$ per group. Statistical comparisons were made using two-way ANOVA. * $p < 0.05$ for AngII values versus saline values in the same genetic group; † $p < 0.05$ for $p47^{phox}$ KO AngII values versus WT AngII values.

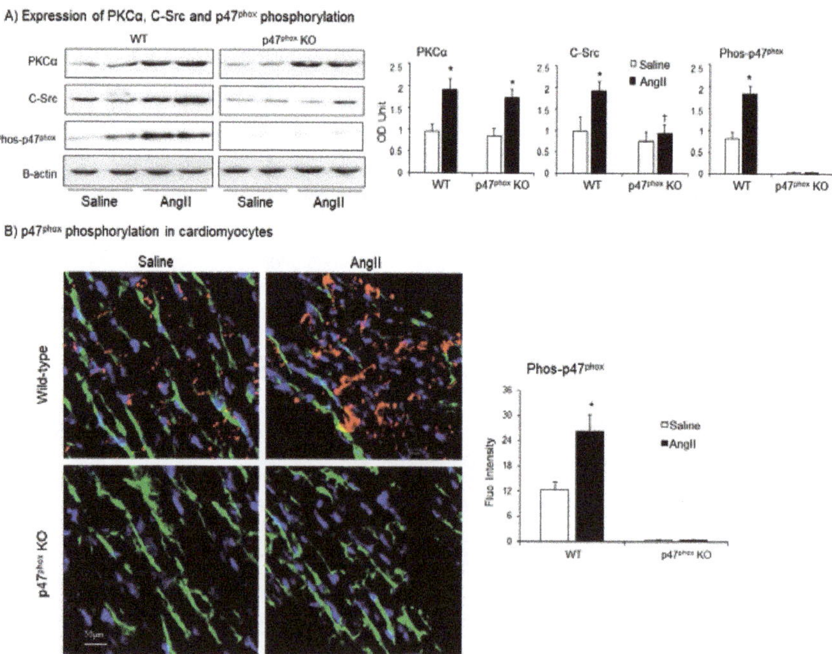

Figure 4. Expressions of protein kinase C alpha (PKCα), Proto-oncogene tyrosine-protein kinase Src (C-Src) and $p47^{phox}$ phosphorylation in AngII-infused murine hearts. (**A**) Western blots. Left panels: Representative images. β-actin detected in the same samples were used as loading controls. Right panels: Optical densities (OD) of Western blot bands were quantified and normalised to the levels of β-actin detected in the same samples. (**B**) Confocal immunofluorescence of cardiac sections. Left panel: Representative immunofluorescence images. Cardiomyocyte cell membrane was labelled by WGA-FITC (green) and $p47^{phox}$ phosphorylation was identified using phos-$p47^{phox}$ specific antibody (Cy3, red). Nuclei were labelled by DAPI (blue). Right panel: Quantification of phos-$p47^{phox}$ fluorescence intensity. $n = 9$ hearts per group. Data were presented as Mean ± SD. Statistical comparisons were made using two-way ANOVA. * $p < 0.05$ for AngII values versus saline values in the same genetic group.

AngII-induced p47^{phox} phosphorylation in the myocardium was further examined by confocal immunofluorescence (Figure 4B). The sarcolemma membranes of cardiomyocytes were labelled with FITC-WGA (green), and the nuclei were labelled with 4′,6-diamidino-2-phenylindole (DAPI, blue) to visualise cardiomyocytes. The phospho-p47^{phox} was labelled by Cy3 (red), and was only detected in WT hearts. AngII infusion significantly increased the levels of p47^{phox} phosphorylation (red) mainly located at the cardiomyocyte gap junctions or at the cell membranes overlapped with FITC-WGA as indicated by the yellow colour (Figure 4B).

3.4. p47^{phox}-Dependent Redox-Signalling through MKK3/6, MAPKs and AKT in AngII-Induced and Cardiac Hypertrophy and Apoptosis

The role of p47^{phox} in modulating AngII signalling in the hearts was examined for the activations of stress-signalling pathways, i.e., mitogen-activated protein kinase kinase (MKK3/6) and down-stream ERK1/2; p38MAPK, JNK and Akt (Figure 5). The total levels of the same protein in the same samples were pre-tested and used as loading controls. In saline-infused control hearts, there was no significant difference in the levels of phosphorylation of these signalling molecules between WT and p47^{phox}KO hearts. Compared to saline-infused controls, AngII-infusion resulted in significant increases in the levels of phosphorylation of MKK3/6, ERK1/2, p38 MAPK, JNK and Akt in WT hearts. However, in the absence of p47^{phox}, AngII failed to induce the phosphorylation of these signalling molecules in the hearts (Figure 5).

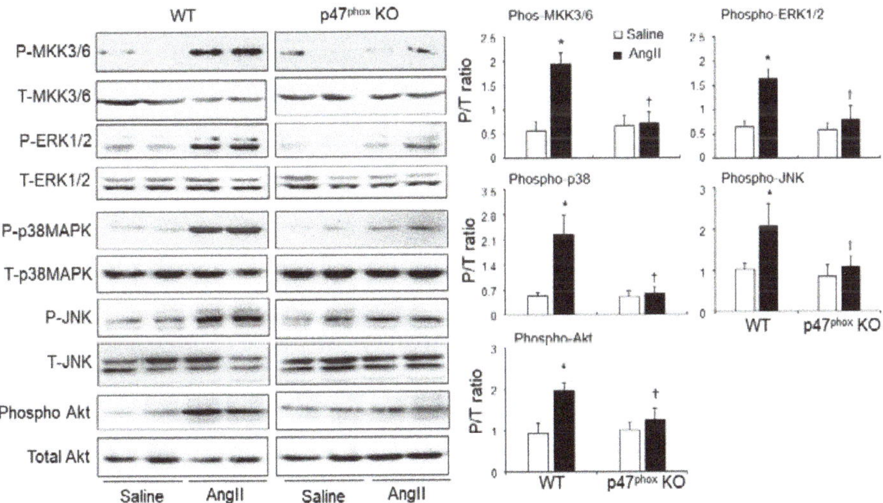

Figure 5. AngII-induced activation of mitogen-activated protein kinase kinase 3/6 (MKK3/6), mitogen-activated protein kinases (i.e., ERK1/2, p38MAPK and JNK) and Akt (also called protein kinase B) in murine hearts. Left panels: Representative immunoblotting images. The total protein bands of each molecule in heart homogenates were pre-tested for equal loading. Right panels: Quantification of the optical densities (OD) of phos-protein bands expressed as phosphorylated/total (P/T) protein ratio. n = 9 mice per group. Data were presented as Mean ± SD. Statistical comparisons were made using two-way ANOVA. * $p < 0.05$ for AngII values versus saline values in the same genetic group. † $p < 0.05$ for p47^{phox} KO AngII values versus WT AngII values.

The effects of genetic knockout of p47^{phox} on AngII-induced oxidative damage of cardiomyocytes and apoptotic death was examine by immunoblotting of apoptosis signal-regulating kinase 1 (ASK1) and biomarkers for DNA double-strand breaks (γH2AX), and apoptosis (p53) (Figure 6A). The levels of β-actin detected in the same sample were used as loading controls. Compared to saline-infused controls, there were remarkable significant increases in the level of expression of phos-ASK1, γH2AX and p53 in AngII-infused WT mice. However, in the absence of p47^{phox}, there was no significant increase in the expression of these markers of cell DNA damage and apoptosis in after two weeks of AngII-infusion.

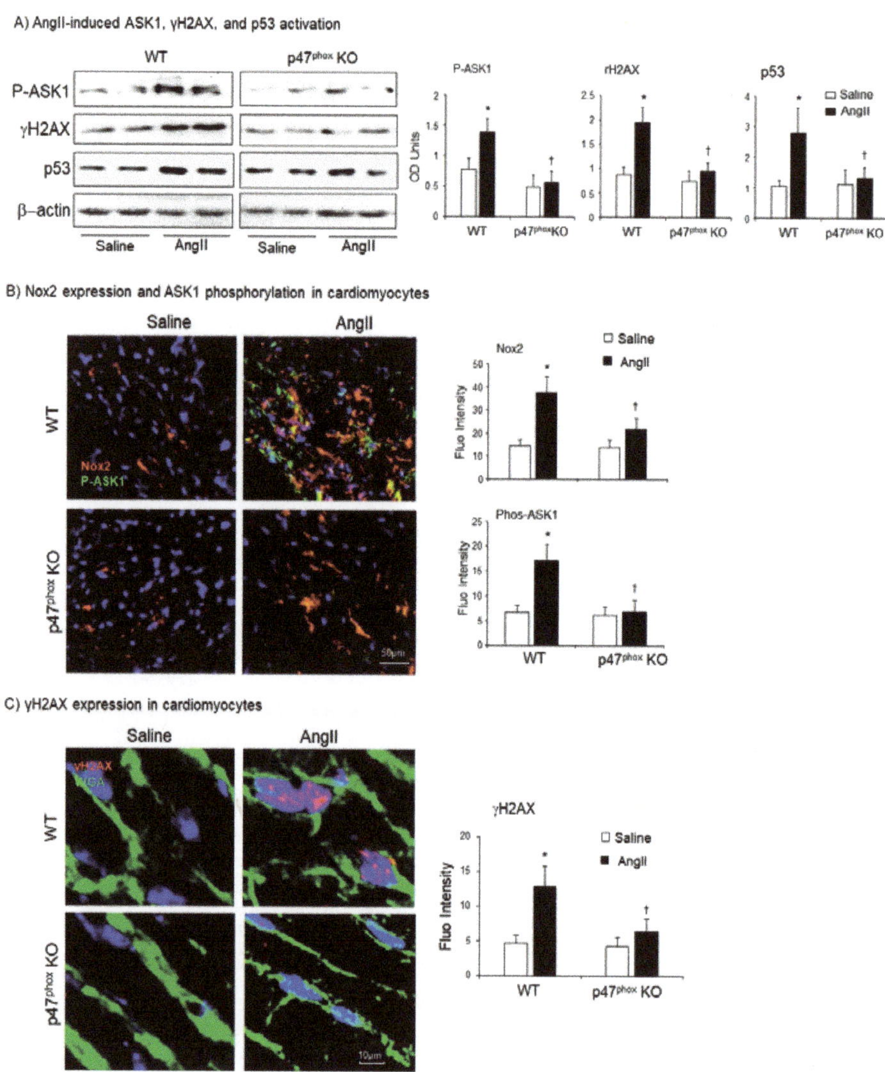

Figure 6. Activation of apoptosis signal-regulating kinase 1 (ASK1), p53 and phosphorylation of H2A histone family member X (γH2AX) in murine hearts. (**A**) Western blots for the expressions of phos-ASK1, p53 and γH2AX. Optical densities of protein bands were quantified and normalised to the levels of β-actin detected in the same samples. (**B**) Confocal immunofluorescence detection of Nox2 (Cy3 labelled, red) and phos-ASK1 expressions (FITC labelled, green) in the cardiac sections. (**C**) Confocal immunofluorescence detection of γH2AX expression (Cy3-labelled, red) in the cardiac sections. The cardiomyocyte membrane was labelled by WGA-FITC (green), and the nuclei were labelled by DAPI (blue). The specific fluorescent densities were quantified. $n = 9$ mice per group. Data were presented as Mean ± SD. Statistical comparisons were made using two-way ANOVA. * $p < 0.05$, for AngII values versus saline values in the same genetic group. † $p < 0.05$ for p47^{phox} KO AngII values versus WT AngII values.

The crucial role of p47^{phox} in regulating AngII-induced Nox2 activation and ASK1 activation in the cardiomyocytes were examined using immunofluorescence confocal microscopy (Figure 6B). Low levels of Nox2 expression (red) could be detected in the control hearts (infused with saline) without significant difference between WT and p47^{phox}KO groups. AngII infusion significantly increased Nox2 expression together with great increases in ASK1 phosphorylation in the WT hearts. AngII-induced Nox2 expression was inhibited in p47^{phox}KO hearts and there was no change in the levels of ASK1 phosphorylation in response to AngII infusion in p47^{phox}KO hearts.

The role of p47^{phox} in modulating AngII-induced DNA damage in cardiomyocytes was further demonstrated using immunofluorescence confocal microscopy (Figure 6C). The cardiomyocyte membranes were labelled with FITC-WAG (green), the nuclei were labelled by DAPI. In saline-infused hearts, there was very low level of γH2AX positive staining. However, in AngII-infused WT hearts, there were clear visible γH2AX foci (red) formation detected in the nuclei (blue) of cardiomyocytes as indicated by the pink colour. AngII-induced nuclear expression of γH2A, seen in WT hearts, was significantly inhibited in p47^{phox}KO hearts. Putting together, our results indicated clearly a key role of p47^{phox} in mediating AngII-induced oxidative stress, activation of stress signalling pathways and oxidative damage of cardiomyocyte DNAs and cell apoptosis.

4. Discussion

AngII is a potent activator of Nox2 enzyme, which by generating ROS is involved in AngII-induced cardiovascular oxidative stress, hypertension, remodelling and organ damage [2,3]. The p47^{phox} is a primary regulatory subunit of Nox2 enzyme, and the phosphorylation of p47^{phox} at multiple serines in the C-terminus is a key step for Nox2 $O_2^{\bullet-}$ production [6,17]. However, the signalling pathways of p47^{phox} in the heart remains unclear. The current study by using a disease model of AngII infusion-induced hypertension and cardiac hypertrophy in WT versus p47^{phox}KO mice, provided novel insights of p47^{phox}-dependent signalling pathways in modulating AngII-induced cardiac hypertrophy and cardiomyocyte apoptosis. We discovered that p47^{phox}-dependent regulation of redox-sensitive signalling cascade through ASK1, MKK3/6 and MAPKs is essential in mediating AngII-induced cardiac hypertrophy and DNA damage in cardiomyocytes. Genetic knockout of p47^{phox} inhibited AngII-induced cardiac ROS production, attenuated ASK1, MKK3/6 and MAPK activation and protected cardiomyocyte from AngII-induced hypertrophic growth, DNA damage and apoptosis.

The mice used in this study were 7-month-old, equivalent to humans at the middle-age, and were more susceptible to AngII-induced cardiovascular damages than mice at younger ages. The crucial role of p47^{phox} in mediating AngII-induced cardiac hypertrophy was properly controlled using age-matched littermates of p47^{phox}KO mice subjected to the same experimental procedures. Despite a mild increase in BP found in p47^{phox}KO mice after two weeks of AngII-infusion, there was no significant cardiac hypertrophy as evaluated using two separate methods, i.e., the changes in HW/BW ratio and cardiomyocyte cross-sectional areas.

NADPH oxidase family contains at least 7 members (Nox1–5 and duox 1–2) [1]. Individual Nox enzyme has distinctive mechanism of activation and functions differently [25,26]. So far, Nox1–2 and Nox4–5 have been found in the hearts [27]. Nox2 relies on p47^{phox} to be active and generates $O_2^{\bullet-}$ involved in many diseased conditions [6]. Whereas, Nox4 is autoactivated and plays a protective role in cardiovascular function [27]. In the current study, we found that AngII-infusion induced a great increase in cardiac Nox2 expression together with increased level of $O_2^{\bullet-}$ production in the WT hearts. $O_2^{\bullet-}$ is short lived and can be quickly converted to H_2O_2 by SOD as a cellular self-protective mechanism [28]. This explained the mild elevation of H_2O_2 production observed in AngII-infused WT hearts. However, AngII-infusion of p47^{phox}KO mice induced a great increase in cardiac Nox4 expression together with a high level of H_2O_2 production indicating Nox4 was the enzymatic source of AngII-induced H_2O_2 production in p47^{phox}KO hearts.

PKCα has been reported to be highly expressed in the hearts [20] and phosphorylates p47^{phox} at multiple serine sites in response to AngII stimulation [21]. C-Src had also been proposed to be an upstream tyrosine kinase of p47^{phox} phosphorylation in response to AngII stimulation [22]. However, a recent study found that C-Src, rather than an upstream kinase of p47^{phox} phosphorylation, was a downstream molecule of p47^{phox}-dependent ROS production in lung inflammation [29]. In the current study, we found that AngII-induced cardiac C-Src activation was oxidant-dependent and was abolished by the knockout of p47^{phox}.

MAPKs belong to a highly conserved family of Ser-Thr protein kinases and have diverse regulatory roles in normal heart development as well as in pathological cardiac hypertrophic growth and remodelling [30]. MAPK activation in response to Nox2-derived oxidative stress is a crucial signalling pathway involved in the development of cardiovascular abnormalities. Akt is also a redox-sensitive Ser-Thr kinase involved in cardiomyocyte hypertrophic growth and survival. In the current study, we showed that knockout of p47^{phox} attenuated Nox2-derived ROS production, inhibited MKK3/6, MAPK and Akt activation in response to AngII challenge and protected murine hearts from AngII-induced cardiac hypertrophy. However, p47^{phox} redox-signalling is a complicated mechanism and we do not know if p47^{phox} is physically a component of these signalling pathways, and how it promotes both the pro- and anti-apoptotic signalling pathways in response to AngII infusion. Further detailed investigation is needed.

ASK1 is a member of the mitogen-activated protein kinase kinase kinase (MAPKKK) family that activates downstream MAPKs, JNKs and p38 MAPKs in response to various stresses, such as ROS [13,31]. H2AX is a variant of the H2A protein family and is a component of the histone octamer in nucleosomes [32]. When DNA is damaged and double stranded DNA breaks, H2AX is phosphorylated to form γH2AX. Therefore, γH2AX has been used as a biomarker of DNA damage [32]. The p53 plays an important role in the regulation of cardiomyocyte hypertrophic growth and apoptosis [33]. An important discovery from this study is that ASK1 links p47^{phox} with the activation of MAPKs and the expression of apoptotic markers, i.e., γH2AX and p53, in AngII-induced cardiac hypertrophy and apoptosis. We showed that ASK1 was phosphorylated in response to AngII-induced oxidative stress in the WT hearts. Knockout of p47^{phox}, inhibited AngII-induced ASK1 phosphorylation and its down-stream signalling pathways. There was no obvious cardiomyocyte hypertrophic growth and no increase in the expression of apoptosis markers in AngII-infused p47^{phox}KO hearts. A schematic illustration of AngII-induced p47^{phox} redox signalling pathways examined in this study is shown in Figure 7.

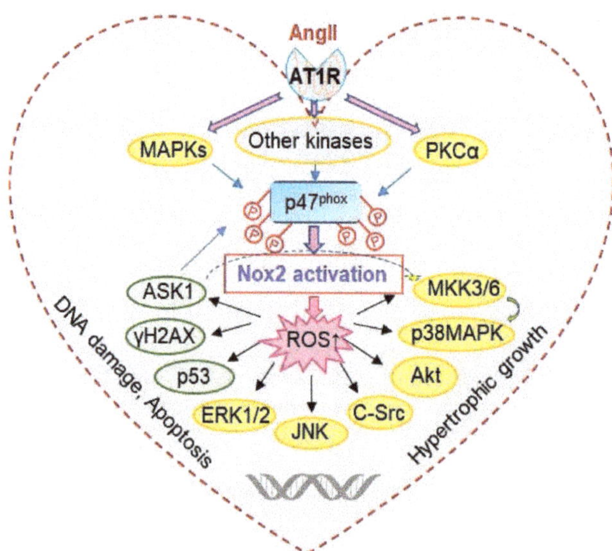

Figure 7. Schematic illustration of p47^{phox} redox-signalling pathways examined.

5. Conclusions

In conclusion, we have reported that p47^{phox} is a key player in mediating AngII-induced oxidative stress signalling cascade from the phosphorylation of ASK1, MKK3/6 and MAPKs to the activation of H2AX and p53 involved in DNA damage and apoptosis of cardiomyocytes. Genetic ablation of p47^{phox} inhibited the cardiac Nox2-derived $O_2^{\bullet-}$ production, attenuated the activation of ASK1 and MAPK signalling pathways and protected hearts from AngII-induced hypertrophic growth and DNA damage. Targeting p47^{phox} has great therapeutic potential in preventing or treating AngII-induced cardiac dysfunction and damages.

Author Contributions: F.L.: data generation and analysis; manuscript drafting; L.M.F.: data analysis and critical revising of the manuscript; L.G.: in vivo sample collection and data generation; J.-M.L.: directed the study and finalised manuscript. All authors have read and agreed to the published version of the manuscript.

Funding: This work was supported by an International PhD Studentship of the University of Reading (F.L.) and the British Heart Foundation (grant number: PG/14/85/31161).

Institutional Review Board Statement: All studies were performed following protocols approved by the Review Board of the Ethics Committees of the Surrey and Reading Universities and the Home Office under the Animals (Scientific Procedures) Act 1986 UK. The animal studies were performed under the Home Office animal work project licences (PPL 70/6729 and PPL 70/7638).

Informed Consent Statement: Not applicable.

Data Availability Statement: Research material sources and all data that support the findings of this study are provided within this paper.

Conflicts of Interest: The authors declare no conflict of interest.

References

1. Lassegue, B.; Martin, A.S.; Griendling, K.K. Biochemistry, physiology, and pathophysiology of NADPH oxidases in the cardiovascular system. *Circ. Res.* **2012**, *110*, 1364–1390. [CrossRef]
2. Forrester, S.J.; Booz, G.W.; Sigmund, C.D.; Coffman, T.M.; Kawai, T.; Rizzo, V.; Scalia, R.; Eguchi, S. Angiotensin II Signal Transduction: An Update on Mechanisms of Physiology and Pathophysiology. *Physiol. Rev.* **2018**, *98*, 1627–1738. [CrossRef]
3. Fan, L.M.; Douglas, G.; Bendall, J.K.; McNeill, E.; Crabtree, M.J.; Hale, A.B.; Mai, A.; Li, J.M.; McAteer, M.A.; Schneider, J.E.; et al. Endothelial cell-specific reactive oxygen species production increases susceptibility to aortic dissection. *Circulation* **2014**, *129*, 2661–2672. [CrossRef]
4. Fan, L.M.; Geng, L.; Cahill-Smith, S.; Liu, F.; Douglas, G.; McKenzie, C.A.; Smith, C.; Brooks, G.; Channon, K.M.; Li, J.M. Nox2 contributes to age-related oxidative damage to neurons and the cerebral vasculature. *J. Clin. Investig.* **2019**, *129*, 3374–3386. [CrossRef] [PubMed]
5. Borchi, E.; Bargelli, V.; Stillitano, F.; Giordano, C.; Sebastiani, M.; Nassi, P.A.; d'Amati, G.; Cerbai, E.; Nediani, C. Enhanced ROS production by NADPH oxidase is correlated to changes in antioxidant enzyme activity in human heart failure. *Biochim. Biophys. Acta* **2010**, *1802*, 331–338. [CrossRef]
6. El-Benna, J.; Dang, P.M.; Gougerot-Pocidalo, M.A.; Marie, J.C.; Braut-Boucher, F. p47^{phox}, the phagocyte NADPH oxidase/NOX2 organizer: Structure, phosphorylation and implication in diseases. *Exp. Mol. Med.* **2009**, *41*, 217–225. [CrossRef]
7. Li, J.M.; Mullen, A.M.; Yun, S.; Wientjes, F.; Brouns, G.Y.; Thrasher, A.J.; Shah, A.M. Essential role of the NADPH oxidase subunit p47^{phox} in endothelial cell superoxide production in response to phorbol ester and tumor necrosis factor-alpha. *Circ. Res.* **2002**, *90*, 143–150. [CrossRef] [PubMed]
8. Patel, V.B.; Wang, Z.; Fan, D.; Zhabyeyev, P.; Basu, R.; Das, S.K.; Wang, W.; Desaulniers, J.; Holland, S.M.; Kassiri, Z.; et al. Loss of p47^{phox} subunit enhances susceptibility to biomechanical stress and heart failure because of dysregulation of cortactin and actin filaments. *Circ. Res.* **2013**, *112*, 1542–1556. [CrossRef] [PubMed]
9. Tamura, M.; Itoh, K.; Akita, H.; Takano, K.; Oku, S. Identification of an actin-binding site in p47^{phox} an organizer protein of NADPH oxidase. *FEBS Lett.* **2006**, *580*, 261–267. [CrossRef]
10. Li, J.M.; Gall, N.P.; Grieve, D.J.; Chen, M.; Shah, A.M. Activation of NADPH oxidase during progression of cardiac hypertrophy to failure. *Hypertension* **2002**, *40*, 477–484. [CrossRef]
11. Cao, M.; Mao, Z.; Peng, M.; Zhao, Q.; Sun, X.; Yan, J.; Yuan, W. Extracellular cyclophilin A induces cardiac hypertrophy via the ERK/p47^{phox} pathway. *Mol. Cell. Endocrinol.* **2020**, *518*, 110990. [CrossRef]
12. Wada, T.; Penninger, J.M. Mitogen-activated protein kinases in apoptosis regulation. *Oncogene* **2004**, *23*, 2838–2849. [CrossRef]

13. Soga, M.; Matsuzawa, A.; Ichijo, H. Oxidative Stress-Induced Diseases via the ASK1 Signaling Pathway. *Int. J. Cell Biol.* **2012**, *2012*, 439587. [CrossRef]
14. Sablina, A.A.; Budanov, A.V.; Ilyinskaya, G.V.; Agapova, L.S.; Kravchenko, J.E.; Chumakov, P.M. The antioxidant function of the p53 tumor suppressor. *Nat. Med.* **2005**, *11*, 1306–1313. [CrossRef] [PubMed]
15. Thomas, M.; Gavrila, D.; McCormick, M.L.; Miller, F.J., Jr.; Daugherty, A.; Cassis, L.A.; Dellsperger, K.C.; Weintraub, N.L. Deletion of $p47^{phox}$ attenuates angiotensin II-induced abdominal aortic aneurysm formation in apolipoprotein E-deficient mice. *Circulation* **2006**, *114*, 404–413. [CrossRef] [PubMed]
16. Doerries, C.; Grote, K.; Hilfiker-Kleiner, D.; Luchtefeld, M.; Schaefer, A.; Holland, S.M.; Sorrentino, S.; Manes, C.; Schieffer, B.; Drexler, H.; et al. Critical role of the NAD(P)H oxidase subunit $p47^{phox}$ for left ventricular remodeling/dysfunction and survival after myocardial infarction. *Circ. Res.* **2007**, *100*, 894–903. [CrossRef] [PubMed]
17. Meijles, D.N.; Fan, L.M.; Howlin, B.J.; Li, J.M. Molecular insights of $p47^{phox}$ phosphorylation dynamics in the regulation of NADPH oxidase activation and superoxide production. *J. Biol. Chem.* **2014**, *289*, 22759–22770. [CrossRef]
18. Du, J.; Fan, L.M.; Mai, A.; Li, J.M. Crucial roles of Nox2-derived oxidative stress in deteriorating the function of insulin receptors and endothelium in dietary obesity of middle-aged mice. *Br. J. Pharmacol.* **2013**, *170*, 1064–1077. [CrossRef] [PubMed]
19. Schipke, J.; Banmann, E.; Nikam, S.; Voswinckel, R.; Kohlstedt, K.; Loot, A.E.; Fleming, I.; Muhlfeld, C. The number of cardiac myocytes in the hypertrophic and hypotrophic left ventricle of the obese and calorie-restricted mouse heart. *J. Anat.* **2014**, *225*, 539–547. [CrossRef] [PubMed]
20. Bensley, J.G.; de Matteo, R.; Harding, R.; Black, M.J. Three-dimensional direct measurement of cardiomyocyte volume, nuclearity, and ploidy in thick histological sections. *Sci. Rep.* **2016**, *6*, 23756. [CrossRef]
21. Csanyi, G.; Cifuentes-Pagano, E.; al Ghouleh, I.; Ranayhossaini, D.J.; Egana, L.; Lopes, L.R.; Jackson, H.M.; Kelley, E.E.; Pagano, P.J. Nox2 B-loop peptide, Nox2ds, specifically inhibits the NADPH oxidase Nox2. *Free Radic. Biol. Med.* **2011**, *51*, 1116–1125. [CrossRef]
22. Dorn, G.W., II; Force, T. Protein kinase cascades in the regulation of cardiac hypertrophy. *J. Clin. Investig.* **2005**, *115*, 527–537. [CrossRef] [PubMed]
23. Fontayne, A.; Dang, P.M.; Gougerot-Pocidalo, M.A.; El-Benna, J. Phosphorylation of $p47^{phox}$ sites by PKC alpha, beta II, delta, and zeta: Effect on binding to $p22^{phox}$ and on NADPH oxidase activation. *Biochemistry* **2002**, *41*, 7743–7750. [CrossRef] [PubMed]
24. Callera, G.E.; Antunes, T.T.; He, Y.; Montezano, A.C.; Yogi, A.; Savoia, C.; Touyz, R.M. c-Src Inhibition Improves Cardiovascular Function but not Remodeling or Fibrosis in Angiotensin II-Induced Hypertension. *Hypertension* **2016**, *68*, 1179–1190. [CrossRef]
25. Altenhofer, S.; Radermacher, K.A.; Kleikers, P.W.; Wingler, K.; Schmidt, H.H. Evolution of NADPH Oxidase Inhibitors: Selectivity and Mechanisms for Target Engagement. *Antioxid. Redox Signal.* **2015**, *23*, 406–427. [CrossRef] [PubMed]
26. Sumimoto, H. Structure, regulation and evolution of Nox-family NADPH oxidases that produce reactive oxygen species. *FEBS J.* **2008**, *275*, 3249–3277. [CrossRef]
27. Zhang, Y.; Murugesan, P.; Huang, K.; Cai, H. NADPH oxidases and oxidase crosstalk in cardiovascular diseases: Novel therapeutic targets. *Nat. Rev. Cardiol.* **2020**, *17*, 170–194. [CrossRef]
28. Fukai, T.; Ushio-Fukai, M. Superoxide dismutases: Role in redox signaling, vascular function, and diseases. *Antioxid. Redox Signal.* **2011**, *15*, 1583–1606. [CrossRef]
29. Lin, C.C.; Lin, W.N.; Cho, R.L.; Yang, C.C.; Yeh, Y.C.; Hsiao, L.D.; Tseng, H.C.; Yang, C.M. Induction of HO-1 by Mevastatin Mediated via a Nox/ROS-Dependent c-Src/PDGFRalpha/PI3K/Akt/Nrf2/ARE Cascade Suppresses TNF-alpha-Induced Lung Inflammation. *J. Clin. Med.* **2020**, *9*, 226. [CrossRef] [PubMed]
30. Wang, Y. Mitogen-activated protein kinases in heart development and diseases. *Circulation* **2007**, *116*, 1413–1423. [CrossRef]
31. Hayakawa, R.; Hayakawa, T.; Takeda, K.; Ichijo, H. Therapeutic targets in the ASK1-dependent stress signaling pathways. *Proc. Jpn. Acad. Ser. B Phys. Biol. Sci.* **2012**, *88*, 434–453. [CrossRef] [PubMed]
32. Kuo, L.J.; Yang, L.X. Gamma-H2AX—A novel biomarker for DNA double-strand breaks. *In Vivo* **2008**, *22*, 305–309. [PubMed]
33. Mak, T.W.; Hauck, L.; Grothe, D.; Billia, F. p53 regulates the cardiac transcriptome. *Proc. Natl. Acad. Sci. USA* **2017**, *114*, 2331–2336. [CrossRef] [PubMed]

MDPI

Article

μ-Opioid Receptor-Mediated AT1R–TLR4 Crosstalk Promotes Microglial Activation to Modulate Blood Pressure Control in the Central Nervous System

Gwo-Ching Sun [1,2,3,4], Jockey Tse [2,3], Yung-Ho Hsu [2,3], Chiu-Yi Ho [4,5], Ching-Jiunn Tseng [4,5] and Pei-Wen Cheng [4,5,*]

1 Department of Anesthesiology, Kaohsiung Veterans General Hospital, Kaohsiung 813414, Taiwan; gcsun39@kmu.edu.tw
2 Department of Anesthesiology, Kaohsiung Medical University Hospital, Kaohsiung 80708, Taiwan; 1030402@kmuh.org.tw (J.T.); 1020366@ms.kmuh.org.tw (Y.-H.H.)
3 Department of Anesthesiology, School of Medicine, College of Medicine, Kaohsiung Medical University, Kaohsiung 80708, Taiwan
4 Department of Medical Education and Research, Kaohsiung Veterans General Hospital, Kaohsiung 813414, Taiwan; cyho@vghks.gov.tw (C.-Y.H.); cjtseng@vghks.gov.tw (C.-J.T.)
5 Department of Biomedical Science, National Sun Yat-Sen University, Kaohsiung 80424, Taiwan
* Correspondence: pwcheng@vghks.gov.tw; Tel.: +886-7-3422121 (ext. 71593); Fax: +886-7-3468056

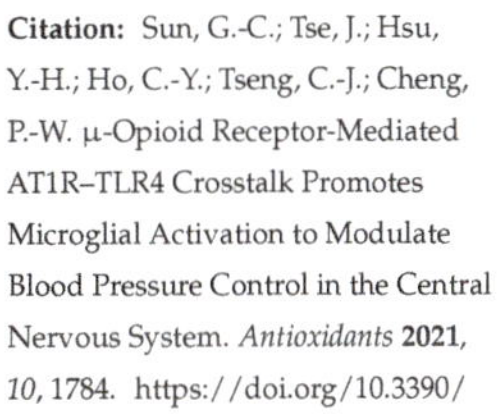

Citation: Sun, G.-C.; Tse, J.; Hsu, Y.-H.; Ho, C.-Y.; Tseng, C.-J.; Cheng, P.-W. μ-Opioid Receptor-Mediated AT1R–TLR4 Crosstalk Promotes Microglial Activation to Modulate Blood Pressure Control in the Central Nervous System. *Antioxidants* **2021**, *10*, 1784. https://doi.org/10.3390/antiox10111784

Academic Editors: Chiara Nediani, Monica Dinu and Francisco J. Romero

Received: 6 August 2021
Accepted: 5 November 2021
Published: 8 November 2021

Abstract: Opioids, a kind of peptide hormone involved in the development of hypertension, cause systemic and cerebral inflammation, and affects regions of the brain that are important for blood pressure (BP) control. A cause-and-effect relationship exists between hypertension and inflammation; however, the role of blood pressure in cerebral inflammation is not clear. Evidence showed that AT1R and μOR heterodimers' formation in the NTS might lead to the progression of hypertension. In this study, we investigated the formation of the μOR/AT1R heterodimer, determined its correlation with μORs level in the NTS, and explored the role of TLR4-dependent inflammation in the development of hypertension. Results showed that Ang II increased superoxide and Iba-1 (microgliosis marker: ionized calcium-binding adaptor molecule (1) levels in the NTS of spontaneously hypertensive rats (SHRs). The AT1R II inhibitor, losartan, significantly decreased BP and abolished superoxide, Iba-1, TLR4 expression induced by Ang II. Furthermore, losartan significantly increased $nNsOS^{S1416}$ phosphorylation. Administration of a μOR agonist or antagonist in the NTS of WKY and SHRs increased endogenous μ-opioids, triggered the formation of μOR/AT1R heterodimers and the TLR4-dependent inflammatory pathway, and attenuated the effect of depressor nitric oxide (NO). These results imply an important link between neurotoxicity and superoxides wherein abnormal increases in NTS endogenous μ-opioids promote the interaction between Ang II and μOR, the binding of Ang II to AT1R, and the activation of microglia. In addition, the interaction between Ang II and μOR enhanced the formation of the AT1R and μOR heterodimers, and inactivated nNOS-derived NO, leading to the development of progressive hypertension.

Keywords: angiotensin II type 1 receptor (AT1R); hypertension; heterodimer; toll like receptor 4; nucleus tractus solitarii; opioids

1. Introduction

Hypertension poses a significant health problem and intensive efforts have been made to elucidate its underlying mechanisms [1]. Approximately one-third of adults have hypertension in the United States and 50% of hypertensive patients show satisfactory results after treatment [2]. The major reason is that most of the existing anti-hypertensive therapies target peripheral mechanisms and are not effective for treating hypertension driven in the central nervous system (CNS) [2]. A better understanding of the CNS mechanisms underpinning the pathogenesis of hypertension can lead to the discovery of novel strategies for the prevention and treatment of hypertension.

The nucleus tractus solitarius (NTS), located in the dorsal medulla of the brainstem, is primarily responsible for integration of cardiovascular (CV) regulation and other autonomic functions of the central nervous system (CNS). In the presence of noxious stimuli, a phenomenon known as "hypertensive hypoalgesia" [3], which is related to decreased pain sensitivity [4], occurs as a result of a homeostatic feedback loop caused by BP stabilization. Numerous neuropeptides such as opioids and angiotensin are implicated in the CV system [5,6]. The renin–angiotensin system (RAS) is an enzyme neuropeptide system in the brain and periphery that has been well-studied, and has served as a neuronal model for peptide regulation. Angiotensin II (Ang II) is a primary effector peptide of the renin–angiotensin–aldosterone system (RAAS) which binds to the G-protein-coupled receptor subtypes AT1R and AT2R with similar affinity in multiple cell and tissue types [7]. Elevated circulating Ang II in the brain was reported to be associated with the genesis of arterial hypertension [8], whereas the overactivation of RAAS is crucially involved in the pathogenesis of hypertension and hypertension-related cardiovascular disorders [9]. Moreover, the beneficial effects of previous studies have shown that RAS blockers can reduce the development of hypertension and its associated neuropathic pain, cognitive impairment and cerebral injury [6,10,11]. By understanding role of the neuropeptides, opioids and angiotensin in CV function, we hope to trace the molecular origin of heart failure during the development of hypertension.

Other experimental evidence has demonstrated significant functional overlapping of RAS components and endogenous opioids (alongside their receptors) in the brain and periphery regions, showing synergistic interaction between angiotensin and opioids [12]. Angiotensin increases opioid levels to induce polydipsia, analgesia, LH secretion and hypertension, which are abolished in the presence of an opioid antagonist, namely naloxone. On the other hand, opioids increase angiotensin II levels by activating renin and angiotensin-converting enzyme (ACE) either directly or indirectly [12]. Furthermore, opioid-induced increases in ACE activity may trigger a negative feedback mechanism that affects the influence of opioids, thereby enhancing the metabolism of endogenous opioids through neutral endopeptidases and dipeptidylcarboxypeptidase [13].

Previous experiments suggest that hypertension is characterized by pro- and antioxidant mechanisms [14], inflammatory disorders [15], GPCR heterodimers [16], and sympathetic/parasympathetic tone imbalances [17]. Accumulating evidence suggests that the Ang II-AT1R axis stimulates innate and adaptive immune systems [18–20]. The blockade or knockdown of toll-like 4 receptor (TLR4), which is required for integral sensing and signaling of the innate system, attenuates Ang II-dependent hypertension, as well as renal and cardiac injury [19]. Nair et al., proposed that Ang II stimulates the AT1R to release high-mobility group protein 1 (HMBG1), a ligand required for TLR4 to evoke inflammation [21]. Direct stimulation of the MD2-TLR4 complex by Ang II is clinically important as Ang II receptor blockers (ARBs) are capable of increasing Ang II through the inhibition of renin release [22]. Thus, the concern is that ARB treatment may cause the unintended consequence of stimulating TLR4-dependent inflammation. This mechanism may potentially diminish the optimal effects of ARBs in the treatment of cardiovascular disease (CVD) [23]. Horvath et al. previously reported that morphine administration results in changes in the microglia and astrocytes, as well as increased cellular hypertrophy, microglial CD11b, Iba1 expression and astrocytic GFAP expression in vitro [24] and in vivo [25]. On the contrary, the inhibition of microglial P2X4 receptors attenuates morphine tolerance, and Iba1, GFAP and μ opioid receptor protein expression [26].

SHRs were used because they show high Ang II and AT1R levels compared tok WKY [27]. During Ang II-induced hypertension, peripheral Ang II infusion increased ROS production and brain inflammation [28]. Evidence indicates that Ang II stimulates the innate system by direct activation of TLR4 through an AT1R-independent mechanism [29]. Opioids increase angiotensin II levels, either directly or indirectly, by activating renin and

ACE [12], and opioids are known to interact with RAS components. A previous report showed that WKY infused with AngII showed increases in BP, and increased AT1R and μOR heterodimers, in the NTS [6], prompting the need to understand the role of μOR in the formation of AT1R-μOR heterodimers and the mechanism of microglial and TLR4 activation. However, the molecular mechanism of this process remains unclear. This study aimed to determine: (1) the interplay between μORs and AT1R in the formation of heterodimers in the NTS when endogeneous opioids are present; (2) the function of Ang II in TLR4-dependent inflammation during high BP in the NTS; and (3) the major factor that leads to the formation of μOR/AT1R heterodimers in the NTS. Our results demonstrated that an increase in endogenous μ-opioids in the NTS induced the formation of μOR/AT1R heterodimers and the TLR4-dependent inflammatory pathway, which attenuated the NO-dependent depressor effect. In summary, endogenous increases in μ-opioid are most likely factor to contribute to the pathogenesis of hypertension through the AT1R–TLR4 axis.

2. Materials and Methods

2.1. Animals

Animal studies were conducted in compliance with the Animal Research: Reporting of In Vivo Experiments (ARRIVE) guidelines as described previously [30,31]. All protocols were approved by Animal Research Committee and the institutional review board at VGHKS (VGHKS-2021-2023-A009; VGHKS-2020-2022-A046) and an affidavit of approval was obtained in with the animal care protocol of Kaohsiung Medical University (109087). Wistar-Kyoto rats (WKY) and SHRs were obtained from the National Science Council Animal Facility (NSCAF; Taipei, Taiwan), and housed in an animal facility at Kaohsiung Veterans General Hospital (VGHKS; Kaohsiung, Taiwan). NSCAF and VGHKS were approved by the Association for Assessment and Accreditation of Laboratory Animal Care (AAALAC). WKY were caged under specific pathogen-free (SPF) conditions at VGHKS, and were free of infectious and pathogenic organisms capable of interfering with research subjects. The WKY were caged in individual cages, provided 12-h light-and-dark cycle, and kept at a temperature between 23 and 24 °C. Animals were provided with normal rat chow (Purina, St. Louis, MO, USA) and tap water ad libitum. The animals were settled in the housing environment for one week for adjustment before being habituated to the indirect blood pressure measurement for another week. The highest and lowest datum were excluded, the remaining six data were averaged for each group ($n = 6$). The SBP of the rats was measured before the start of the WKY, SHR, and losartan treatments (week 0) using a tail-cuff monitor (Noninvasive Blood Pressure System, SINGA, Taipei, Taiwan). The rats were placed in the fixer for 30 min at a constant temperature of 37 °C. During measurement, six individual readings were obtained. The highest and lowest readings were discarded, and the averages of the remaining eight readings were obtained. The SBP was measured at the same time on a daily basis. The rats were randomly assigned to five groups, with six in each group—(1) WKY: 6 wo WKY; (2) SHR: 6 wo SHR; (3) WKY: 20 wo WKY; (4) SHR: 20 wo SHR (saline); (5) SHR + losartan: 20 wo SHR + losartan (week-old abbreviated as wo). Losartan (30 mg/kg per day) was administered by gavage to 20 wo SHRs for 2 weeks. Animals were euthanized using 100% CO_2, a procedure that was in accordance with the 2013 American Veterinary Medical Association (AVMA) guidelines.

2.2. Intra-NTS Microinjection

Twenty-week-old WKY were anesthetized using urethane (1.0 g kg^{-1} intraperitoneally (i.p.), supplemented with 300 mg kg^{-1} intravenously (i.v.) when required). Blood pressure and heart rate were measured via femoral-artery cannula using a pressure transducer and polygraph (Gould, Cleveland, OH, USA), and a tachograph preamplifier (Gould, Cleveland, OH, USA), respectively. Tracheostomy was performed to maintain airway patency. For brainstem nuclei microinjection, animals were placed in a stereotaxic instrument (Kopf, Tujunga, CA, USA), with the head positioned 45°downward to expose dorsal surface of the medulla with limited craniotomy, followed by a 1h resting period. Single-barrel glass catheters (0.031-inch outer diameter (OD), 0.006-inch internal diameter (ID); Richland Glass Co, Vineland, NJ, USA) with external tips of 40 μm in diameter were used. L-glutamate (0.154 nmol 60 nL^{-1}) was microinjected to induce the characteristic decrease in BP (BP ≥ -35 mmHg), in order to verify that needle tip was located in the medial site, in one-third of the NTS. The precise coordinates were as follows: anteroposterior, 0.0 mm; mediolateral, 0.5 mm; and vertical, 0.4 mm (with the obex as a reference) [32]. For microinjections, 0.3 nmol of the DAMGO μ opioid agonist (Sigma, St. Louis, MO, USA), or 0.3 nmol of the guanfacine α2A agonist (Sigma, St. Louis, MO, USA), were prepared in 0.9% saline for use.

2.3. *In Situ Detection of Superoxide in the NTS*

Endogenous in vivo superoxide production in the NTS was determined via dihydroethidium staining (DHE; Invitrogen, Carlsbad, CA, USA). The NTS was dissected, quickly frozen, embedded in OCT, and immersed in liquid nitrogen. Cryostat slices (30 μm) were stained with 1 μM DHE in the dark for 30 min at 37 °C. The samples were analyzed using a confocal microscope (Carl Zeiss LSM 5 PASCAL, Göttingen, Germany).

2.4. *Immunofluorescence Staining Analysis*

The rats were perfused using 0.9% saline and 4% formaldehyde, followed by 30% sucrose solution. Brainstems were cut into 20 μm-thick sections, incubated in anti-endomorphin-2, anti-IBA1, anti-AT1R (ab124505) and anti-$nNOS^{S1416}$ primary antibodies at a dilution ratio of 1:100. After washing with PBS, sections were incubated in Alexa Fluor 488 or 588-conjugated donkey anti-rabbit IgG (1:200; Invitrogen, Carlsbad, CA, USA) at 25 °C for 2 h, and analyzed using a fluorescence microscope and Zeiss LSM Image software (Carl Zeiss MicroImaging).

2.5. *Proximity Ligation Assay (PLA)*

The Duolink in situ proximity ligation assay (PLA; OLINK Bioscience, Uppsala, Sweden) was utilized to detect the formation of AT1R/μOR heterodimers. The NTSs of SHRs and WKY were examined to detect the formation of AT1R (sc-515884) and μOR (bs-3623R) heterodimers in situ. The rats were first perfused in saline, then 4% formaldehyde, and finally 30% sucrose solution. Brainstem's microsections of 5 μm thickness were obtained. Primary antibody diluent (OLINK Bioscience), containing two primary antibodies (1:100 for goat anti-μ receptor antibodies and 1:100 rabbit anti-AT1R antibodies), was added to the sections, and incubated overnight at 4 °C. Next, PLA secondary antibody with specific oligonucleotides, anti-rabbit plus and anti-goat minus (OLINK Bioscience), were applied to the sections and incubated for 2 h at 37 °C. Samples underwent ligation to allow nearby oligonucleotide probe pairs to form closed circles, and signals were amplified in the amplification solution. Images were acquired using a confocal laser scanning microscope (Carl Zeiss LSM 5 PASCAL), and were further processed by LSM 5 PASCAL software (Version 3.5, Carl Zeiss), which automatically counted the number of spots per unit of surface area.

2.6. *Measurement of NO in the NTS*

The NTS (10 mg) was deproteinized using Microcon YM-30 centrifugal filter units (Millipore, Bedford, MA, USA). The total content of NO in the samples was determined through a procedure that is based on the purge system, using the Sievers Nitric Oxide Analyzer (NOA 280i) (Sievers Instruments, Boulder, CO, USA) to assess chemiluminescence. The samples (10 μL) were injected into a reflux column containing 0.1 mol/L VCl_3 in 1 mol/L HCl at 90 °C to reduce any existing nitrate and nitrite (NOx) to NO. NO reacted with the O_3 produced by the analyzer to form NO_2. The resulting emission from the excited NO_2 was detected by a photomultiplier tube and recorded digitally (mV). The standard curve was determined for the $NaNO_3$ concentrations, and NO levels were corrected for the rats' NTS.

2.7. Immunoblotting Analysis

Proteins (20 µg per sample) were quantified using a BCA protein assay (Pierce Chemical Co., Rockford, IL, USA), resolved in 6% polyacrylamide gel, and transferred to the PVDF membrane (GE Healthcare, Buckinghamshire, UK). The membranes were incubated at 4 °C overnight using the following primary antibodies: mouse anti-P-eNOS^{S1177}, mouse anti-DDAH1, and mouse anti-eNOS (BD Biosciences, San Jose, CA, USA); mouse anti-actin and mouse anti-nNOS (Millipore); and mouse anti-P-nNOS (Abcam, Cambridge, UK) (dilution at 1:1000).

2.8. Statistical Analysis

All measurements were repeated at least three times under independent conditions. The results shown are the mean ± the standard error of the mean (SEM). Statistics were analyzed using the Mann–Whitney U-test. One-way analysis of variance (ANOVA) with Scheffé post-hoc comparison was used to compare differences between groups. SPSS version 20.0 (SPSS Inc, Chicago, IL, USA) was applied for analyzing the raw data. * $p < 0.05$ and ** $p < 0.01$ indicate significance.

3. Results

3.1. Ang II Elevates the ROS-Microglial Activation and Reduces the Systemic Vasodepressor Effect of NO by Impairing the nNOS Pathway in the NTS of Spontaneously Hypertensive Rats

To determine the effect of ROS-dependent NO release on systemic BP (SBP) in the NTS, we examined the SBP, nitrate, IBA1 level, and ROS production in the NTS of WKY controls, prehypertensive 6 wo SHRs, and hypertensive 20 wo SHRs. A significant age-dependent increase in BP occurred in SHRs between 6 and 20 wo, and NTS NO levels were significantly decreased ($p < 0.05$, $n = 6$; Figure 1A,B). Superoxide levels in the NTS were significantly high in the SHRs between 6 and 20 wo ($p < 0.05$, $n = 6$; Figure 1C). Immunoblotting analyses demonstrated that the phosphorylation of nNOS^{S1416} was significantly decreased in 20 wo SHRs ($p < 0.05$, $n = 6$; Figure 1D). The α2A-AR and µOR heterodimers were determined using PLA. At 6 wo, SHRs and WKY exhibited normal systolic BP, and no differences in the levels of the α2A-AR and µOR heterodimers were observed. Interestingly, the levels of NTS AT1R and IBA1 were significantly elevated in adult SHRs compared to WKY (Figure 1E), indicating that Ang II enhanced ROS production and microglial activation, which impaired the nNOS pathway in the NTS of spontaneously hypertensive rats.

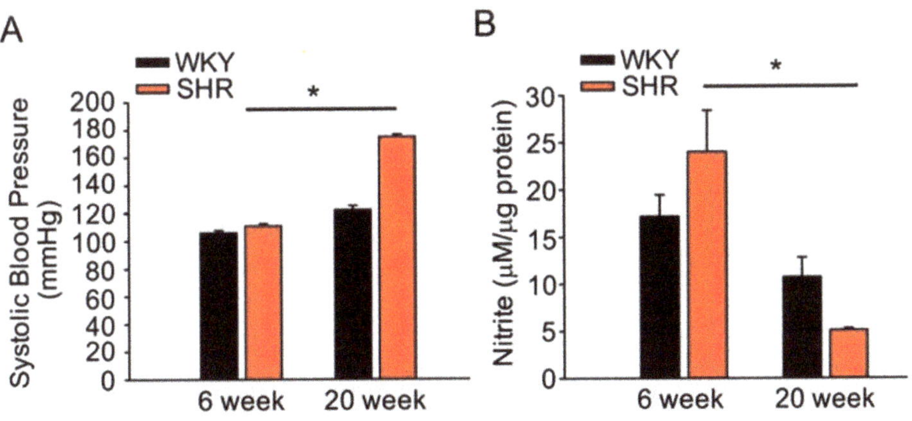

Figure 1. *Cont.*

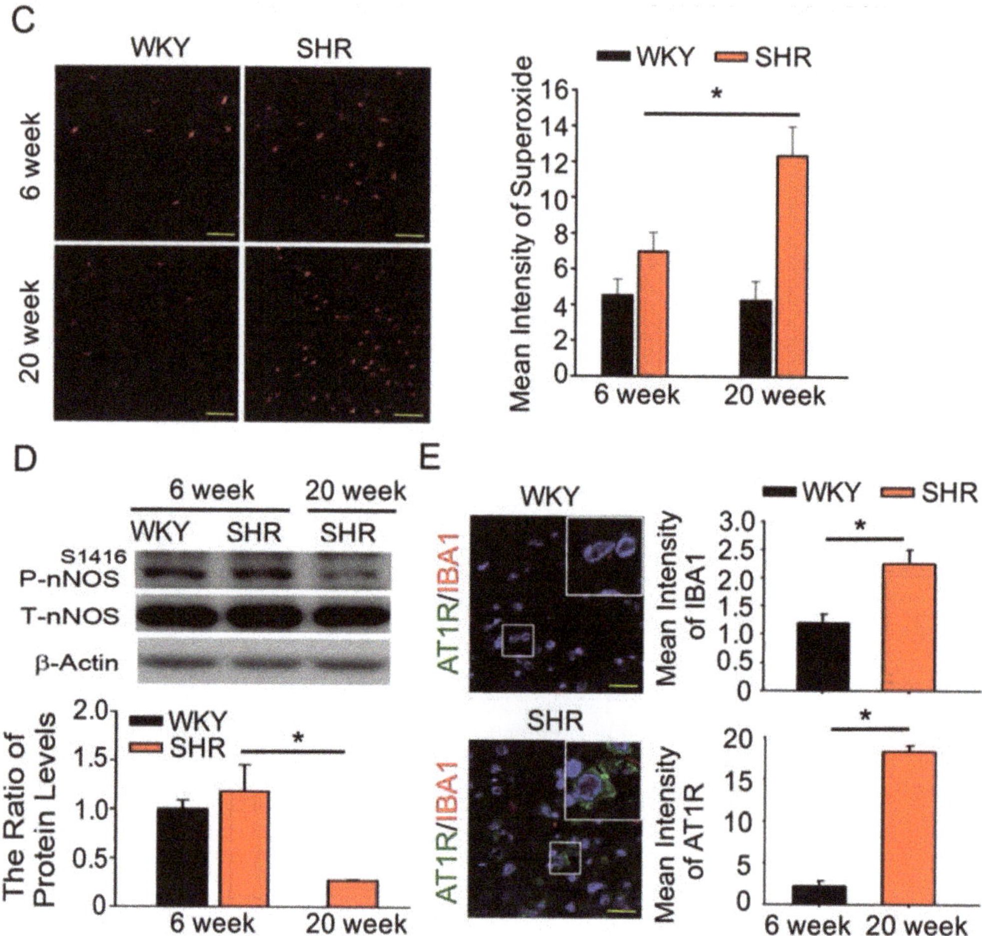

Figure 1. Ang II supports superoxide generation, increases the activation of microglia and restores the nNOS pathway in the NTS of spontaneously hypertensive rats. (**A**,**B**) Graph showing systemic blood pressure (SBP) and nitric oxide (NO) concentrations in the nucleus tractus solitarius (NTS) of normotensive Wistar Kyoto rats (WKY; 6-, 20-week-old) and spontaneously hypertensive rats (SHRs; 6-, 20-week-old). The bar graph shows the NO concentration as μM nitrate per μg of total protein. (**C**) Representative images of DHE-treated brain sections, images photographed at ×280 magnification. ROS index in the NTS of the SHRs groups (20 weeks old) was compared to SHRs (6 weeks old). The ROS index is the relative mean fluorescence intensity for dihydroethidium. Sections including the NTS of SHRs rats displayed significant increase in DHE fluorescence compared with the SHRs (6 weeks old) group sections. (**D**) Quantitative immunoblotting analysis of $nNOS^{S1416}$ phosphorylation in the NTS of 6-week-old WKY, SHRs and 20-week-old SHRs. One-way analysis of variance (ANOVA) with Scheffé post-hoc was performed for statistical analysis in Figure 1A–D. (**E**) Representative images for immunofluorescence staining of AT1R-positive cells (green) and microglial marker IBA-1 (red) in NTS sections of WKY and SHRs, counterstained with 4′,6-diamidino-2-phenylindole (DAPI for blue color). The images were photographed at ×400 and 1000 magnification. The Mann–Whitney U-test was performed for statistical analysis in Figure 1E. The values are presented as mean ± SEM. * $p < 0.05$ indicates significant difference from 6-week-old SHRs (n = 6~8 per group).

3.2. AT1R Inhibitors Decrease BP via Inhibition of AT1R-Induced Superoxide to Enhance Microglia Activity in the NTS

In this section, we show that the activation of AT1R increased microglia and ROS production in the NTS, which further reduced nNOS^{S1416} phosphorylation during the development of ANG II-induced hypertension. Therefore, we investigated the progression of hypertension after the activation of the microglia and TLR4 resulted from the increase in AT1R, further studied the effects of the AT1R inhibitor losartan on SBP, superoxide production, and activity of the microglia. SBP was found to be significantly lower, and superoxide levels in the NTS were significantly lower in the losartan-treated SHRs as compared to the untreated SHRs (Figure 2A and Figure 4B, lane 2 and lane 3, respectively; $^{\#} p < 0.05$, $n = 6$). Immunofluorescence staining demonstrated that losartan-treated SHRs showed significantly reduced AT1R and TLR4 activation of microglial cells in the NTS (Figure 2C,D, lane 2 and lane 3, respectively; $^{\#}p < 0.05$, $n = 6$). These results indicated that AT1R-induced superoxide generation led to an increase in the activation of microglia and microglial TLR4, thereby inducing progressive hypertension.

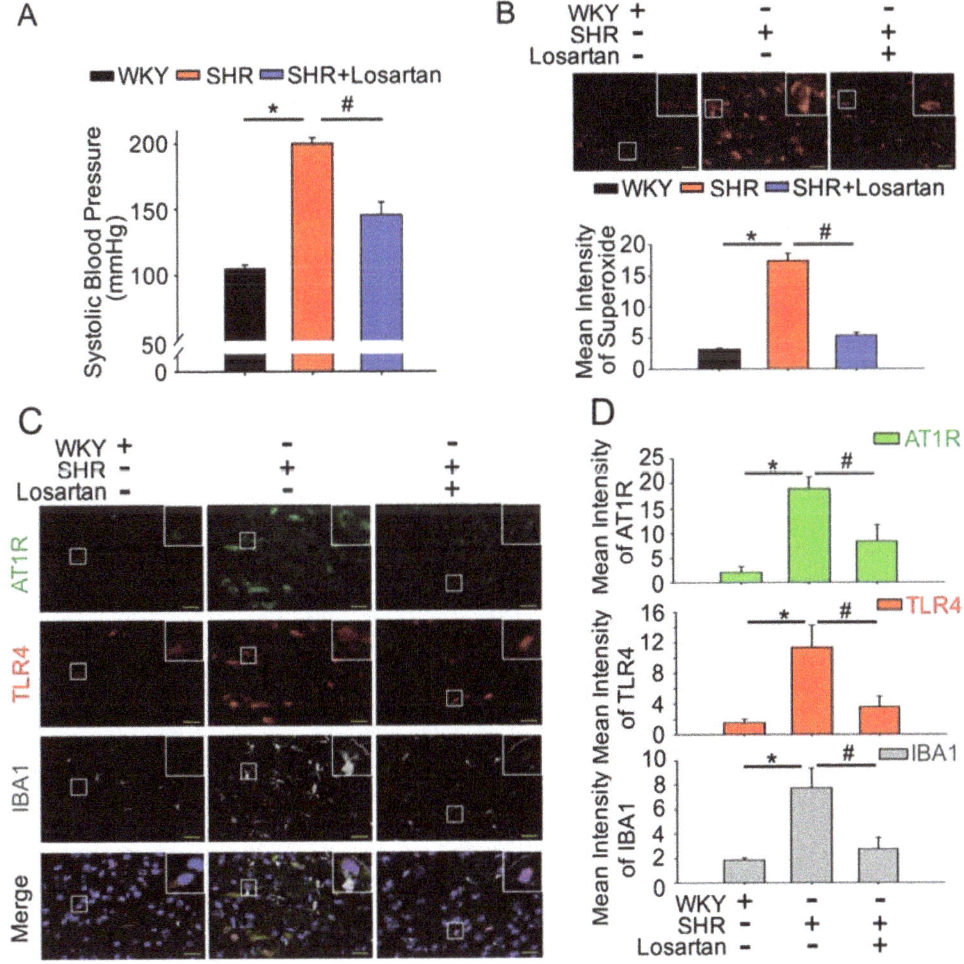

Figure 2. The downregulation of AT1R-induced superoxide generation is associated with the activation of microglia and

the expression of TLR4 in the NTS of spontaneously hypertensive rats. (**A**) SBP after losartan administration for 2 weeks. (**B**) Representative images of DHE-treated brain sections, photographed at ×280 magnification. Bar graph representation of ROS index in the NTS of WKY, SHRs, and SHRs treated with losartan. (**C**,**D**) Representative fluorescence images for AT1R (green), TLR4 (red) and microglial marker IBA-1 (white)-positive cells in the NTS sections of WKY, SHRs, and SHRs treated with losartan. The images were photographed at ×400 and 1000 magnification. The Mann–Whitney U-test was performed for statistical analysis in Figure 1E. One-way analysis of variance (ANOVA) with Scheffé post-hoc was performed for statistical analysis. The values are presented as mean ± SEM. * $p < 0.05$ indicates significant difference from 20-week-old WKY. # $p < 0.05$ versus SHR. All data are presented as means ± SEM ($n = 6$ per group).

3.3. *The Potential Role of μOR in AT1R-Induced Microglia Activation and TLR4 Expression in the NTS Explains Its Function in Cardiovascular Regulation*

Figure 3A shows that the unilateral microinjection of DAMGO, a μOR-specific agonist, into the NTS elicited a depressor effect as compared to the control (73.9 ± 2.03 mmHg and 93.3 ± 1.42 bpm, $p < 0.05$, paired t test; $n = 3$, Figure 3C). In addition, the endomorphin-2 level in the NTS of WKY was significantly increased (Figure 3E,F). Hypertensive SHRs had the highest endomorphin-1/2 level in their NTSs. Endogenous μ-opioids were blocked by a μOR-specific antagonist [D-Phe-Cys-Tyr-D-Trp-Arg-Thr-Pen-Thr-NH_2] (CTAP). In Figure 3B, the BP of the hypertensive SHRs began to gradually decrease, reaching a minimum at approximately 30 min after intra-NTS CTAP microinjection (105.97 ± 5.05 mmHg and 59.77 ± 4.67 bpm, $p < 0.05$, paired t test; $n = 4$, Figure 3D) and showed significantly lower endomorphin-2 levels in the NTS of SHRs after CTAP injection (Figure 3E,F).

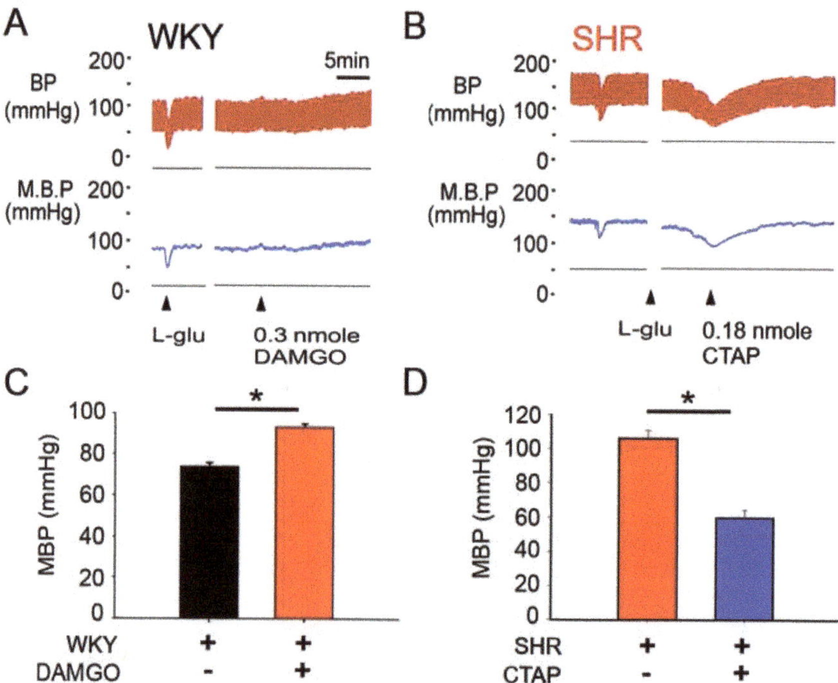

Figure 3. *Cont.*

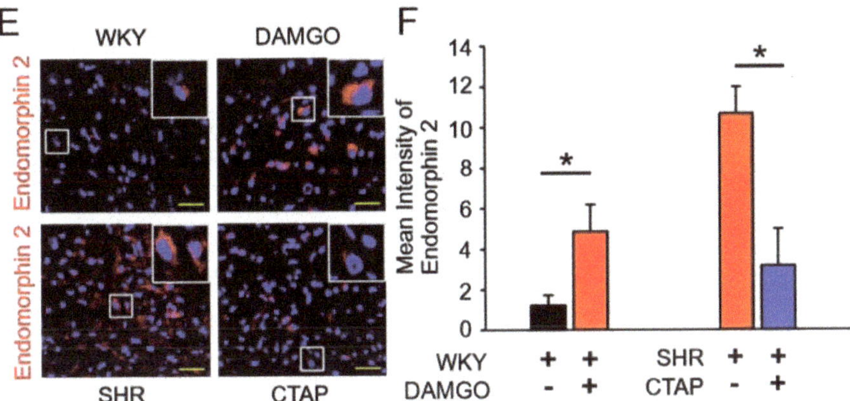

Figure 3. Activation of μOR in the NTS and cardiovascular regulation of WKY rats and SHRs. (**A**) Representative tracings demonstrating the cardiovascular effects of [D-Ala^2, MePhe^4, Gly^5-ol]-enkephalin (DAMGO) (0.3 nmol) injected into the unilateral NTS of anesthetized WKY (▲ time of injection). (**B**) Representative BP recordings of an intra-NTS microinjection of the μOR antagonist CTAP in hypertensive SHRs (▲ time of injection). (**C**) Bar graphs showing the effects of 10-min treatment with DAMGO on the mean blood pressure (MBP) of anesthetized WKY. (**D**) Bar graphs showing the effects of 10-min treatment with DAMGO on the mean blood pressure (MBP) of anesthetized SHRs. (**E**,**F**) Representative red fluorescence images and the statistical analysis for Endomorphin-2-positive cells after DAMGO or CTAP treatment. The images were photographed at ×400 and 1000 magnification. The Mann–Whitney U-test was used for statistical analysis. Bar values are shown as mean ± SEM (n = 6); * $p < 0.05$ versus control.

The regulation of BP by AT1R/μOR heterodimers was further investigated. The DAMGO-induced formation of AT1R/μOR heterodimers peaked at 10 min after DAMGO microinjection, and a reduction in CTAP was observed in the AT1R/μOR heterodimers (Figure 4A, n = 6). Furthermore, we found that AT1R, TLR4, activated microglial cells and nNOS^{S1416} phosphorylation were significantly increased in the DAMGO group compared to the control group. However, AT1R and TLR4 levels, activated microglial cells and nNOS^{S1416} phosphorylation were significantly lower in the CTAP group compared to the control group (Figure 4B,E, * $p < 0.05$, n = 6). These results indicate that the activation of μOR in the NTS may elevate AT1R to increase the activation of microglia and microglial TLR4.

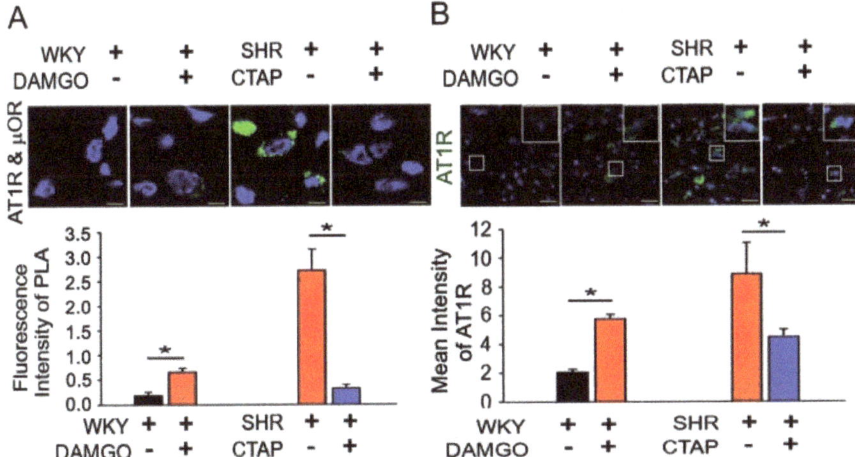

Figure 4. *Cont.*

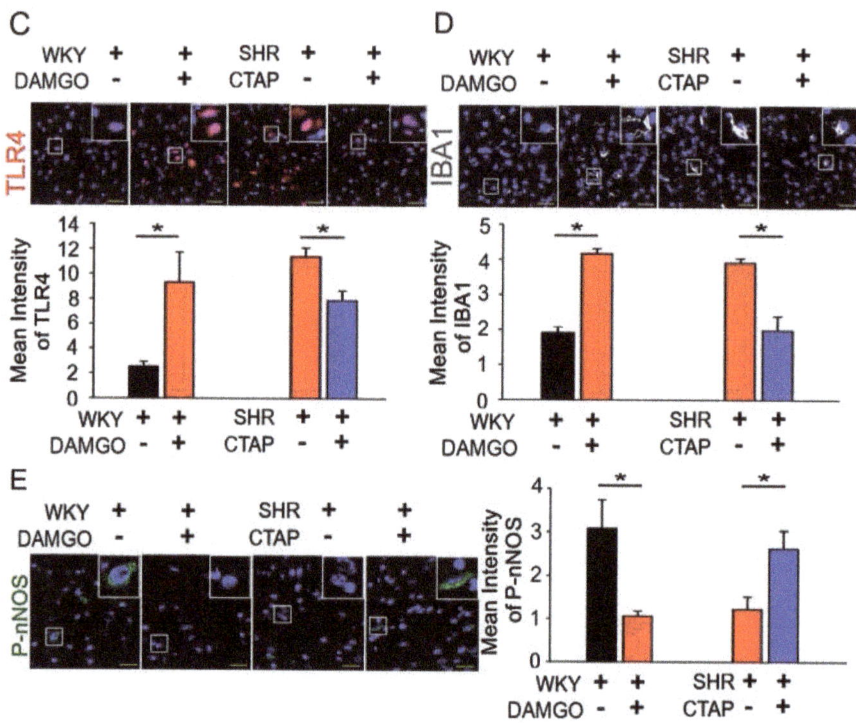

Figure 4. μOR elevates the formation of AT1R and μOR heterodimers, and induces the activation of microglia and the expression of TLR4 to impair the nNOS pathway in the NTS. (**A**) The in situ PLA (Proximity Ligation Assay) was used to confirm the formation of AT1R and μOR (μ-opioid receptors) heterodimers after intra-NTS DAMGO or CTAP microinjection. Green color indicates AT1R and μOR heterodimers; the nuclei were counterstained with DAPI. The images were photographed at 1000 magnification. (**B–E**) Representative fluorescence images of AT1R (green), TLR4 (red) microglial marker IBA-1 (white) and nNOS^{S1416} (green)-positive cells after intra-NTS DAMGO or CTAP microinjection. The images were photographed at ×400 and 1000 magnification. The Mann–Whitney U-test was used for statistical analysis. Bar values are shown as mean ± SEM (n = 6); * $p < 0.05$ versus control.

4. Discussion

Despite recent advances in treatment options, approximately one-third of the adult population are affected by hypertension in the United States [33]. Essential hypertension, which is a rise in BP due to undetermined causes, includes 90% of all hypertensive cases and is estimated to cause 13% of all deaths [2]. Most patients with essential hypertension are obese and suffer from increased RAAS and AT1R activity, which exacerbates their risk for cardiovascular disease. A century of discoveries has established the importance of the RAAS in maintaining BP, fluid volume and electrolyte homeostasis through autocrine, paracrine and endocrine signaling. While research continues to yield novel functions for Ang II, angiotensin (1–7), angiotensin-converting enzyme inhibitors and Ang II receptor blockers, the gap between basic research and actual clinical application is yet to be solved [23].

Microglia is the major player in the brain innate immune system. Recent studies indicate that microglia and astrocytes in the brainstem and hypothalamus are involved in cardiovascular and metabolic events [34]. Our previous studies found that C-X3-C motif chemokine receptor 1 (CX3CR1) functions as a microglia biomarker, and that microglia suppresses the nNOS signaling pathway and promotes chronic inflammation in fructose-induced hypertension [15]. Recent studies also demonstrated that microglial activation in the paraventricular hypothalamic nucleus (PVN) and elevated proinflammatory cytokines (PICs) are found in Ang II-induced hypertension and SHR with high BP [35]. Previous reports showed that the blockade of brain microglia or the targeted depletion of activated microglia in the PVN attenuated Ang II-induced hypertension, decreased PVN cytokines and reduced cardiac hypertrophy, strongly demonstrating the important role of Ang II in microglial activation and the release of PICs in the pathogenesis of hypertension. Previous findings demonstrated that TLR4s modulated inflammatory responses implicated in the development of hypertension. As a prototypic TLR4 ligand, the acute administration of LPS activates microglia in the brain, and this response is attenuated by the blockade of AT1R [19]. In addition, Okechukwu et al. investigated the ability of Ang II to induce the release of the TLR4 ligand and high-mobility group protein 1 (HMBG1), and to augment TLR4 expression, which represents an alternative mechanism for Ang II stimulation of the innate system in the renal cells [20]. Therefore, we investigated the progression of hypertension after the activation of microglia and TLR4, the effects of TLR4 inhibitor TAK242 on SBP, the phosphorylation of $nNOS^{S1416}$, and the activity of the microglia (Supplementary Figure S1). The present result showed that TAK-242-treated SHRs' SBP was found to be significantly lower and the phosphorylation of $nNOS^{S1416}$ was significantly higher in the NTS (Supplementary Figure S1A,B). Furthermore, TAK242-treated SHRs showed significantly reduced AT1R and TLR4 activation of microglial cells in the NTS (Supplementary Figure S1C,D). NO, the gas involved in sympathetic activity and blood pressure regulation in the NTS, was elevated through the inhibition of TLR4 microglia. Our results demonstrated that the increase in endogenous μ-opioid in the NTS induced the formation of μOR/AT1R heterodimers and the TLR4-dependent inflammatory pathway, which attenuated the NO-dependent depressor effect.

Oxidative stress and inflammation are essential for hypertension-induced renal injury. Toll-like receptors (TLR) are key regulators of the innate immune system, and TLR-4 deficiency reduces Ang-II-induced real injury and fibrosis via the attenuation of reactive oxygen species (ROS) production and inflammation in hypertensive kidneys [36]. Here, we show that AT1R-induced superoxide generation led to the activation of microglia and an increase in microglial TLR4, which were abolished by losartan treatment (Figure 2B–D). As previously noted, the regulation of Ang II and TLR4 involves the body–brain communication between afferent neural and humoral pathways that activate the central neural network to control cardiovascular function [36]. Neither RAS nor inflammatory mediators can contribute individually to the pathogenesis of hypertension; thus, the interactions between RAS components and inflammation mediators are likely synergistic [37]. Our findings concluded that the downregulation of AT1R-induced superoxide generation is associated with the activation of microglia and the expression of TLR4 in the NTS of SHR.

Opioids have been widely applied in clinics for centuries as one of the most potent pain relievers, but their abuse has deleterious physiological effects that are difficult to predict. In earlier studies, hypertension was characterized by pro- and antioxidant mechanisms [14], inflammatory disorders [15], GPCR heterodimers [16], and sympathetic/-parasympathetic tone imbalances [17]. Nevertheless, the mechanisms of the GPCRs involved in essential hypertension are not fully understood. A previous study showed that the formation of μOR/α_{2A}-AR heterodimers in the NTS contributed to hypertension by disrupting the BP-lowering function of α_{2A}-ARs [16]. Our previous results revealed that Ang II-induced stimulation generated the formation of AT1R and μOR heterodimers in the NTS, and downregulated the activity of the ERK1/2-RSK-nNOS pathway and the production of NO. In addition, opioids were previously shown to activate renin and ACE to increase the level of angiotensin II through either direct or indirect pathways [12]. A previous report showed that WKY that were ICV-infused with AngII showed increased BP, and that the AT1R and μOR heterodimers were also increased in the NTS [6].

Microglia cells express the three subtypes of opioid receptors, μ, δ and κ [38]. Studies have shown that glial and neuronal μ opioid receptors have similar morphine binding affinities; however, glial cells express five times lower μ opioid receptors compared to neurons [39]. Wong et al. indicated that TLR9, not TLR2 or TLR4, plays a role in the morphine inhibition of *S. pneumoniae*-induced NF-κB activity in the early stage of infection [40]. He et al. also showed that TLR9 was required for the morphine-induced apoptosis of microglia through the p38 MAPK signaling pathway; in addition, he suggested that the inhibition of TLR9 and/or the blockage of μOR prevented opioid-induced brain damage [41]. Surprisingly, we observed that unilateral microinjection of a μOR-specific agonist, DAMGO, in the NTS led to the formation of the AT1R/μOR heterodimer, and that TLR4 expression was involved in the progression of hypertension. However, μOR specific antagonists [D-Phe-Cys-Tyr-D-Trp-Arg-Thr-Pen-Thr-NH_2] (CTAP) have the reverse effect (Figure 4A,C). The μOR activated the AT1R, increased the number of microglial cells, and downregulated the phosphorylation of $nNOS^{S1416}$ in the NTS (Figure 4B,D,E). These results indicate that μOR activation elevated AT1R to augment the activation of microglia and cause an increase in microglial TLR4, thereby leading to the progression of hypertension. These results also suggest that a reduction in GPCR stimulation through μOR is required to impair the formation of GPCR heterodimers and the depressor's response. Given the role of Ang II in the maintenance of renal homeostasis, any novel inhibitor ought to possess improved selectivity for the targeting of pathogenic Ang II signaling to enable better hypertension treatment. Most importantly, we observed that the upregulation of endogenous μ-opioids in the NTS led to the interaction of Ang II and μOR, which promoted the binding of Ang II to AT1R, thereby activating the microglia and triggering superoxide production, and finally leading to neurotoxicity (Figure 5).

This study provides novel evidence that: (1) TLR4-dependent inflammatory levels were upregulated in the NTS of hypertensive SHRs; (2) AT1R inhibitors decreased BP and abolished TLR4-dependent inflammation in the NTS; (3) high μ-opioids levels triggered the formation of μOR/AT1R heterodimers in the NTS, which contributed to the development of hypertension; (4) the formation of the μOR/AT1R heterodimers enhanced TLR4-dependent inflammation, which impaired the NO-dependent depressor effect in the NTS; and (5) the TLR4-dependent inflammatory pathway also attenuated the NO-dependent depressor effect. Previous reports support these findings, indicating that μORs tend to form heterodimers with the formation of the α_{2A}-ARs [42], and that μOR/α_{2A}-AR heterodimers impair the function of α_{2A}-ARs, which is consistent with our previous study [43]. G protein-coupled receptors (GPCRs) play an important role in drug therapy and are one of the largest families for drug targets. Similarly to other GPCRs, the μ-opioid receptor (μOR) carries out its function by stimulating the heterotrimeric G protein [44]. The formation of—or changes in—these ligands can alter dimer binding and receptor activation, and cause desensitization and trafficking, leading to pathophysiological processes. Further studies of these heterodimers, including AT1R and μOR or α2A-AR and μOR, will provide new insights into therapies against hypertensive conditions.

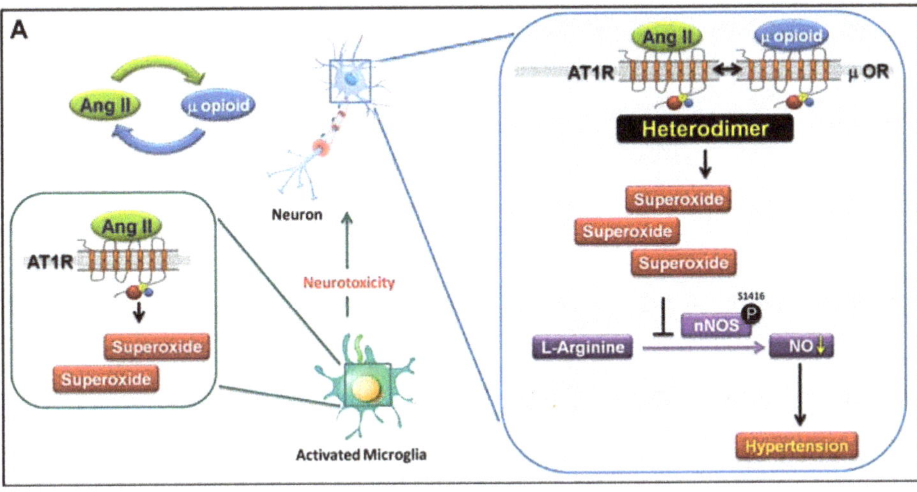

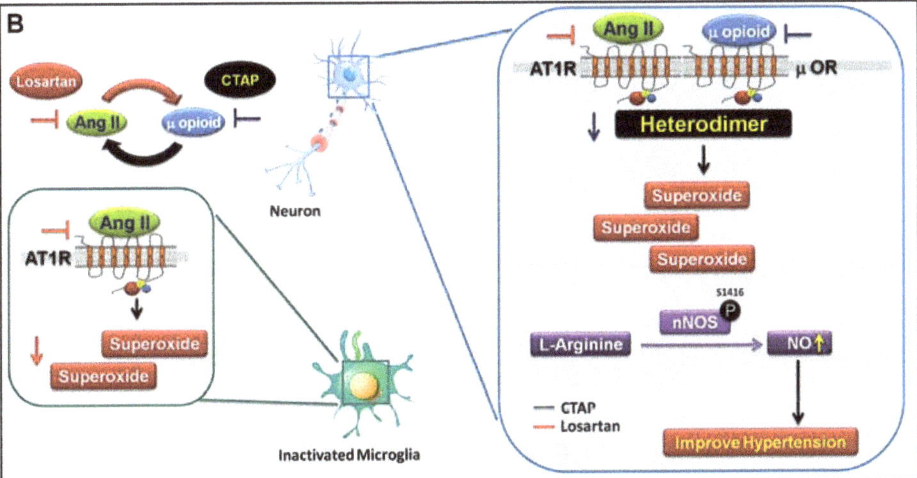

Figure 5. Proposed pathogenic mechanism for neurogenic hypertension. (**A**) The interaction of Ang II and μOR enhances the binding of Ang II to the AT1R receptor, activates the microglia and promotes the formation of AT1R-μOR heterodimers in the NTS, leading to superoxide production. This inactivates nNOS-derived NO, causing systemic elevations in blood pressure. (**B**) AT1R inhibitors (such as losartan) decrease superoxide production, abolish the activation of microglia and AT1R, and decrease TLR4 expression, which ultimately leads to improved hypertension (red line). Interestingly, μOR inhibitors (such as CTAP) decrease BP and abolish μOR-induced formation of AT1R and μOR heterodimers. CTAP also significantly inactivates the activation of microglia and AT1R, and reduces TLR4 expression, which can lead to an increase nNOS-derived NO levels, thereby improving hypertension (black line). In summary, this study shows how the interaction between Ang II and μOR enhances the binding of Ang II to AT1R, thereby causing microglia activation and inducing superoxide production, which, in turn, leads to neurotoxicity. Furthermore, the interaction of Ang II and μOR also enhances the formation of the AT1R and μOR heterodimers and inactivates nNOS-derived NO, leading to the development of progressive hypertension.

5. Conclusions

In conclusion, an abnormal increase in endogenous μ-opioid in the NTS not only induces a neurotoxicity cascade with enhanced Ang II binding to the AT1R receptor, and activates the microglia (which induces superoxide production), but also induces the formation of μOR/AT1R heterodimers and the TLR4-dependent inflammatory response, which attenuate the NO-dependent depressor effect. These findings deepen our understanding of μOR as a novel candidate for intervention in hypertensive conditions.

Supplementary Materials: The following are available online at https://www.mdpi.com/article/10.3390/antiox10111784/s1, Figure S1: Downregulation of TLR4-induced neurotoxicity is associated with nNOSS1416 phosphorylation in the NTS of spontaneously hypertensive rats.

Author Contributions: The study was conceived and designed by G.-C.S., P.-W.C. conducted most of the experiments with assistance from J.T., Y.-H.H., C.-Y.H. and C.-J.T. The paper was written by P.-W.C. All authors have read and agreed to the published version of the manuscript.

Funding: This work was supported by the Kaohsiung Medical University Chung-Ho Memorial Hospital (KMUH 109-9R85), the Ministry of Science and Technology (MOST108-2314-B-037-044-MY2, MOST110-2314-B-037-127-MY3) and the Kaohsiung Veterans General Hospital KSVGH110-141) (to P.-W.C).

Institutional Review Board Statement: Animal studies were conducted in compliance with the Animal Research: Report-ing of In Vivo Experiments (ARRIVE) guidelines as described previously [30,31]. All protocols were approved by Animal Research Committee and the institutional review board at VGHKS (VGHKS-2021-2023-A009; VGHKS-2020-2022-A046) and an affida-vit of approval was obtained in with the animal care protocol of Kaohsiung Medical University (109087).

Informed Consent Statement: Not applicable.

Data Availability Statement: All data generated or analysed during this study are included in this published article.

Acknowledgments: The authors gratefully acknowledge technical assistance and the invaluable input and support from Yu-Ju Hsiao and Jui-Hsiang Tseng.

Conflicts of Interest: The authors declare no conflict of interest. The funders had no role in the design of the study; in the collection, analyses, or interpretation of data; in the writing of the manuscript, or in the decision to publish the results.

Abbreviations

AT1R, angiotensin II type 1 receptor; Ang II, angiotensin II; AR, adrenergic receptors; α2A-AR, α2A-adrenergic receptor; BP, blood pressure; CNS, central nervous system; DAMGO, [D-Ala^2, MePhe^4, Gly^5-ol]-enkephalin; GPCRs, G protein-coupled receptors; IBA1, ionized calcium binding adaptor molecule 1; PLA, Proximity Ligation Assay; NTS, nucleus tractus solitarii; NO, nitric oxide; μORs, μ-opioid receptors; RAAS, renin–angiotensin–aldosterone system; SHRs, spontaneous hypertensive rats; TLR4, toll-like receptor 4; WKY, Wistar-Kyoto rats.

References

1. Gheorghe, A.; Griffiths, U.; Murphy, A.; Legido-Quigley, H.; Lamptey, P.; Perel, P. The economic burden of cardiovascular disease and hypertension in low- and middle-income countries: A systematic review. *BMC Public Health* **2018**, *18*, 975. [CrossRef]
2. Go, A.S.; Mozaffarian, D.; Roger, V.L.; Benjamin, E.J.; Berry, J.D.; Blaha, M.J.; Dai, S.; Ford, E.S.; Fox, C.S.; Franco, S.; et al. Heart disease and stroke statistics—2014 update: A report from the American Heart Association. *Circulation* **2014**, *129*, e28–e292. [CrossRef] [PubMed]
3. Campbell, T.S.; Ditto, B.; Seguin, J.R.; Sinray, S.; Tremblay, R.E. Adolescent pain sensitivity is associated with cardiac autonomic function and blood pressure over 8 years. *Hypertension* **2003**, *41*, 1228–1233. [CrossRef]
4. Bruehl, S.; Burns, J.W.; Chung, O.Y.; Magid, E.; Chont, M.; Gilliam, W.; Matsuura, J.; Somar, K.; Goodlad, J.K.; Stone, K.; et al. Hypoalgesia associated with elevated resting blood pressure: Evidence for endogenous opioid involvement. *J. Behav. Med.* **2010**, *33*, 168–176. [CrossRef]
5. Nicolo, C.; Perier, C.; Prague, M.; Bellera, C.; MacGrogan, G.; Saut, O.; Benzekry, S. Machine Learning and Mechanistic Modeling for Prediction of Metastatic Relapse in Early-Stage Breast Cancer. *JCO Clin. Cancer Inform.* **2020**, *4*, 259–274. [CrossRef] [PubMed]
6. Sun, G.C.; Wong, T.Y.; Chen, H.H.; Ho, C.Y.; Yeh, T.C.; Ho, W.Y.; Tseng, C.J.; Cheng, P.W. Angiotensin II inhibits DDAH1-nNOS signaling via AT1R and muOR dimerization to modulate blood pressure control in the central nervous system. *Clin. Sci.* **2019**, *133*, 2401–2413. [CrossRef]

7. Paul, M.; Poyan Mehr, A.; Kreutz, R. Physiology of local renin-angiotensin systems. *Physiol. Rev.* **2006**, *86*, 747–803. [CrossRef]
8. Munoz-Durango, N.; Fuentes, C.A.; Castillo, A.E.; Gonzalez-Gomez, L.M.; Vecchiola, A.; Fardella, C.E.; Kalergis, A.M. Role of the Renin-Angiotensin-Aldosterone System beyond Blood Pressure Regulation: Molecular and Cellular Mechanisms Involved in End-Organ Damage during Arterial Hypertension. *Int. J. Mol. Sci.* **2016**, *17*, 797. [CrossRef] [PubMed]
9. Tan, X.; Li, J.K.; Sun, J.C.; Jiao, P.L.; Wang, Y.K.; Wu, Z.T.; Liu, B.; Wang, W.Z. The asymmetric dimethylarginine-mediated inhibition of nitric oxide in the rostral ventrolateral medulla contributes to regulation of blood pressure in hypertensive rats. *Nitric Oxide* **2017**, *67*, 58–67. [CrossRef]
10. Balogh, M.; Aguilar, C.; Nguyen, N.T.; Shepherd, A.J. Angiotensin receptors and neuropathic pain. *Pain Rep.* **2021**, *6*, e869. [CrossRef]
11. Lin, Y.T.; Wu, Y.C.; Sun, G.C.; Ho, C.Y.; Wong, T.Y.; Lin, C.H.; Chen, H.H.; Yeh, T.C.; Li, C.J.; Tseng, C.J.; et al. Effect of Resveratrol on Reactive Oxygen Species-Induced Cognitive Impairment in Rats with Angiotensin II-Induced Early Alzheimer's Disease (dagger). *J. Clin. Med.* **2018**, *7*, 329. [CrossRef]
12. Bali, A.; Randhawa, P.K.; Jaggi, A.S. Interplay between RAS and opioids: Opening the Pandora of complexities. *Neuropeptides* **2014**, *48*, 249–256. [CrossRef] [PubMed]
13. Shepherd, A.J.; Copits, B.A.; Mickle, A.D.; Karlsson, P.; Kadunganattil, S.; Haroutounian, S.; Tadinada, S.M.; De Kloet, A.D.; Valtcheva, M.V.; McIlvried, L.A.; et al. Angiotensin II Triggers Peripheral Macrophage-to-Sensory Neuron Redox Crosstalk to Elicit Pain. *J. Neurosci.* **2018**, *38*, 7032–7057. [CrossRef] [PubMed]
14. Billingsley, H.E.; Carbone, S. The antioxidant potential of the Mediterranean diet in patients at high cardiovascular risk: An in-depth review of the PREDIMED. *Nutr. Diabetes* **2018**, *8*, 13. [CrossRef]
15. Ho, C.Y.; Lin, Y.T.; Chen, H.H.; Ho, W.Y.; Sun, G.C.; Hsiao, M.; Yeh, T.C.; Hsiao, M.; Lu, P.J.; Tseng, C.J. CX3CR1-microglia mediates neuroinflammation and blood pressure regulation in the nucleus tractus solitarii of fructose-induced hypertensive rats. *J. Neuroinflam.* **2020**, *17*, 185. [CrossRef] [PubMed]
16. Sun, G.C.; Ho, W.Y.; Chen, B.R.; Cheng, P.W.; Cheng, W.H.; Hsu, M.C.; Lu, P.J.; Tseng, C.J. GPCR dimerization in brainstem nuclei contributes to the development of hypertension. *Br. J. Pharmacol.* **2015**, *172*, 2507–2518. [CrossRef]
17. Shen, M.J.; Zipes, D.P. Role of the autonomic nervous system in modulating cardiac arrhythmias. *Circ. Res.* **2014**, *114*, 1004–1021. [CrossRef]
18. Jaen, R.I.; Val-Blasco, A.; Prieto, P.; Gil-Fernandez, M.; Smani, T.; Lopez-Sendon, J.L.; Delgado, C.; Boscá, L.; Fernández-Velasco, M. Innate Immune Receptors, Key Actors in Cardiovascular Diseases. *JACC Basic Transl. Sci.* **2020**, *5*, 735–749. [CrossRef]
19. Singh, M.V.; Cicha, M.Z.; Nunez, S.; Meyerholz, D.K.; Chapleau, M.W.; Abboud, F.M. Angiotensin II-induced hypertension and cardiac hypertrophy are differentially mediated by TLR3- and TLR4-dependent pathways. *Am. J. Physiol. Heart Circ. Physiol.* **2019**, *316*, H1027–H1038. [CrossRef]
20. Okechukwu, C.C.; Pirro, N.T.; Chappell, M.C. Evidence that angiotensin II does not directly stimulate the MD2-TLR4 innate inflammatory pathway. *Peptides* **2021**, *136*, 170436. [CrossRef]
21. Nair, A.R.; Ebenezer, P.J.; Saini, Y.; Francis, J. Angiotensin II-induced hypertensive renal inflammation is mediated through HMGB1-TLR4 signaling in rat tubulo-epithelial cells. *Exp. Cell. Res.* **2015**, *335*, 238–247. [CrossRef] [PubMed]
22. Te Riet, L.; van Esch, J.H.; Roks, A.J.; van den Meiracker, A.H.; Danser, A.H. Hypertension: Renin-angiotensin-aldosterone system alterations. *Circ. Res.* **2015**, *116*, 960–975. [CrossRef] [PubMed]
23. Ferrario, C.M.; Mullick, A.E. Renin angiotensin aldosterone inhibition in the treatment of cardiovascular disease. *Pharmacol. Res.* **2017**, *125 Pt A*, 57–71. [CrossRef]
24. Horvath, R.J.; DeLeo, J.A. Morphine enhances microglial migration through modulation of P2X4 receptor signaling. *J. Neurosci.* **2009**, *29*, 998–1005. [CrossRef]
25. Tawfik, V.L.; LaCroix-Fralish, M.L.; Nutile-McMenemy, N.; DeLeo, J.A. Transcriptional and translational regulation of glial activation by morphine in a rodent model of neuropathic pain. *J. Pharmacol. Exp. Ther.* **2005**, *313*, 1239–1247. [CrossRef]
26. Horvath, R.J.; Romero-Sandoval, A.E.; De Leo, J.A. Inhibition of microglial P2X4 receptors attenuates morphine tolerance, Iba1, GFAP and mu opioid receptor protein expression while enhancing perivascular microglial ED2. *Pain* **2010**, *150*, 401–413. [CrossRef] [PubMed]
27. McGrath, J.C.; Lilley, E. Implementing guidelines on reporting research using animals (ARRIVE etc.): New requirements for publication in BJP. *Br. J. Pharmacol.* **2015**, *172*, 3189–3193. [CrossRef]
28. Kilkenny, C.; Browne, W.; Cuthill, I.C.; Emerson, M.; Altman, D.G.; Group NCRRGW. Animal research: Reporting in vivo experiments: The ARRIVE guidelines. *Br. J. Pharmacol.* **2010**, *160*, 1577–1579. [CrossRef] [PubMed]
29. Tseng, C.J.; Liu, H.Y.; Lin, H.C.; Ger, L.P.; Tung, C.S.; Yen, M.H. Cardiovascular effects of nitric oxide in the brain stem nuclei of rats. *Hypertension* **1996**, *27*, 36–42. [CrossRef]
30. Cheng, W.H.; Lu, P.J.; Ho, W.Y.; Tung, C.S.; Cheng, P.W.; Hsiao, M.; Tseng, C.J. Angiotensin II inhibits neuronal nitric oxide synthase activation through the ERK1/2-RSK signaling pathway to modulate central control of blood pressure. *Circ. Res.* **2010**, *106*, 788–795. [CrossRef]
31. Dange, R.B.; Agarwal, D.; Masson, G.S.; Vila, J.; Wilson, B.; Nair, A.; Francis, J. Central blockade of TLR4 improves cardiac function and attenuates myocardial inflammation in angiotensin II-induced hypertension. *Cardiovasc. Res.* **2014**, *103*, 17–27. [CrossRef]

32. Han, J.; Zou, C.; Mei, L.; Zhang, Y.; Qian, Y.; You, S.; Pan, Y.; Xu, Z.; Bai, B.; Huang, W.; et al. MD2 mediates angiotensin II-induced cardiac inflammation and remodeling via directly binding to Ang II and activating TLR4/NF-kappaB signaling pathway. *Basic Res. Cardiol.* **2017**, *112*, 9. [CrossRef] [PubMed]
33. Ong, K.L.; Tso, A.W.; Lam, K.S.; Cheung, B.M. Gender difference in blood pressure control and cardiovascular risk factors in Americans with diagnosed hypertension. *Hypertension* **2008**, *51*, 1142–1148. [CrossRef]
34. Dampney, R.A. Central neural control of the cardiovascular system: Current perspectives. *Adv. Physiol. Educ.* **2016**, *40*, 283–296. [CrossRef]
35. Shen, X.Z.; Li, Y.; Li, L.; Shah, K.H.; Bernstein, K.E.; Lyden, P.; Shi, P. Microglia participate in neurogenic regulation of hypertension. *Hypertension* **2015**, *66*, 309–316. [CrossRef]
36. Pushpakumar, S.; Ren, L.; Kundu, S.; Gamon, A.; Tyagi, S.C.; Sen, U. Toll-like Receptor 4 Deficiency Reduces Oxidative Stress and Macrophage Mediated Inflammation in Hypertensive Kidney. *Sci. Rep.* **2017**, *7*, 6349. [CrossRef] [PubMed]
37. Xue, B.; Zhang, Y.; Johnson, A.K. Interactions of the Brain Renin-Angiotensin-System (RAS) and Inflammation in the Sensitization of Hypertension. *Front. Neurosci.* **2020**, *14*, 650. [CrossRef] [PubMed]
38. Turchan-Cholewo, J.; Dimayuga, F.O.; Ding, Q.; Keller, J.N.; Hauser, K.F.; Knapp, P.E.; Bruce-Keller, A.J. Cell-specific actions of HIV-Tat and morphine on opioid receptor expression in glia. *J. Neurosci. Res.* **2008**, *86*, 2100–2110. [CrossRef] [PubMed]
39. Maderspach, K.; Solomonia, R. Glial and neuronal opioid receptors: Apparent positive cooperativity observed in intact cultured cells. *Brain Res.* **1988**, *441*, 41–47. [CrossRef]
40. Wang, J.; Barke, R.A.; Charboneau, R.; Schwendener, R.; Roy, S. Morphine induces defects in early response of alveolar macrophages to Streptococcus pneumoniae by modulating TLR9-NF-kappa B signaling. *J. Immunol.* **2008**, *180*, 3594–3600. [CrossRef]
41. He, L.; Li, H.; Chen, L.; Miao, J.; Jiang, Y.; Zhang, Y.; Xiao, Z.; Hanley, G.; Li, Y.; Zhang, X.; et al. Toll-like receptor 9 is required for opioid-induced microglia apoptosis. *PLoS ONE* **2011**, *6*, e18190. [CrossRef] [PubMed]
42. Jordan, B.A.; Gomes, I.; Rios, C.; Filipovska, J.; Devi, L.A. Functional interactions between mu opioid and alpha 2A-adrenergic receptors. *Mol. Pharmacol.* **2003**, *64*, 1317–1324. [CrossRef]
43. Vilardaga, J.P.; Nikolaev, V.O.; Lorenz, K.; Ferrandon, S.; Zhuang, Z.; Lohse, M.J. Conformational cross-talk between alpha2A-adrenergic and mu-opioid receptors controls cell signaling. *Nat. Chem. Biol.* **2008**, *4*, 126–131. [CrossRef] [PubMed]
44. Koehl, A.; Hu, H.; Maeda, S.; Zhang, Y.; Qu, Q.; Paggi, J.M.; Latorraca, N.R.; Hilger, D.; Dawson, R.; Matile, H.; et al. Structure of the micro-opioid receptor-Gi protein complex. *Nature* **2018**, *558*, 547–552. [CrossRef] [PubMed]

Article

Oleuropein-Rich Leaf Extract as a Broad Inhibitor of Tumour and Macrophage iNOS in an Apc Mutant Rat Model

Jessica Ruzzolini [1,†], Sofia Chioccioli [2,†], Noemi Monaco [1], Silvia Peppicelli [1], Elena Andreucci [1], Silvia Urciuoli [3], Annalisa Romani [3], Cristina Luceri [2], Katia Tortora [2], Lido Calorini [1,4], Giovanna Caderni [2], Chiara Nediani [1,*] and Francesca Bianchini [1,*]

1 Department of Experimental and Clinical Biomedical Sciences "Mario Serio", University of Florence, 50134 Florence, Italy; jessica.ruzzolini@unifi.it (J.R.); noemi.monaco@unifi.it (N.M.); silvia.peppicelli@unifi.it (S.P.); e.andreucci@unifi.it (E.A.); lido.calorini@unifi.it (L.C.)
2 NEUROFARBA Department of Neurosciences, Psychology, Drug Research and Child Health, Pharmacology and Toxicology Section, University of Florence, 50139 Florence, Italy; sofia.chiccoli@unifi.it (S.C.); cristina.luceri@unifi.it (C.L.); katia.tortora@unifi.it (K.T.); giovanna.caderni@unifi.it (G.C.)
3 PHYTOLAB (Pharmaceutical, Cosmetic, Food Supplement Technology and Analysis)-DiSIA, Department of Statistics, Informatics, Applications "Giuseppe Parenti", Scientific and Technological Pole, University of Florence, 50019 Sesto, Italy; silvia.urciuoli@unifi.it (S.U.); annalisa.romani@unifi.it (A.R.)
4 Center of Excellence for Research, Transfer and High Education DenoTHE, School of Medicine, University of Florence, 50134 Florence, Italy
* Correspondence: chiara.nediani@unifi.it (C.N.); francesca.bianchini@unifi.it (F.B.)
† These authors contributed equally to this work.

Citation: Ruzzolini, J.; Chioccioli, S.; Monaco, N.; Peppicelli, S.; Andreucci, E.; Urciuoli, S.; Romani, A.; Luceri, C.; Tortora, K.; Calorini, L.; et al. Oleuropein-Rich Leaf Extract as a Broad Inhibitor of Tumour and Macrophage iNOS in an Apc Mutant Rat Model. *Antioxidants* **2021**, *10*, 1577. https://doi.org/10.3390/antiox10101577

Academic Editors: Ferdinando Nicoletti and José M. Matés

Received: 28 July 2021
Accepted: 1 October 2021
Published: 6 October 2021

Abstract: Oleuropein, the major compound found in olive leaves, has been reported to exert numerous pharmacological properties, including anti-inflammatory, anti-diabetic and anti-cancer effects. The purpose of this study was to evaluate, for the first time, the effect of oleuropein-rich leaf extracts (ORLE) in already-developed colon tumours arising in *Apc* (adenomatous polyposis coli) mutated PIRC rats (F344/NTac-Ap^{cam1137}). Here, we were able to investigate in parallel the anti-cancer effect of ORLE, both in vivo and in vitro, and its anti-inflammatory effect on macrophages, representing a critical and abundant population in most solid tumour microenvironment. We found that in vivo ORLE treatment promoted apoptosis and attenuated iNOS activity both in colon tumours as in peritoneal macrophages of PIRC rats. We this confirmed in vitro using primary RAW264.7 cells: ORLE reduced iNOS activity in parallel with COX-2 and pro-inflammatory cytokines, such as IL-1β, IL-6 and TGF-β. These findings suggest that ORLE possess a strong anti-inflammatory activity, which could be crucial for dampening the pro-tumourigenic activity elicited by a chronic inflammatory state generated by either tumour cells or tumour-associated macrophages.

Keywords: oleuropein; colon tumours; PIRC rats; activated macrophages; chronic inflammation; inducible nitric oxide synthetase (iNOS); cyclooxygenase-2 (COX-2); nitric oxide (NO)

1. Introduction

Cancer is currently the second leading cause of death worldwide and highly efficient anti-cancer drugs are currently used to counteract the uncontrolled proliferative activity of neoplastic cells. The effectiveness of most chemotherapeutic agents is accompanied by systemic toxicity, since anti-cancer agents discriminate poorly between normal and cancerous cells. In addition, the efficacy of these treatments is still limited, due to the adverse side effects and the frequent development of resistance.

Among alternative therapies for cancer treatment, there is a growing interest in the anti-cancer action of natural substances, that are non-toxic, affordable, readily accessible, and, some of which, present in large amounts in byproducts from agro-food chains [1].

Natural products are well-established to have pharmacological or biological activities that can be of therapeutic benefits for cancer therapy. Accumulating evidence has revealed that natural products can modulate a series of key signalling pathways displaying therapeutic effects, such as pro-apoptotic, anti-proliferative, anti-angiogenic effects, in different types of human cancers [2]. More recently the use of natural compounds as differentiation inducing agents leading maturation of low differentiated cancer cells rendering them less aggressive and more sensitive to conventional treatments has emerged [3].

The beneficial effects of olive leaves or different preparations (e.g., infusions, extracts) have a several-century-long tradition and have been used for the treatment or to alleviate the symptoms of many diseases (such as diabetes mellitus, arterial hypertension, and bronchial asthma), and are currently contemplated in the Ph. Eur. 5 pharmacopoeia [4].

In particular, *Olea europaea* L. leaves are rich in oleuropein (Ole), a secoiridoid compound that exhibits a wide range of anti-oxidant, anti-inflammatory, anti-diabetic, neuro- and cardio-protective, anti-microbic and immunomodulatory activities [5–8]. Recently, preclinical studies have provided convincing evidence that Ole has, also, peculiar properties as autophagic, and pro-apoptotic inducer and amyloid fibril growth inhibitor [9–13]. In our experience, we found that Ole might exert an anti-cancer activity, alone or in combination with conventional treatments, through different mechanism in different cancer cell lines [13,14].

Colorectal cancer (CRC), one of most common type of cancer in the Western world of both men and women [15], is one of the solid tumours that may take advantage of a nutritional intervention. Indeed, in addition to a complex genetic susceptibility, the key environmental factors for colon cancer include the diet. Preclinical evidences have demonstrated that olive oil-derived substances have a beneficial effect against colorectal cancer through the modulation of gut microbiota composition or activity [16]. In particular, Ole was able to reduce crypt dysplasia in a rat short-term colon carcinogenesis experiment [17] and to show protective effects in colitis-associated CRC in mice, suggesting, together with the results obtained in cancer cells in vitro, that this molecule may decrease colon tumorigenesis. Whether these protective effects can be extended also to already-developed colon tumours is not known.

Based on these considerations, the aim of the present study is to explore, for the first time, whether oleuropein-rich leaf extracts (ORLE), exerts anti-tumoural and anti-inflammatory activity in colon tumours and peritoneal activated macrophages of PIRC rats carrying a heterozygous germline mutation in the *Apc* gene. The *APC* mutation is the first event triggering colon carcinogenesis both in the majority of sporadic cases and in familial adenomatous polyposis (FAP) syndrome, a hereditary form of colon cancer [18]. Accordingly, PIRC rat spontaneously develops multiple tumours in the colon and small intestine, thus standing as a robust model to study the protective effect of ORLE, derived from olive leaves, on colon cancer progression.

We found that an ORLE enriched diet reduces cell proliferation and increases cell apoptosis in tumours and reduces nitric oxide synthase (iNOS) in colon tumour lesions and peritoneal macrophages of PIRC rats. We confirm that ORLE inhibits the pro-inflammatory features of activated murine macrophages through the reduction of iNOS, cyclooxygenase-2 (COX-2), interleukin (IL)-1β, IL-6 and TGF-β expression, both in acute as in a chronic exposure. We suggest that an ORLE-enriched diet contributes to switching-off the pro-inflammatory signal released either by tumour cells or by inflammatory cells of tumour microenvironment critical for colon cancer progression.

2. Materials and Methods

2.1. Olive Leaf Extract's Preparation and Toxicity

Organic olive (Leccino cultivar) leaves were harvested in Tuscany (Vinci, Florence, Italy) and immediately processed to obtain a powder extract rich in active compounds, as previously described in Romani et al. [19]. The characterization of the minor polar compounds and the phenolic profile of olive leaves extract was carried out by HPLC-DAD-MS (high-performance liquid chromatography coupled with diode-array detection and mass spectrometry). The total polyphenol content of dry extract is about 400 mg/g, of which oleuropein was about 379 mg/g. For in vitro experiments ORLE was reconstituted to a final concentration of 14 mM in PBS.

2.2. Cell Lines and Culture Conditions

HCT-116, colorectal carcinoma cells were purchased from European Collection of Authenticated Cell Cultures (ECACC, Porton Down, SP4 0JG Salisbury, UK). The murine macrophage RAW 264.7 cell line was purchased from the American Type Culture Collection (ATCC, Manassas, VA, USA. Cells were cultured in Dulbecco's Modified Eagle Medium high glucose (DMEM 4500, EuroClone, Milan, Italy) supplemented with 10% fetal bovine serum (FBS, EuroClone) and maintained at 37 °C in humidified atmosphere containing 90% air and 10% CO_2 and they harvested from subconfluent cultures by incubation with a trypsin-EDTA solution (EuroClone) and propagated every three days. Viability of the cells was determined by trypan blue exclusion test. Cultures were periodically monitored for mycoplasma contamination using Chen's fluorochrome test. HCT116 or RAW cells were exposed for 24 h or 72 h to 50μM ORLE in complete medium according to different experimental procedures. This concentration was tested on colon and macrophage cell lines in preliminary experiments and chosen because resulted non-toxic (see Figure 3a).

2.3. MTT Assay

HCT116 cell viability was assessed using MTT (3-(4,5-dimethylthiazol-2-yl)-2,5-diphenyltetrazolium bromide) tetrazolium reduction assay (Sigma Aldrich, Milan, Italy) as described in [13]. Cells (2.5×10^3) were plated into 96-multiwell plates in complete medium without red phenol. The ORLE treatment was added to the medium culture at different dose for 72 h. Then the MTT reagent was added to the medium, and plates were incubated at 37 °C. After 2 h, MTT was removed and the blue MTT–formazan product was solubilized with dimethyl sulfoxide (DMSO, Sigma Aldrich). The absorbance of the formazan solution was read at 595 nm using the microplate reader (Bio-Rad, Milan, Italy).

2.4. Animal Maintenance and Ex-Vivo Analysis

PIRC rats (F344/NTac-Apcam1137) and wild type (wt) (Fisher F344) rats were originally obtained by the National Institutes of Health (NIH), Rat Resource and Research Center (RRRC) (University of Missouri, Columbia, MO, USA) and bred in Ce.S.A.L. (Housing Center for Experimental Animals of the University of Florence, Florence, Italy) in accordance with the Commission for Animal Experimentation of the Italian Ministry of Health (EU Directive 2010/63/EU for animal experiments), as described [20]; rats were maintained in polyethylene cages and fed with a standard AIN-76 diet (Laboratorio Dottori Piccioni, s.r.l., Gessate MI, Italy). Eight PIRC rats aged 12 months were randomly assigned to the AIN-76 diet (Control group: two males, two females) or to the same diet containing ORLE (2.7 g/kg of diet) (ORLE group: one male, three females) as reported [13]. The number of rats in each treatment was based on the expected number of tumours/animal in which to carry out our analyses (considering tumour apoptosis as primary endpoint). Accordingly, to observe a significant increase in apoptosis (expected increase of at least 60%, as we previously documented for other cancer-preventive plant compounds [21]), we considered that 6–7 tumours/group, as we actually found, would be sufficient, as calculated with an "a priori "analysis [22]. Considering that rats eat about 11 g of diet/day, and a mean body-weight of 300 g, we administered a dose of ORLE of about 100 mg/kg b.w [13]. Rats were euthanized by CO_2 asphyxia after one week of treatment, in line with the experimental protocol approved by the Commission for Animal Experimentation of the Italian Ministry of Health. The entire colon and small intestine were flushed with saline solution and opened to check for the presence of tumours which were collected and processed for histological procedure and RNA-extraction as reported [20].

Expression of CD68, as a measure of macrophage infiltration, and expression of proliferating cell nuclear antigen (PCNA), as a measure of proliferative activity, were determined in the tumour lesions of PIRC rats fed with different diets. Longitudinal colon sections (4 µm) were mounted on electrostatic-treated slides (Superfrost® Plus, Medite, Wollenweberstrasse 12 31303 Burgdorf Germany) and processed as described [23] using as primary antibodies: mouse monoclonal antibodies against PCNA (PC-10, Santa Cruz, CA, USA) and rat CD68 (AbD Serotec, Oxford, UK). Both antibodies were diluted in PBS 1:200. CD68 reactivity was quantified as number of labeled cells/areas scored evaluated with the ACT-2U software program (Nikon, Instruments Europe, Badhoevedorp, The Netherlands) connected via a camera to the microscope (Optiphot-2, Nikon, Tokyo, Japan). Evaluation was performed at 400× magnification.

2.5. Peritoneal Macrophages Isolation

Macrophage cultures were established from peritoneal exudates collected from rats fed with different diets. Briefly, 20 mL of ice-cold PBS were injected in the peritoneal cavity of the rats and collected immediately. Peritoneal exudates were washed by centrifugation and macrophage monolayers allowed to adhere to plastic dishes in DMEM4500 medium (without phenol red) containing 250 µg/mL bovine serum albumin (BSA), at the density of 125×10^3 cells/cm^2 in a 24-well dishes. After adhesion, macrophages cultures were washed with PBS and then incubated in DMEM 4500 (w BSA, *w/o* PhR), at 37 °C in a 10% CO_2 humidified atmosphere and exposed to rIFNγ (50 U/mL) (Immunotools, Friesoythe, Germany) and LPS (10 ng/mL) (Sigma) [24] for 48 h.

2.6. Mucosal Samples Collection and RT-PCR Analysis

Mucosal samples were collected in RNAlater and stored at −80 °C until extraction of nucleic acids. DNA quality was assessed by gel electrophoresis and spectrophotometry, measuring OD 260/280.

2.7. Determination of Apoptosis

Apoptosis was evaluated in histological sections (4 µm thick) of tumours stained with hematoxylin eosin as recommended by Femia et al. [25], determining cells with the following characteristics of apoptosis: cell shrinkage, loss of normal contact with the adjacent cells of the crypt, chromatin condensation, or formation of round or oval nuclear fragments. Apoptosis was quantified as the number of apoptotic cells/area measured using the ACT-2U software program (Nikon, Instruments Europe) connected via a camera to a microscope (Nikon Optiphot-2). The evaluation was performed at 1000× magnification.

2.8. Nitric Oxide Assay

NO concentration was measured in the culture medium of rat peritoneal macrophages or RAW cells using the Griess reaction. Namely, NO production was measured in rat peritoneal macrophages cultures after 48 h of an in vitro treatment with IFNγ and LPS, while in RAW264.7 cells (1.6×10^5 cells/well) was measured after 24 h treatment with 1 µg/mL LPS [26]. In particular, RAW264.7 cells were exposed for 24 h to a co-treatment with LPS and 50 µM ORLE, or for 72 h to 50 µM ORLE pre-treatment and a sequential 24 h treatment with LPS. Briefly, 100 µL of cell culture medium from RAW264.7 cells or 250 µL from peritoneal macrophages cultures were mixed with an equal volume of Griess reagent (1% sulfanilamide, 0.1% N-1-naphthalenediamine dihydrochloride, and 2.5% H_3PO_4) and transferred to 96-well plates. Plates were incubated at room temperature for 10 min. Then the absorbance was measured at 540 nm in a microplate reader (BioTek, Winooski, VT, USA). The amount of nitrite in the media was calculated from sodium nitrite ($NaNO_2$) standard curve. Results were normalized to protein concentration. For RAW264.7 cells NO production was expressed referred to LPS as 100%.

2.9. Cytofluorimetric Assay of iNOS in HCT116 Cells

The expression of intracellular iNOS in HCT116 cells was assessed by flow cytometry using iNOS monoclonal antibody recommended for the detection of NOS2 of mouse, rat and human origin. HCT116 cells were exposed to a standard medium of a medium containing 50 μM ORLE for 72 h. At the end of the incubation, cells were harvested, fixed in ethanol 70%, permeabilized (Triton-X100, 0.01%), and then stained using anti-human iNOS mouse IgG1 monoclonal antibody (sc-7271, Santa Cruz, 1 μg/10^5 cells). At the end of the incubation (45′, 4 °C), cells were exposed to PE-labeled goat anti-mouse IgG as secondary antibody (#22549814, Immunotools). Stained cells were analyzed on a fluorescence-activated cell sorting (FACS) flow cytometer (FACScan, Becton Dickinson, BD Biosciences Torreyana Rd - 92121 San Diego, CA, USA).

2.10. Western Blotting Analysis

RAW cells were exposed to 1 μg/mL LPS alone, or 50 μM ORLE, or LPS/ORLE in complete medium as previously described. After incubation, cells and supernatants were lysed together and separated using electrophoresis [13]. Cells were washed with ice cold PBS containing 1 μM Na_4VO_3, and lysed in 100 μL of cell RIPA lysis buffer (Merck Millipore). PMSF (1 mM final concentration), sodium orthovanadate (100 μM final concentration) and protease inhibitor cocktail set III (from Sigma-Aldrich), have been added to RIPA buffer. Aliquots of supernatants containing equal amounts of protein (40 μg) in Laemmli buffer were separated on Bolt® Bis-Tris Plus gels 4–12 and fractionated proteins were transferred to a PVDF membrane using iBlot 2 system (Life Technologies, Carlsbad, CA, USA). Membranes were blocked for 1 h (RT) using Odyssey blocking buffer. Subsequently, the membranes were probed (4 °C O/N) with primary antibodies. The primary antibodies were diluted 1:1000 in a solution of 1:1 Odyssey blocking buffer/T-PBS buffer. The following antibodies were used: COX-2 rabbit anti h/m/r mAb (#4842S, Cell Signaling, Danvers, MA, USA, 74 kDa), iNOS rabbit anti mouse mAb (#13120S, Cell Signaling, 130 kDa), and IL-1βRI rabbit anti mouse mAb (sc-689, Santa Cruz, 80 kDa). The membranes were washed in T-PBS buffer and incubated for 1 h (RT) with goat anti-rabbit IgG Alexa Flour 750 antibody or with goat anti-mouse IgG Alexa Fluor 680 antibody (Invitrogen, Waltham, MA, USA, 1:10000). Bands were then visualized using Odyssey Infrared Imaging System (LI-COR® Bioscience, Lincoln, NE, USA). The vinculin rabbit anti h/m/r/mk mAb (#13901, Cell Signaling, 124 kDa) (1:1000) and anti h/m/r/mk mAb α-tubulin (#3873, Cell Signaling, 52 kDa) were used to assess equal amount of protein loaded in each lane.

2.11. RNA Extraction and Quantitative PCR

Total RNA from rat tumours or RAW264.7 cells was extracted using the Trizol reagent. The concentration and purity of RNA samples were determined using the 260 to 280 ratio (A260/A280) and the 260 to 230 ratio (A260/A230) readings of an ultraviolet spectrophotometer. The first-strand cDNA was synthesized, from 1 μg of total RNA, using the iScript cDNA Synthesis Kit (#1708891, BioRad, Hercules, CA, USA) following the provided instructions. Quantitative real-time polymerase chain reaction (RT-PCR) was carried out using SYBR Green PCR Master Mix (#4309155, Applied Biosystems, Waltham, MA, USA). The results were analyzed on the (BioRad CFX96 qPCR Instrument). Values were normalized to 18S, and all the results were obtained from at least three experiments independently. The sequences of primers used in this study are listed in Table 1.

Table 1. Primer sequences for qRT-PCR.

Gene Name (Accession nr.)	Forward Sequence (5′-3′)	Reverse Sequence (5′-3′)
IL-1β (NM_008361)	CCT GCA GCT GGA GAG TGT GGA	CCC ATC AGA GGC AAG GAG GAA
IL-6 (NM_031168)	CTT CCA TCC AGT TGC CTT CT	TGC ATC ATC GTT GTT CAT AC
TGF-β (NM_011577.2)	GGC TTC TAG TGC TGA CG	GGG TGC TGT TGT ACA AAG
18S (NR_003278)	CGC CGC TAG AGG TGA AAT TCT	CGA ACC TCC GAC TTT CGT TCT

2.12. Statistical Analysis

All data were obtained based on at least on three independent experiments and expressed as mean ± SEM. Statistical analysis between two groups was performed using unpaired Student's *t*-tests. When comparing three or more groups for one or two independent variables one-way or two-way analysis of variance (ANOVA) were performed followed by Tukey's post hoc test. P-values were calculated using GraphPad Prism version 6.04 for Windows (GraphPad Software, La Jolla, CA, USA) and are provided in the figure legends. Band intensities in Western blot analysis were quantified using the computer-based ImageJ software.

3. Results

3.1. In Vivo Effect of ORLE on PIRC Rats

One-year old PIRC rats with consolidated colon tumours were fed for one week with ORLE diet or standard diet. At the end of treatment, no differences in body weight were found between the two groups, indicating an unchanged caloric intake. At sacrifice, colon tumours were present in both controls and ORLE treated groups and were processed and analyzed for cell proliferation and apoptosis level. ORLE induced a significant reduction in cell proliferation and augmented the levels of apoptotic bodies in the tumour lesions of PIRC rats, when compared to those of control group fed with a standard diet (Figure 1a).

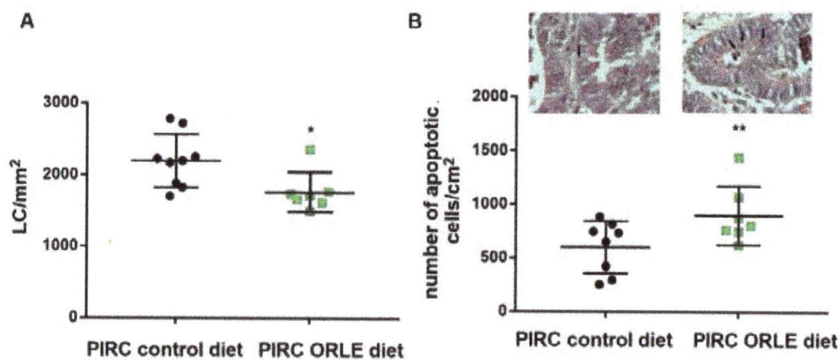

Figure 1. (**A**) Proliferative activity ((PCNA-Labelled cells (LC))/mm^2) in colon tumours), (**B**) apoptosis (apoptotic cells/mm^2) in colon tumours) in PIRC rats fed with different diets determined as described in the Methods section. Representative images of either group were shown (arrows indicate apoptotic cells; original magnification 1000×). Statistical significances were determined using two-tailed unpaired student's *t*-test corresponding to * $p = 0.022$ and ** $p = 0.041$.

To explore the in vivo effect of ORLE, we also evaluated the response of ORLE enriched diet on macrophages. Macrophage recruitment to neoplastic mucosa was determined with the evaluation of CD68 antigen expression in the colon tumours. As showed in Figure 2a, ORLE intake did not alter the number of macrophages infiltrating the lesions. In addition, as NO is often associated with cancer aggressiveness, we determined mRNA level of iNOS in the colon tumours and compared to that expressed in normal mucosa of the same rats. We found that iNOS expression was significantly augmented in tumour lesions compared to their normal mucosa, while in rats fed with an ORLE diet, iNOS mRNA over-expression was not significant compared to their normal mucosa (Figure 2b). The effect of the ORLE diet was also explored in peritoneal macrophages collected upon the sacrifice of rats fed with the two diets. In particular, we examined the ex-vivo NO production by collected macrophages after in vitro treatment with IFNγ and LPS. Macrophages recovered from PIRC rats fed with ORLE diet were unresponsive to the exposure to IFNγ/LPS. Indeed, the release of NO was comparable to that of relative untreated macrophages. Conversely, peritoneal macrophages collected from PIRC rats fed with a control diet were extremely responsive to the exposure to IFNγ/LPS. Indeed, the NO release was remarkably increased compared to relative untreated macrophages (Figure 2c).

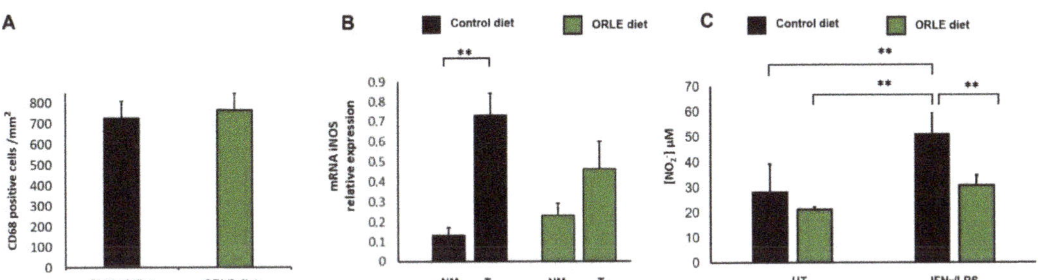

Figure 2. Effect of ORLE on rat macrophages. (**A**) CD-68 expression (positive cells/mm^2) in colon tumours from PIRC rats fed with a control diet (black column) or a diet enriched in ORLE (green column). (**B**) iNOS expression in colon tumours (T) from PIRC rats fed with a control diet (black columns) or a diet enriched in ORLE (green columns) compared to iNOS expression in normal mucosa (NM). Data are expressed as means + SEM; two-way-ANOVA of the data shows a statistical significance (** $p = 0.0016$) for the effect of tissue (i.e., tumour expression being higher than that in the normal mucosa), while dietary treatment was not significant. Post-hoc analysis of the differences between different groups (Tukey's multiple comparisons test) shows a significant difference between NM and tumours in the Control diet (** $p = 0.0047$), while the difference between NM and tumours in the ORLE diet is not significant. (**C**) Effect of ORLE on NO production by peritoneal macrophages recovered from tumour bearing rats fed a control diet (black columns) or a diet enriched in ORLE (green columns) and untreated or exposed, in vitro, to IFNγ/LPS. Two-way-ANOVA of the data shows a statistical significance for the effect of ORLE diet ($p = 0.0001$), and for the effect of treatment ($p = 0.0002$). Post-hoc analysis of the differences between different groups (Tukey's) shows a significant difference as indicated by the asterisks ($p < 0.001$).

Finally, since a reduction of iNOS mRNA was observed in rat colon tumours, we aimed to determine whether the expression of iNOS protein by colon cancer cells changed after the treatment with ORLE. By using flow cytometry analysis, we found a decreased iNOS protein expression in HCT116 cells treated with 50 μM ORLE for 72 h.

HCT116 cells exposed to 50 μM ORLE for 72 h treatment did not show, compared to untreated cells, a significant reduction of cell viability. Significant reduction on cell viability was found at higher doses (100 μM and 200 μM ORLE), this resistance is probably related to their high aggressiveness (Figure 3a,b).

3.2. *Effect of ORLE on Primed Murine RAW264.7 Cells*

To further explore ORLE effect on macrophages pro-inflammatory activity, we evaluated the release of NO in RAW264.7 cells treated with 50 μM ORLE for 24 h in the presence of LPS (acute exposure) or pretreated with 50 μM ORLE for 72 h (chronic exposure) and next exposed to LPS. NO production, in quiescent RAW264.7 cells, was not affected by the treatment with ORLE, neither after an acute or chronic modality. NO production in RAW264.7 cells activated with ORLE acute exposure was significantly inhibited compared to that measured in absence of it (50% reduction). Interestingly, NO production by LPS-activated macrophages, after ORLE chronic exposure, was strongly inhibited (70% compared to LPS-treated) (Figure 4a). In parallel, ORLE treatment strongly inhibited the expression of iNOS after ORLE acute or chronic exposure (Figure 4B).

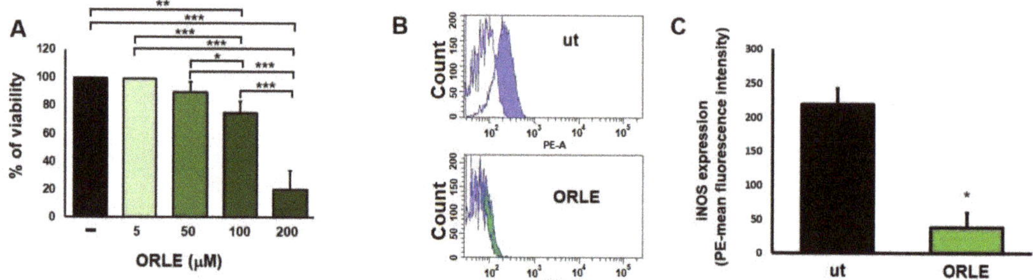

Figure 3. MTT assay of HTC116 cells exposed for 72 h to different concentration of ORLE (**A**). Data are expressed as means ± SEM of the percentage of viability and are representative of three independent experiments (n = 4). Statistical evaluation of the effect of ORLE on tumour cell viability was analyzed by one-way ANOVA with post-hoc Tukey's test, * $p < 0.001$, ** $p < 0.0001$, and *** $p < 0.00001$. (**B**) Inhibition of intracellular protein expression of iNOS in HCT116 cells. Representative histogram plots of untreated HCT116 cells (blu istogram) or HCT116 cells exposed for 72 h to ORLE (50 μM) (green istograms). Coloured istograms represents HCT116 cells exposed to primary and secondary antibody, while white istograms represents cells exposed to secondary antibody only. (**C**) Quantitative analysis of mean florescence intensity. Data were expressed as means ± SEM (n = 3). Statistical significance was determined using two-tailed unpaired student's *t*-test corresponding to * $p < 0.0001$.

Further, ORLE inhibits the expression of COX-2 by an acute treatment and completely abolished it after a chronic exposure (Figure 5). It has to be noted that the expression of iNOS and COX-2 was evaluated on the same set of samples, and, necessarily, the same tubulin reference was used. We hypothesize a possible cooperation of COX-2 with iNOS abrogation during tumour lesion regression. Indeed, over-expression of iNOS may generate reactive mutagenic agents causing as DNA damage or impairment of DNA repair, and COX-2 stimulation leads to sustain tumour growth [27].

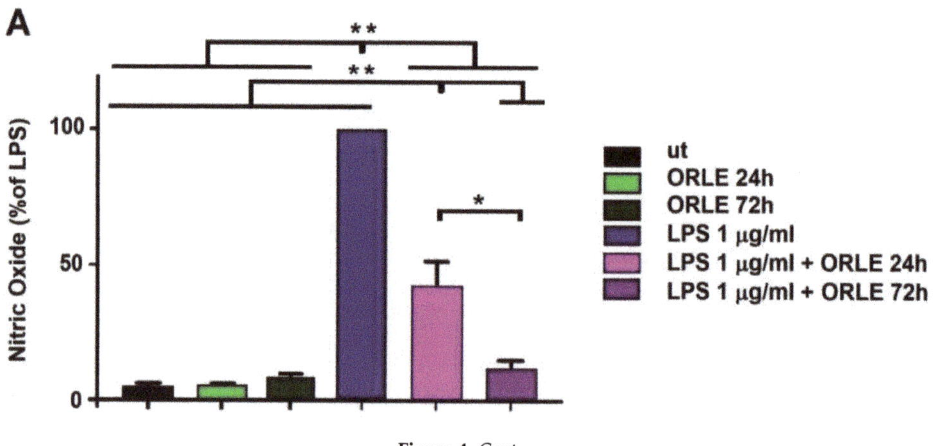

Figure 4. *Cont.*

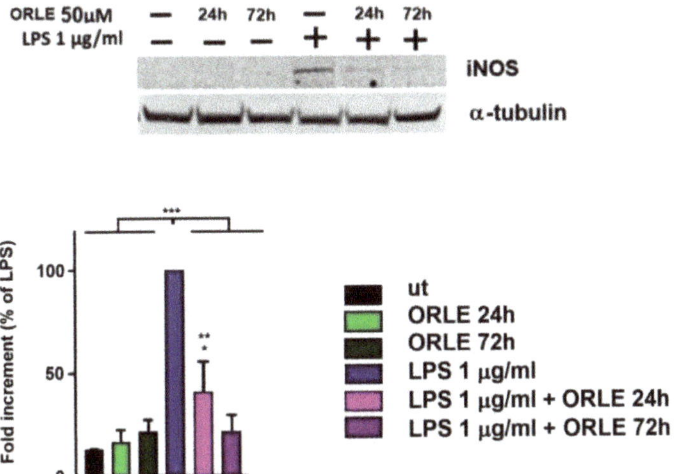

Figure 4. Inhibition of NO production in RAW264.7 macrophages: (**A**) exposed to 50 µM ORLE and 1 µg/mL LPS. Data are expressed as means ± SEM of the percentage of inhibition compared to LPS treated cells. Data are representative of three independent experiments ($n = 4$). Two-way-ANOVA of the data shows a statistical significance for the effect of ORLE and for the effect of LPS ($p < 0.0001$). Post-hoc analysis of the differences between different groups (Tukey's) shows a significant difference as indicated by the asterisks (* $p < 0.0001$, and ** $p < 0.01$). (**B**) Upper panel, representative Western blot of inducible nitric oxide synthase (iNOS) protein expression, lower panel, densitometric analysis of iNOS expression in RAW.264.7 cells. Data are expressed as means ± SEM of percentage compared to LPS-stimulated cells from at least three independent experiments. Two-way-ANOVA of the data shows a statistical significance for the effect of ORLE and for the effect of LPS ($P < 0.001$). Post-hoc analysis of the differences between different groups (Tukey's) shows a significant difference as indicated by the asterisks (* $p < 0.01$, ** $p < 0.05$ vs. UT, and *** $p < 0.0001$).

The evaluation of anti inflammatory effect of ORLE, either acutely or chronically, on RAW 264.7 macrophages exposed to LPS was finally explored through the analysis of the relative mRNA expression of IL-1β and IL-6 cytokines and TGF-β. The results indicated that ORLE decreased the mRNA expression of IL-1β, and IL-6, in LPS-induced RAW264.7 cells in time-dependent manner, the more prolonged was the exposure to ORLE, the more significant was the inhibitory effect (Figure 6a,b). In addition to this inhibitory effect, ORLE promotes, in an acute exposure, a reduction of IL-1βR protein expression and a reduction of mRNA expression for TGF-β. (Figure 6c,d).

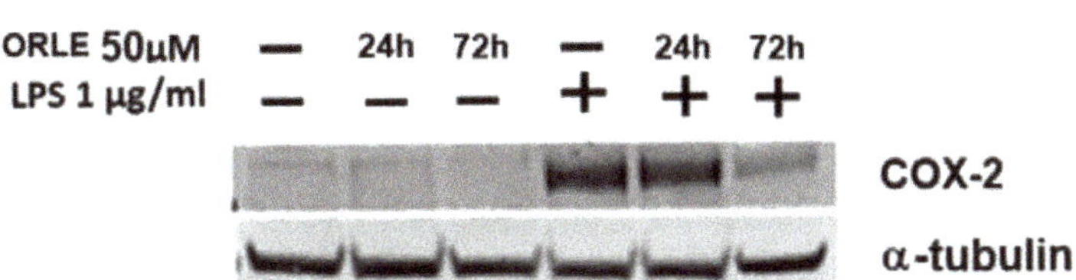

Figure 5. *Cont.*

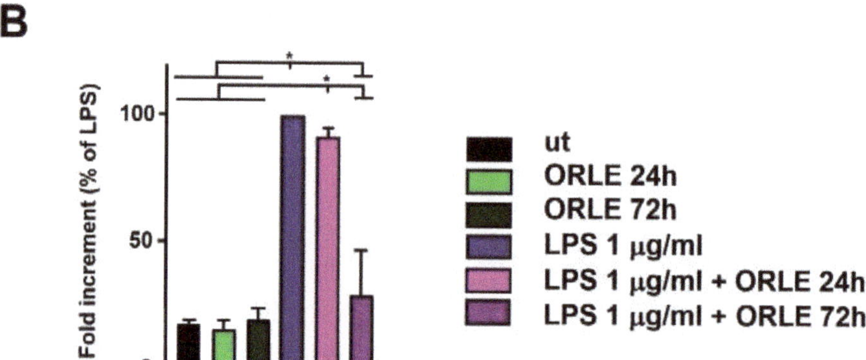

Figure 5. Inhibition of COX-2 expression. (**A**) Representative Western blot of inducible cyclooxygenase-2 (COX-2) protein expression; (**B**) densitometric analysis of COX-2 expression in RAW.264.7 cells. Data are expressed as means ± SEM of percentage compared to LPS-stimulated cells from at least three independent experiments. Two-way-ANOVA of the data shows a statistical significance for the effect of LPS ($p < 0.005$). Post-hoc analysis of the differences between different groups (Tukey's) shows a significant difference as indicated by the asterisks (* $p < 0.0001$).

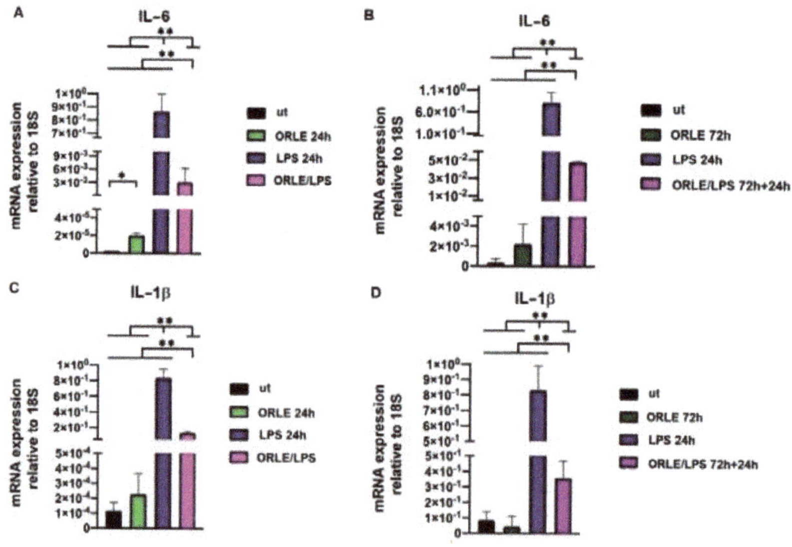

Figure 6. *Cont.*

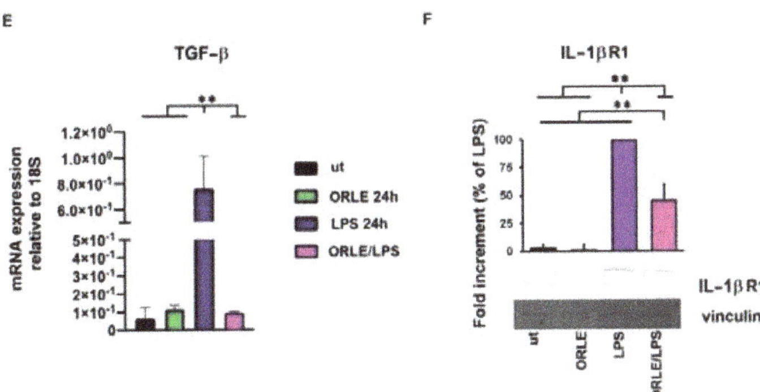

Figure 6. Evaluation by quantitative real-time PCR of IL-6 mRNA (panel (**A**) for ORLE acute, and panel (**B**) for ORLE chronic exposure), and IL-1β mRNA (panel (**C**) for ORLE acute, and panel (**D**) for ORLE chronic exposure) in RAW264.7 cells. Quantitative real-time PCR of TGF-β mRNA in RAW264.7 cells exposed to ORLE acute treatment (**E**). mRNA levels were normalized to 18S as an endogenous control. Two-way-ANOVA of the data shows a statistical significance for the effect of ORLE treatment (p: 0.001), and for the effect of LPS (P: 0.002). Post-hoc analysis of the differences between different groups (Tukey's) shows a significant difference as indicated by the asterisks (* $p < 0.01$ and ** $p < 0.0001$). IL-1R1 protein expression and relative densitometric analysis in RAW264.7 cells exposed to ORLE acute treatment (**F**). Data are expressed as reduction relative to LPS. Western blot images are representative of at least three independent experiments. Two-way-ANOVA of the data shows a statistical significance for the effect of ORLE treatment, and for the effect of LPS (p: 0.001). Post-hoc analysis of the differences between different groups (Tukey's) shows a significant difference as indicated by the asterisks (** $p < 0.0001$).

4. Discussion

The anti-inflammatory properties of phenolic secoiridoids have been recognized since a long time, and the reduction of oxidative stress and inflammatory cells recruitment, have been clearly demonstrated [28]. Ole is one of the most intriguing members of secoiridoids family, and is able to dampen systemic inflammation through the modulation of pro-inflammatory cell recruitment [29,30]. Given the close correlation between the perseverance of chronic inflammation and tumour progression, Ole and the other phenolic compounds have been studied for their beneficial effect on different models of in vitro and in vivo cancer progression.

The aim of the present study is to evaluate, for the first time, the impact of an ORLE-enriched diet, in an animal model with already-developed colon tumours, exploring its effects on different characteristics of colon cancer and associated systemic inflammation [31]. We focused our interest on in vivo model of colon cancer, using the PIRC rats that spontaneously develop tumours in the colon [32,33]. We choose one-week treatment of ORLE-enriched diet equivalent to the consumption of a Ole dose (100 mg/kg of ORLE) according to previous studies [13,17], to better highlight whether this low-dose treatment might exert a beneficial effect against established cancer lesions and local and systemic inflammation. Secondly, we confirm the anti-inflammation activity of ORLE in murine activated peritoneal macrophages.

The novelty of our study is that one year old PIRC rats, bearing consolidated tumours, were fed with ORLE-enriched diet for a short period of time (one week). On the contrary, previous studies explored the chemoprevention effects of Ole/ORLE during the first phases of the development of the tumours [17,34]. We focused our attention on one of the key mediators released in tumour microenvironment, namely NO. Accordingly, it is well known that the expression of inducible NO synthase in cancer correlates with a patient poor prognosis [35]. However, it should be considered that, due to the highly sophisticated network of interactions, in which the released NO is involved in tumor microenvironment, the role of NO in colon cancer progression remains controversial [36,37]. We demonstrated that ORLE-enriched diet decreased cell proliferation and increased apoptosis in colon tumours in vivo; we also documented that ORLE diet was able to counteract the tumour-associated iNOS over-expression present in the tumours of control rats.

It has to be noted that peritoneal macrophages not only monitor and maintain local homeostasis, but also play the role of sentinel cells against threats such as infections, tissue damages, and tumours [38]. Thus, in some instances, the evaluation of peritoneal macrophages activation could be used as a measure of local and systemic responsiveness to pro-inflammatory stimuli [39]. We found that ORLE-enriched diet dampened the pro-inflammatory behavior of peritoneal macrophages from tumour-bearing rats. Indeed, one-week exposure to ORLE-enriched diet was sufficient to clearly reduce peritoneal macrophages responsiveness to the in vitro treatment with IFNγ/LPS. Thereafter, we demonstrate that ORLE was also effective in reducing in vitro iNOS expression in human adenocarcinoma cells. It has to be noted that we observed an in vitro anti-proliferative effect on colon carcinoma cells only at higher ORLE concentrations. Similar results have been found by others using ORLE [13] or Ole at higher concentrations [40,41].

The upregulation of iNOS and the elevated NO release is an unavoidable aspect of tumour microenvironment. The NO unpaired electron rapidly reacts with other radical species present in the tumour microenvironment driving to an increased DNA damage and the acquisition of additional mutations on surrounding tumor cells contributing in the maintenance of an aggressive tumour phenotype [42–44].

Further, in colon cancer the release of cytokines as IL-1β, IL-6 and the activation of COX-2, collectively, support neoplastic transformation and malignant progression [45,46]. Thus, we investigated whether ORLE treatment might modulate COX-2 and cytokine release of activated macrophages. There is an intense debate regarding the role of tumour-associated macrophages and cancer outcomes, mostly due to the highly plastic behavior of these immune cell population [47,48]. Despite this, clinical evidence suggests that, in colon cancer, the presence of a rich macrophage infiltrate accounts for a better prognosis [47]. Interestingly, we found that the number of macrophages infiltrating tumour lesions of ORLE enriched diet fed rats was similar to that of macrophages infiltrating tumour lesions in rats fed a regular diet. Conversely, we found that ORLE treatment significantly inhibited LPS activated macrophages, in term of NO release and iNOS expression, consistent with our in vivo results and other studies [26]. In addition to this, we found that COX-2 expression in LPS-activated macrophages was completely abolished after ORLE chronic exposure. We also found that ORLE treatment significantly downregulates LPS-macrophage activation in term of IL-1β, IL-6, and TGF-β mRNA expression. These cytokines and growth factor strongly cooperate in the maintenance of a pro-inflammatory and toxic microenvironment, that correlates with disease progression and drug resistance [39,49–51].

5. Conclusions

The present study assesses whether one week-low-dose treatment with an ORLE-enriched diet exerts a beneficial effect against established colon cancer lesions of PIRC rats and local and systemic inflammation. Although in vivo experiments were performed with a limited number of PIRC rats fed with ORLE, the overall results disclose a significant increase in tumour apoptosis together with a downregulation of proliferation associated with the inhibition of NO and relative pro-inflammatory mediators expressed by tumour cells and inflammatory cells of tumour microenvironment. These findings suggest the possibility to test ORLE as a complementary therapy in combination with standard anti-cancer drugs.

Author Contributions: Conceptualization, F.B.; Data curation, J.R., S.C., S.P. and E.A.; Formal analysis, J.R., C.L., S.C. and S.U.; Funding acquisition, L.C., G.C., C.N. and F.B.; Investigation, J.R., S.C., N.M., S.P., E.A., K.T. and F.B.; Methodology, N.M., E.A., S.U., A.R. and K.T.; Resources, A.R., L.C. and G.C.; Supervision, A.R., L.C., C.L., C.N. and F.B.; Writing—original draft, J.R., S.C. and F.B.;

Writing—review & editing, A.R., L.C., G.C., C.N. and F.B. All authors have read and agreed to the published version of the manuscript.

Funding: This research was supported by the Italian Ministry of Education, University and Research (MIUR) (PRIN 2015 contract nr. 2015RNWJAM and 20157WW5EH), by the University of Florence (Fondo ex-60%), by Ente Cassa di Risparmio di Firenze (BIANCRF181014) and Regione Toscana (BIOSYNOL PSGO 52 2017).

Institutional Review Board Statement: The study was conducted according to the guidelines of the Declaration of Helsinki and approved by the Commission for Animal Experimentation of the Italian Ministry of Health (Authorization number 323/2016-PR).

Informed Consent Statement: Not applicable.

Data Availability Statement: Data is contained within the article.

Conflicts of Interest: The authors declare no conflict of interest.

References

1. Romani, A.; Ieri, F.; Urciuoli, S.; Noce, A.; Marrone, G.; Nediani, C.; Bernini, R. Health Effects of Phenolic Compounds Found in Extra-Virgin Olive Oil, By-Products, and Leaf of Olea europaea L. *Nutrients* **2019**, *11*, 1776. [CrossRef] [PubMed]
2. Taylor, W.F.; Moghadam, S.E.; Farimani, M.M.; Ebrahimi, S.N.; Tabefam, M.; Jabbarzadeh, E. A multi-targeting natural compound with growth inhibitory and anti-angiogenic properties re-sensitizes chemotherapy resistant cancer. *PLoS ONE* **2019**, *14*, e0218125. [CrossRef] [PubMed]
3. Mijatović, S.; Bramanti, A.; Nicoletti, F.; Fagone, P.; Kaluđerović, G.N.; Maksimović-Ivanić, D. Naturally occurring compounds in differentiation based therapy of cancer. *Biotechnol. Adv.* **2018**, *36*, 1622–1632. [CrossRef] [PubMed]
4. El, S.N.; Karakaya, S. Olive tree (*Olea europaea*) leaves: Potential beneficial effects on human health. *Nutr. Rev.* **2009**, *67*, 632–638. [CrossRef] [PubMed]
5. Nediani, C.; Ruzzolini, J.; Romani, A.; Calorini, L. Oleuropein, a Bioactive Compound from Olea europaea L., as a Potential Preventive and Therapeutic Agent in Non-Communicable Diseases. *Antioxidants* **2019**, *8*, 578. [CrossRef]
6. Dinu, M.; Pagliai, G.; Scavone, F.; Bellumori, M.; Cecchi, L.; Nediani, C.; Maggini, N.; Sofi, F.; Giovannelli, L.; Mulinacci, N. Effects of an Olive By-Product Called Pâté on Cardiovascular Risk Factors. *J. Am. Coll. Nutr.* **2020**, *40*, 617–623. [CrossRef]
7. Bulotta, S.; Celano, M.; Lepore, S.M.; Montalcini, T.; Pujia, A.; Russo, D. Beneficial effects of the olive oil phenolic components oleuropein and hydroxytyrosol: Focus on protection against cardiovascular and metabolic diseases. *J. Transl. Med.* **2014**, *12*, 219. [CrossRef] [PubMed]
8. Lee, O.-H.; Lee, B.-Y. Antioxidant and antimicrobial activities of individual and combined phenolics in Olea europaea leaf extract. *Bioresour. Technol.* **2010**, *101*, 3751–3754. [CrossRef]
9. Miceli, C.; Santin, Y.; Manzella, N.; Coppini, R.; Berti, A.; Stefani, M.; Parini, A.; Mialet-Perez, J.; Nediani, C. Oleuropein Aglycone Protects against MAO-A-Induced Autophagy Impairment and Cardiomyocyte Death through Activation of TFEB. *Oxid. Med. Cell. Longev.* **2018**, *2018*, 8067592. [CrossRef]
10. Rigacci, S.; Guidotti, V.; Bucciantini, M.; Parri, M.; Nediani, C.; Cerbai, E.; Stefani, M.; Berti, A. Oleuropein aglycon prevents cytotoxic amyloid aggregation of human amylin. *J. Nutr. Biochem.* **2010**, *21*, 726–735. [CrossRef]
11. Rigacci, S.; Miceli, C.; Nediani, C.; Berti, A.; Cascella, R.; Pantano, D.; Nardiello, P.; Luccarini, I.; Casamenti, F.; Stefani, M. Oleuropein aglycone induces autophagy via the AMPK/mTOR signalling pathway: A mechanistic insight. *Oncotarget* **2015**, *6*, 35344–35357. [CrossRef]
12. Luccarini, I.; Pantano, D.; Nardiello, P.; Cavone, L.; Lapucci, A.; Miceli, C.; Nediani, C.; Berti, A.; Stefani, M.; Casamenti, F. The Polyphenol Oleuropein Aglycone Modulates the PARP1-SIRT1 Interplay: An In Vitro and In Vivo Study. *J. Alzheimer's Dis.* **2016**, *54*, 737–750. [CrossRef]
13. Ruzzolini, J.; Peppicelli, S.; Bianchini, F.; Andreucci, E.; Urciuoli, S.; Romani, A.; Tortora, K.; Caderni, G.; Nediani, C.; Calorini, L. Cancer Glycolytic Dependence as a New Target of Olive Leaf Extract. *Cancers* **2020**, *12*, 317. [CrossRef] [PubMed]
14. Ruzzolini, J.; Peppicelli, S.; Andreucci, E.; Bianchini, F.; Scardigli, A.; Romani, A.; La Marca, G.; Nediani, C.; Calorini, L. Oleuropein, the Main Polyphenol of Olea europaea Leaf Extract, Has an Anti-Cancer Effect on Human BRAF Melanoma Cells and Potentiates the Cytotoxicity of Current Chemotherapies. *Nutrients* **2018**, *10*, 1950. [CrossRef] [PubMed]
15. Armaghany, T.; Wilson, J.D.; Chu, Q.; Mills, G. Genetic Alterations in Colorectal Cancer. *Gastrointest Cancer Res.* **2012**, *5*, 19–27. [PubMed]
16. Borzì, A.M.; Biondi, A.; Basile, F.; Luca, S.; Vicari, E.S.D.; Vacante, M. Olive Oil Effects on Colorectal Cancer. *Nutrients* **2018**, *11*, 32. [CrossRef] [PubMed]
17. Sepporta, M.V.; Fuccelli, R.; Rosignoli, P.; Ricci, G.; Servili, M.; Fabiani, R. Oleuropein Prevents Azoxymethane-Induced Colon Crypt Dysplasia and Leukocytes DNA Damage in A/J Mice. *J. Med. Food* **2016**, *19*, 983–989. [CrossRef]
18. Tortora, K.; Vitali, F.; De Filippo, C.; Caderni, G.; Giovannelli, L. DNA damage in colon mucosa of Pirc rats, an Apc-driven model of colon tumorigenesis. *Toxicol. Lett.* **2020**, *324*, 12–19. [CrossRef] [PubMed]

19. Romani, A.; Campo, M.; Urciuoli, S.; Marrone, G.; Noce, A.; Bernini, R. An Industrial and Sustainable Platform for the Production of Bioactive Micronized Powders and Extracts Enriched in Polyphenols From Olea europaea L. and Vitis vinifera L. Wastes. *Front. Nutr.* **2020**, *7*, 120. [CrossRef]
20. Luceri, C.; Femia, A.P.; Tortora, K.; D'Ambrosio, M.; Fabbri, S.; Fazi, M.; Caderni, G. Supplementation with phytoestrogens and insoluble fibers reduces intestinal carcinogenesis and restores ER-β expression in Apc-driven colorectal carcinogenesis. *Eur. J. Cancer Prev.* **2020**, *29*, 27–35. [CrossRef]
21. Navarra, M.; Femia, A.P.; Romagnoli, A.; Tortora, K.; Luceri, C.; Cirmi, S.; Ferlazzo, N.; Caderni, G. A flavonoid-rich extract from bergamot juice prevents carcinogenesis in a genetic model of colorectal cancer, the Pirc rat (F344/NTac-Apcam1137). *Eur. J. Nutr.* **2019**, *59*, 885–894. [CrossRef]
22. Faul, F.; Erdfelder, E.; Lang, A.-G.; Buchner, A. G*Power 3: A flexible statistical power analysis program for the social, behavioral, and biomedical sciences. *Behav. Res. Methods* **2007**, *39*, 175–191. [CrossRef]
23. Femia, A.P.; Dolara, P.; Luceri, C.; Salvadori, M.; Caderni, G. Mucin-depleted foci show strong activation of inflammatory markers in 1,2-dimethylhydrazine-induced carcinogenesis and are promoted by the inflammatory agent sodium dextran sulfate. *Int. J. Cancer* **2009**, *125*, 541–547. [CrossRef]
24. Calorini, L.; Bianchini, F.; Mannini, A.; Mugnai, G.; Ruggieri, S. Enhancement of Nitric Oxide Release in Mouse Inflammatory Macrophages Co-cultivated with Tumor Cells of a Different Origin. *Clin. Exp. Metastasis* **2005**, *22*, 413–419. [CrossRef]
25. Femia, A.P.; Luceri, C.; Dolara, P.; Giannini, A.; Biggeri, A.; Salvadori, M.; Clune, Y.; Collins, K.J.; Paglierani, M.; Caderni, G. Antitumorigenic activity of the prebiotic inulin enriched with oligofructose in combination with the probiotics Lactobacillus rhamnosus and Bifidobacterium lactis on azoxymethane-induced colon carcinogenesis in rats. *Carcinogenesis* **2002**, *23*, 1953–1960. [CrossRef]
26. Ryu, S.-J.; Choi, H.-S.; Yoon, K.-Y.; Lee, O.-H.; Kim, K.-J.; Lee, B.-Y. Oleuropein Suppresses LPS-Induced Inflammatory Responses in RAW 264.7 Cell and Zebrafish. *J. Agric. Food Chem.* **2015**, *63*, 2098–2105. [CrossRef] [PubMed]
27. Sarkar, F.H.; Li, Y. Cell signaling pathways altered by natural chemopreventive agents. *Mutat. Res. Mol. Mech. Mutagen.* **2004**, *555*, 53–64. [CrossRef]
28. Huang, Y.-L.; Oppong, M.B.; Guo, Y.; Wang, L.-Z.; Fang, S.-M.; Deng, Y.-R.; Gao, X.-M. The Oleaceae family: A source of secoiridoids with multiple biological activities. *Fitoterapia* **2019**, *136*, 104155. [CrossRef] [PubMed]
29. Sun, W.; Wang, X.; Hou, C.; Yang, L.; Li, H.; Guo, J.; Huo, C.; Wang, M.; Miao, Y.; Liu, J.; et al. Oleuropein improves mitochondrial function to attenuate oxidative stress by activating the Nrf2 pathway in the hypothalamic paraventricular nucleus of spontaneously hypertensive rats. *Neuropharmacology* **2017**, *113*, 556–566. [CrossRef] [PubMed]
30. Simsek, T.; Erbas, M.; Buyuk, B.; Pala, C.; Sahin, H.; Altinisik, B. Prevention of rocuronium induced mast cell activation with prophylactic oleuropein rich diet in anesthetized rabbits. *Acta Cir. Bras.* **2018**, *33*, 954–963. [CrossRef]
31. E Tuomisto, A.; Mäkinen, M.J.; Väyrynen, J. Systemic inflammation in colorectal cancer: Underlying factors, effects, and prognostic significance. *World J. Gastroenterol.* **2019**, *25*, 4383–4404. [CrossRef]
32. Lori, G.; Paoli, P.; Femia, A.P.; Pranzini, E.; Caselli, A.; Tortora, K.; Romagnoli, A.; Raugei, G.; Caderni, G. Morin-dependent inhibition of low molecular weight protein tyrosine phosphatase (LMW-PTP) restores sensitivity to apoptosis during colon carcinogenesis: Studies in vitro and in vivo, in anApc-driven model of colon cancer. *Mol. Carcinog.* **2018**, *58*, 686–698. [CrossRef]
33. Amos-Landgraf, J.; Kwong, L.N.; Kendziorski, C.M.; Reichelderfer, M.; Torrealba, J.; Weichert, J.; Haag, J.D.; Chen, K.-S.; Waller, J.L.; Gould, M.N.; et al. A target-selected Apc-mutant rat kindred enhances the modeling of familial human colon cancer. *Proc. Natl. Acad. Sci. USA* **2007**, *104*, 4036–4041. [CrossRef]
34. Giner, E.; Recio, M.C.; Ríos, J.L.; Cerdá-Nicolás, J.M.; Giner, R.M. Chemopreventive effect of oleuropein in colitis-associated colorectal cancer in c57bl/6 mice. *Mol. Nutr. Food Res.* **2015**, *60*, 242–255. [CrossRef]
35. Liao, W.; Ye, T.; Liu, H. Prognostic Value of Inducible Nitric Oxide Synthase (iNOS) in Human Cancer: A Systematic Review and Meta-Analysis. *BioMed Res. Int.* **2019**, *2019*, 1–9. [CrossRef]
36. Lala, P.K.; Chakraborty, C. Role of nitric oxide in carcinogenesis and tumour progression. *Lancet Oncol.* **2001**, *2*, 149–156. [CrossRef]
37. Mijatović, S.; Savić-Radojević, A.; Plješa-Ercegovac, M.; Simić, T.; Nicoletti, F.; Maksimović-Ivanić, D. The Double-Faced Role of Nitric Oxide and Reactive Oxygen Species in Solid Tumors. *Antioxidants* **2020**, *9*, 374. [CrossRef]
38. Cassado, A.D.A.; Lima, M.R.D.; Bortoluci, K. Revisiting Mouse Peritoneal Macrophages: Heterogeneity, Development, and Function. *Front. Immunol.* **2015**, *6*, 225. [CrossRef]
39. Machado, M.C.C.; Coelho, A.M.M. Role of Peritoneal Macrophages on Local and Systemic Inflammatory Response in Acute Pancreatitis. *Acute Pancreat.* **2012**, *101*. [CrossRef]
40. Vanella, L.; Acquaviva, R.; Di Giacomo, C.; Sorrenti, V.; Galvano, F.; Santangelo, R.; Cardile, V.; Gangia, S.; D'Orazio, N.; Abraham, N.G. Antiproliferative effect of oleuropein in prostate cell lines. *Int. J. Oncol.* **2012**, *41*, 31–38. [CrossRef] [PubMed]
41. Bulotta, S.; Corradino, R.; Celano, M.; D'Agostino, M.; Maiuolo, J.; Oliverio, M.; Procopio, A.; Iannone, M.; Rotiroti, D.; Russo, D. Antiproliferative and antioxidant effects on breast cancer cells of oleuropein and its semisynthetic peracetylated derivatives. *Food Chem.* **2011**, *127*, 1609–1614. [CrossRef]
42. Watanabe, K.; Kawamori, T.; Nakatsugi, S.; Wakabayashi, K. COX-2 and iNOS, good targets for chemoprevention of colon cancer. *BioFactors* **2000**, *12*, 129–133. [CrossRef] [PubMed]

43. Xu, W.; Liu, L.Z.; Loizidou, M.; Ahmed, M.; Charles, I.G. The role of nitric oxide in cancer. *Cell Res.* **2002**, *12*, 311–320. [CrossRef] [PubMed]
44. Luanpitpong, S.; Chanvorachote, P. Nitric Oxide and Aggressive Behavior of Lung Cancer Cells. *Anticancer. Res.* **2015**, *35*, 4585–4592.
45. Li, J.; Huang, L.; Zhao, H.; Yan, Y.; Lu, J. The Role of Interleukins in Colorectal Cancer. *Int. J. Biol. Sci.* **2020**, *16*, 2323–2339. [CrossRef] [PubMed]
46. Cianchi, F.; Cortesini, C.; Fantappiè, O.; Messerini, L.; Sardi, I.; Lasagna, N.; Perna, F.; Fabbroni, V.; Di Felice, A.; Perigli, G.; et al. Cyclooxygenase-2 Activation Mediates the Proangiogenic Effect of Nitric Oxide in Colorectal Cancer. *Clin. Cancer Res.* **2004**, *10*, 2694–2704. [CrossRef]
47. Hang, Q.; Liu, L.; Gong, C.-Y.; Shi, H.-S.; Zeng, Y.-H.; Wang, X.-Z.; Zhao, Y.-W.; Wei, Y.-Q. Prognostic Significance of Tumor-Associated Macrophages in Solid Tumor: A Meta-Analysis of the Literature. *PLoS ONE* **2012**, *7*, e50946. [CrossRef]
48. Ma, J.; Liu, L.; Che, G.; Yu, N.; Dai, F.; You, Z. The M1 form of tumor-associated macrophages in non-small cell lung cancer is positively associated with survival time. *BMC Cancer* **2010**, *10*, 112. [CrossRef]
49. Ping, P.H.; Bo, T.F.; Li, L.; Hui, Y.N.; Hong, Z. IL-1β/NF-kb signaling promotes colorectal cancer cell growth through miR-181a/PTEN axis. *Arch. Biochem. Biophys.* **2016**, *604*, 20–26. [CrossRef]
50. Masjedi, A.; Hashemi, V.; Hojjat-Farsangi, M.; Ghalamfarsa, G.; Azizi, G.; Yousefi, M.; Jadidi-Niaragh, F. The significant role of interleukin-6 and its signaling pathway in the immunopathogenesis and treatment of breast cancer. *Biomed. Pharmacother.* **2018**, *108*, 1415–1424. [CrossRef]
51. Toyoshima, Y.; Kitamura, H.; Xiang, H.; Ohno, Y.; Homma, S.; Kawamura, H.; Takahashi, N.; Kamiyama, T.; Tanino, M.; Taketomi, A. IL6 Modulates the Immune Status of the Tumor Microenvironment to Facilitate Metastatic Colonization of Colorectal Cancer Cells. *Cancer Immunol. Res.* **2019**, *7*, 1944–1957. [CrossRef] [PubMed]

 *antioxidants* 

Article

Columbianadin Dampens In Vitro Inflammatory Actions and Inhibits Liver Injury via Inhibition of NF-κB/MAPKs: Impacts on •OH Radicals and HO-1 Expression

Thanasekaran Jayakumar [1], Shaw-Min Hou [1,2,3], Chao-Chien Chang [1,2,3], Tsorng-Harn Fong [4], Chih-Wei Hsia [1], Yen-Jen Chen [1,5], Wei-Chieh Huang [1], Periyakali Saravanabhavan [6], Manjunath Manubolu [7], Joen-Rong Sheu [1,5,*] and Chih-Hsuan Hsia [1,8,*]

1 Graduate Institute of Medical Sciences, College of Medicine, Taipei Medical University, Taipei 110, Taiwan; jayakumar@tmu.edu.tw (T.J.); houshawmin@gmail.com (S.-M.H.); change@seed.net.tw (C.-C.C.); d119106003@tmu.edu.tw (C.-W.H.); m120104004@tmu.edu.tw (Y.-J.C.); m120107013@tmu.edu.tw (W.-C.H.)
2 Department of Cardiovascular Center, Cathay General Hospital, Taipei 106, Taiwan
3 Division of Cardiovascular Surgery, Department of Surgery, School of Medicine, College of Medicine, Fu Jen Catholic University, New Taipei City 242, Taiwan
4 Department of Anatomy and Cell Biology, School of Medicine, College of Medicine, Taipei Medical University, Taipei 110, Taiwan; thfong@tmu.edu.tw
5 Department of Pharmacology, School of Medicine, College of Medicine, Taipei Medical University, Taipei 110, Taiwan
6 Department of Zoology, Bharathiar University, Coimbatore 641046, Tamil Nadu, India; bhavan@buc.edu.in
7 Department of Evolution, Ecology and Organismal Biology, Ohio State University, Columbus, OH 43212, USA; manubolu.1@osu.edu
8 Translational Medicine Center, Shin Kong Wu Ho-Su Memorial Hospital, Taipei 111, Taiwan
* Correspondence: sheujr@tmu.edu.tw (J.-R.S.); T014913@ms.skh.org.tw (C.-H.H.)

Citation: Jayakumar, T.; Hou, S.-M.; Chang, C.-C.; Fong, T.-H.; Hsia, C.-W.; Chen, Y.-J.; Huang, W.-C.; Saravanabhavan, P.; Manubolu, M.; Sheu, J.-R.; et al. Columbianadin Dampens In Vitro Inflammatory Actions and Inhibits Liver Injury via Inhibition of NF-κB/MAPKs: Impacts on •OH Radicals and HO-1 Expression. *Antioxidants* **2021**, *10*, 553. https://doi.org/10.3390/antiox10040553

Academic Editor: Han Moshage

Received: 26 February 2021
Accepted: 30 March 2021
Published: 2 April 2021

Abstract: Columbianadin (CBN), a natural coumarin isolated from *Angelica decursiva,* is reported to have numerous biological activities, including anticancer and platelet aggregation inhibiting properties. Here, we investigated CBN's anti-inflammatory effect in lipopolysaccharide (LPS)-stimulated RAW 264.7 cell activation and deciphered the signaling process, which could be targeted by CBN as part of the mechanisms. Using a mouse model of LPS-induced acute liver inflammation, the CBN effects were examined by distinct histologic methods using trichrome, reticulin, and Weigert's resorcin fuchsin staining. The result showed that CBN decreased LPS-induced expressions of TNF-α, IL-1β, and iNOS and NO production in RAW 264.7 cells and mouse liver. CBN inhibited LPS-induced ERK and JNK phosphorylation, increased IκBα levels, and inhibited NF-κB p65 phosphorylation and its nuclear translocation. Application of inhibitors for ERK (PD98059) and JNK (SP600125) abolished the LPS-induced effect on NF-κB p65 phosphorylation, which indicated that ERK and JNK signaling pathways were involved in CBN-mediated inhibition of NF-κB activation. Treatment with CBN decreased hydroxyl radical (•OH) generation and increased HO-1 expression in RAW 264.7 cells. Furthermore, LPS-induced liver injury, as indicated by elevated serum levels of liver marker enzymes (aspartate aminotransferase (AST) and alanine aminotransferase (ALT)) and histopathological alterations, were reversed by CBN. This work demonstrates the utility of CBN against LPS-induced inflammation, liver injury, and oxidative stress by targeting JNK/ERK and NF-κB signaling pathways.

Keywords: columbianadin; LPS; hydroxyl radicals; HO-1 expression; liver-injury; NF-κB/MAPK signaling pathways

1. Introduction

Inflammation is one of the most common clinical indicators of many disease conditions, including Alzheimer's, gout, arthritis, and obesity [1]. Therefore, suppression of the inflammatory process is the first significant treatment step in almost all pathological conditions. At present, various chronic inflammatory diseases are controlled by treatment with non-steroidal anti-inflammatory drugs (NSAIDs). However, severe side effects of these treatments are commonly reported [2], necessitating the development of safer anti-inflammatory substances. The molecular mechanisms connected with the inflammatory process are intricate. Among them, the nuclear factor kappa B (NF-κB) and mitogen-activated protein kinase (MAPK) signaling pathways have been found to play significant roles in the inflammatory response. Lipopolysaccharides (LPSs), also known as endotoxins, are found in the outer membrane of Gram-negative bacteria and have been extensively used in models of both systemic and local inflammation. LPS activates macrophage inflammation via activation of NF-κB [3], and controls acute inflammatory and innate immunity functional genes. Rapid phosphorylation of inhibitory κB (IκB) protein results in NF-κB activation by the IκB kinase (IKK) signaling process, which enhances degradation of the IκBα protein [4]. Increased degradation of IκBα subsequently initiates the translocation of active IκB-free NF-κB from the cytoplasm to the nucleus, which results in an activation of promoter regions of target genes and transcription of pro-inflammatory markers, such as tumor necrosis factor (TNF-α), interleukin-6 (IL-6), inducible nitric oxide synthase (iNOS), and cyclooxygenase-2 (COX-2). [5].

NF-κB is closely related to MAPK signaling (the extracellular signaling kinases (ERK1 and 2), the c-jun N-terminal kinases (JNK1-3), and the p38 MAPK (p38α, β, γ, and δ)) pathway, which is responsible for the release of various inflammatory cytokines [6]. In addition, a previous study examined the role of the MAPKs in the induction of iNOS and cytokine expression in activated macrophages [7]. LPS-mediated induction of NF-κB also activates iNOS and inflammatory liver injury [8]. Moreover, the pro-oxidative effect of LPS is generated via the stimulation of reactive oxygen species (ROS) production. ROS act as cellular messengers and provoke an inflammatory response. Due to the imbalance between ROS and antioxidant defense systems, oxidative stress occurs [9]. ROS are reported to induce cell and tissue injury through pro-inflammatory cytokine production and triggering NF-κB [10]. LPS induces iNOS and nitric oxide (NO) at sites of inflammation via activation of p38MAPK and NF-κB in RAW 264.7 cells [11]. Inhibition of JNK has been identified as an important anti-inflammatory mechanism through the suppression of inflammatory genes in several diseases [12]. Evidence shows that inflammation induces ERK activation. Therefore, the reduction of pro-inflammatory mediators and tissue injury via regulating NF-κB/MAPK and free radicals should be effective for treating inflammatory diseases.

Coumarins, the most common secondary metabolites in plants, have numerous physiological activities [13]. Columbianadin (CBN), 1-[(8S)-8, 9-dihydro-2-oxo-2Hfuro [2, 3-h]-1-benzopyran-8-yl]-1-methylethyl-[(2Z)-2-methylbutenoic acid] ester, is a principle component extracted from the root of *Angelica pubescens* Maxim [14]. As an angular dihydrofurocoumarin, it has shown several activities, including antiplatelet activity [15], cytotoxicity against various cancer cells, and in vivo palliative actions in mice [16]. Transportation and absorption kinetics studies revealed that CBN easily entered into the blood and circulated mainly in hepatic tissue [17]. Our recent study also reported the antiplatelet effect of CBN via different molecular mechanisms and suggested that CBN could act as a potential drug to treat thromboembolic disorders [18]. Although several basic biological and anti-inflammatory properties of CBN have been explored, the underlying mechanisms of the anti-inflammatory and liver protective effects are unclear. The present study may provide an improved scientific motivation for its clinical use as a candidate drug for the treatment of inflammatory liver diseases.

2. Materials and Methods

2.1. Materials

Columbianadin (CBN, >98%) was purchased from ChemFaces Biochem, Wuhan, Hubei, China. Dimethyl sulfoxide (DMSO), PD98059, SP600125, BAY11-7082, and DMPO were purchased from Sigma (St Louis, MO, USA). The antibodies against phospho-p38 MAPK Ser^{182} (pAb), IκBα (44D4), phospho-c-JNK (Thr^{183}/Tyr^{185}), p38 MAPK, NF-κB p65 (mAb), phospho-p44/p42 ERK (Thr^{202}/Tyr^{204}), and phospho-NF-κB p65 (Ser^{536}) pAb were all purchased from Cell Signaling (Beverly, MA, UAS). Anti-HO-1 pAb was purchased from Enzo (Farmingdale, New York, USA). The monoclonal antibody against α-tubulin was derived from NeoMarkers (Fremont, CA, USA). CBN was dissolved in 0.1% DMSO. All other chemicals and reagents used in this study were commercially purchased from Sigma until specified otherwise.

2.2. Cell Cultivation and MTT Assay for Cell Viability

RAW 264.7 cells were procured from the American Type Culture Collection (ATCC, Manassas, VA, USA, TIB-71) and cultivated in DMEM at 37 °C under 5% CO_2 and 95% air. Cells (2×10^5 cells/well) were pretreated with CBN (10–180 μM) for 20 min, followed by stimulation with LPS (1 μg/mL) for 24 h. An MTT assay was utilized for cell viability. Concisely, a 5 mg/mL MTT working solution was added into each well. After 4 h incubation at 37 °C, the culture medium was collected and 300 μL of DMSO were added to dissolve the crystals. The cell viability index was measured by calculating the absorbance of treated cells/absorbance of control cells ×100%.

2.3. Detection of Hydroxyl Radicals

According to our previous study, electron spin resonance (ESR) spectrometry analysis was done [19]. Briefly, RAW 264.7 cells (5×10^5 cells/mL) were exposed to 1 μg/mL LPS after a 20 min incubation with 20 and 40 μM CBN. After a 5 min incubation, the suspensions were added 100 μM DMPO before the ESR analysis was performed. The spectrometer functioned at 20 mW of power, 9.78 GHz of frequency, 100 G of scan range, and 5×10^4 of receiver gain. Variation amplitudes, 1 G; time constant, 164 ms; and scanning for 42 s with 3 scans accumulated.

2.4. Immunoblotting Study

For Western blotting analysis, cells and liver tissues were lysed and homogenized using lysis buffer. Fifty micrograms of the extracted proteins were electrophoretically separated using 12% SDS-PAGE and transferred to PVDF membranes and then blocked with 5% skimmed milk. The membranes were subjected to various desired primary antibodies for 2 h. After washing with PBS, they were incubated with HRP-conjugated donkey antirabbit IgG or sheep anti-mouse IgG for 1 h. The immunoreactive bands were identified with an enhanced chemiluminescent (ECL) system. The density of protein bands was quantified by using ImageJ software (NIH, Bethesda, MD, USA). The results were evaluated as relative units determined by normalization of the density of each band to that of the corresponding α-tubulin protein band.

2.5. Confocal Microscopy Assay

Cells at a density of 5×10^4 cells per well in 6-well plates on cover slips were pretreated with 40 μM CBN for 20 min followed by stimulation with 1 μg/mL of LPS for 30 min. Cells were fixed with 4% paraformaldehyde for 10 min. After double washing with PBS, cells were permeated with 0.1% Triton X-100 for 10 min and then blocked with 5% BSA for 1 h. Further, the primary p65 antibody was added to cells and kept at 4 °C for overnight. After that, cells were subjected to conjugated goat anti-rabbit IgG for 1 h and then stained with 4,6-diamidino-2 phenylindole (DAPI). Nuclear translocation of NF-κB p65 was examined by using a laser scanning confocal microscope (Nikon Ti-E A1, Tokyo, Japan) and images were captured for further analysis.

2.6. Measurement of NO Production

Cells were treated with CBN (20–40 μM) in the presence or absence of LPS (1 μg/mL) for 24 h. Briefly, 100 μL of each culture suspension were incubated with 100 μL Griess reagent for 10 min, and an microplate absorbance (MRX) reader measured the optical density at 550 nm. The NO production was measured with reference to the standard curve of sodium nitrite.

2.7. Animals and Experimental Design

Sixty healthy male C57BL/6 mice (weighing 25–30 g) were obtained from BioLasco Taiwan Co., Ld., Taipei, Taiwan. The Institutional Animal Care and Use Committee, Taipei Medical University, Taiwan (LAC-2016-0395) approved the study. After a week of acclimatization in a laboratory condition, the animals were separated into four groups (n = 12 per group): (i) Normal saline group (control), (ii) LPS control group (2.5 mg/kg, LPS), and (iii and iv) CBN + LPS groups (10 and 20 mg/kg, respectively). In the drug pretreatment groups, mice were intraperitoneally treated with CBN (10 and 20 mg/kg) for 2 h and then injected with LPS for 6 h. After anesthetizing with isoflurane (induced at 4% and maintained at 3%), blood was collected for biochemical analyses and liver tissues were dissected out for histology study. All mice were in fasting condition prior to collection of blood and tissue.

2.8. Histological Analysis

The liver tissues from all animals were fixed in 4% phosphate-buffered formalin for histopathological analysis. The fixed tissues were desiccated and inserted in paraffin and cut into 7 µm thin sections. After overnight drying, sections were dewaxed, rehydrated, and stained using trichrome, reticulin, and Weigert's resorcin fuchsin. The alterations were observed under a Nikon (Eclipse Ci-L) light microscope (Nikon Co., Tokyo, Japan).

2.9. Measurement of Liver Function Enzymes

Markers of liver function, serum aspartate transaminase (AST), and alanine transaminase (ALT) were determined using the Vet-Test® chemistry analyzer (IDEXX, Westbrook, ME, USA) with the results being expressed as international units per liter (U/L).

2.10. Statistical Evaluation

The results are given as mean ± standard error (S.E.M). Data were tested using SAS (version 9.2; SAS Inc., Cary, NC, USA). Statistical difference was determined by one-way analysis of variance (ANOVA). If significant variation was identified with multiple comparisons, a Student–Newman–Keuls test was performed. $p < 0.05$ indicated statistical significance. * $p < 0.05$; ** $p < 0.01$; *** $p < 0.001$. The statistical power was calculated with G * Power software (v.3.1) (Heinrich-Heine-Universität Düsseldorf, Düsseldorf, Germany) for animal experimental data. The statistical power was ranged from 0.8–0.9 in all the variables analyzed.

3. Results

3.1. Effects of CBN on Cytotoxicity, •OH Radical Production, and HO-1 Expression

To find whether CBN (Figure 1A) produces inhibition of •OH radical production without affecting cell viability, its cytotoxicity was detected by an MTT assay in RAW 264.7 cells. As shown in Figure 1B,C, CBN (10–180 µM) alone or with LPS (1 µg/mL) treatment had no noticeable cytotoxicity up to the concentration of 60 µM. The observed cell viabilities were >90%. Despite CBN showing no toxicity up to 60 µM in RAW cells, we used more safe and effective concentrations of 20 and 40 µM in follow-up experiments. The capability of CBN to scavenge free radicals was measured using the ESR radical scavenging assay and the results are shown in Figure 1D. CBN significantly (*** $p < 0.001$) and concentration (20 and 40 µM) dependently scavenges •OH radicals induced by LPS. Heme oxygenase-1 (HO-1) has been reported as an essential molecular target for anti-inflammatory activity [20] and several natural products have been found to exhibit anti-inflammatory activity via HO-1-mediated NF-E2-related factor 2 (Nrf2) activation [21]. Accordingly, this study tested the effect of CBN on HO-1 protein expression. As shown in Figure 1E, CBN treatment increased HO-1 expression in a concentration-dependent manner. The results may indicate that CBN scavenges LPS-induced •OH radical formation via at least partially increasing HO-1 expression.

A

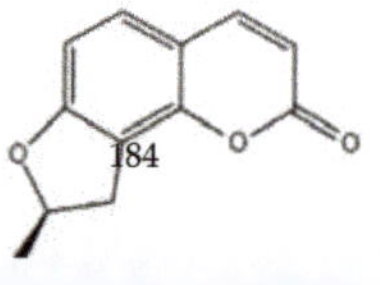

3.2. CBN Suppresses LPS-Induced MAPK Activation

Abnormal activation of MAPK signaling results in excessive inflammatory actions, since these molecules are important players in regulating inflammatory-mediated macrophage activation. Inhibition of MAPK activation has been reported to block the activation of inflammatory cytokines and molecules, reducing the severity of inflammatory disease. In this study, as shown in Figure 2, LPS could cause a significant elevation of ERK, p38 MAPK, and JNK phosphorylation after exposure for 30 min. CBN treatment significantly reduced ERK and JNK, but not p38 MAPK phosphorylation. Studies with pharmacological inhibitors demonstrated that MAPK is associated with regulating the transactivation function of NF-κB in macrophages [22]. This result evidenced that ERK and JNK play major roles as regulators of CBN-mediated anti-inflammatory effects in LPS-induced macrophages.

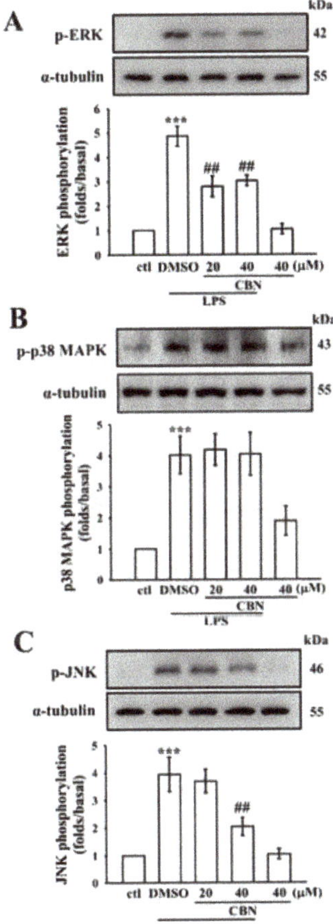

Figure 2. Effects of CBN on ERK1/2, p38MAPK, and JNK1/2 phosphorylation in LPS-induced RAW cells. Cells were treated with 0.1% DMSO or CBN (20–40 μM) for 20 min, and then induced by LPS (1 μg/mL) for 30 min. The (**A**) ERK, (**B**) p38MAPK, and (**C**) JNK phosphorylation was detected by immunoblotting. Data are expressed as the means ± S.E.M. (n = 4). *** $p < 0.001$, LPS vs. control cells; ## $p < 0.01$, LPS vs. treatment groups.

3.3. CBN Regulates NF-κB Signaling Pathways

Studies have stated that LPS could activate NF-κB in addition to activating various transcription factors, which regulate numerous inflammatory signals due to its role in stimulating the generation of NO, TNF-α, IL-6, and other inflammatory mediators in activated macrophages [23]. These mediators are also related to the modulation of iNOS. Consequently, in order to investigate the effect of CBN in LPS-induced NF-κB signaling, the levels of IκBα degradation, p-p65 expression, and its nuclear translocation were analyzed. As described in Figure 3A–D, IκBα degradation (Figure 3A), p65 phosphorylation (Figure 3B,C) and its nuclear translocation (Figure 3D) were increased in LPS-induced cells when compared with normal cells. Interestingly, CBN reversed IκBα degradation and inhibited the phosphorylation and nuclear translocation of p65 in LPS-induced cells. Moreover, similar to CBN, specific inhibitors for ERK (PD98059), JNK (SP600125), and NF-κB (BAY117082) significantly reversed LPS-induced p65 phosphorylation (Figure 3C). These results suggest that ERK and JNK1/2 are involved in the upstream signal pathways, which mediate LPS-induced NF-κB activation. Overall, these findings indicate the crucial role of ERK and JNK, but not p38 MAPK, in the upstream regulators of NF-κB activation and play a vital role in the CBN-mediated anti-inflammatory effects.

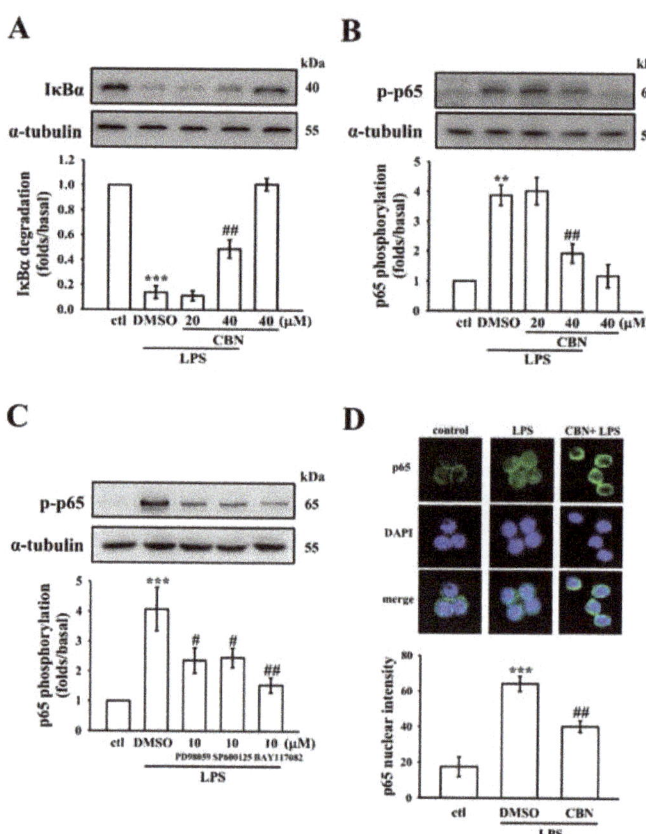

Figure 3. Effects of CBN on LPS-induced NF-κB signaling pathway in RAW cells. RAW cells were treated with CBN (20 and 40 μM) for 20 min with or without LPS (1 μg/mL) for 30 min. Immunoblotting assay was performed to detect (**A**) IκBα degradation and (**B**) p65 phosphorylation. (**C**) Effect of the BAY117082, PD98059, or SP600125 on p65 phosphorylation in LPS-induced RAW cells. (**D**) CBN reversed LPS-induced NF-κB p65 nuclear translocation. Data were graphed by pooling multiple images, with each individual data point corresponding to the mean fluorescence intensity of each individual cell nucleus. ** $p < 0.01$ and *** $p < 0.001$, LPS vs. control cells; # $p < 0.05$, and ## $p < 0.01$, LPS vs. CBN-treated cells ($n = 4$).

3.4. CBN Reduced the Expression of Inflammatory Mediators

3.4.1. Effects on NO and iNOS

Elevated levels of NO are considered to be an index for inflammatory disorders and a suitable target to find potent anti-inflammatory agents [24]. The increased production of NO can be attributed to the overexpression of iNOS, which is an important inflammatory mechanism. The NO production and iNOS protein expression were effectively augmented in LPS-induced RAW 264.7 cells (Figure 4A,B), however, co-treatment with CBN significantly reduced the extent of augmentation. This result suggests that the inhibition of NO production by CBN may be connected with its suppressive effect on LPS-induced iNOS protein.

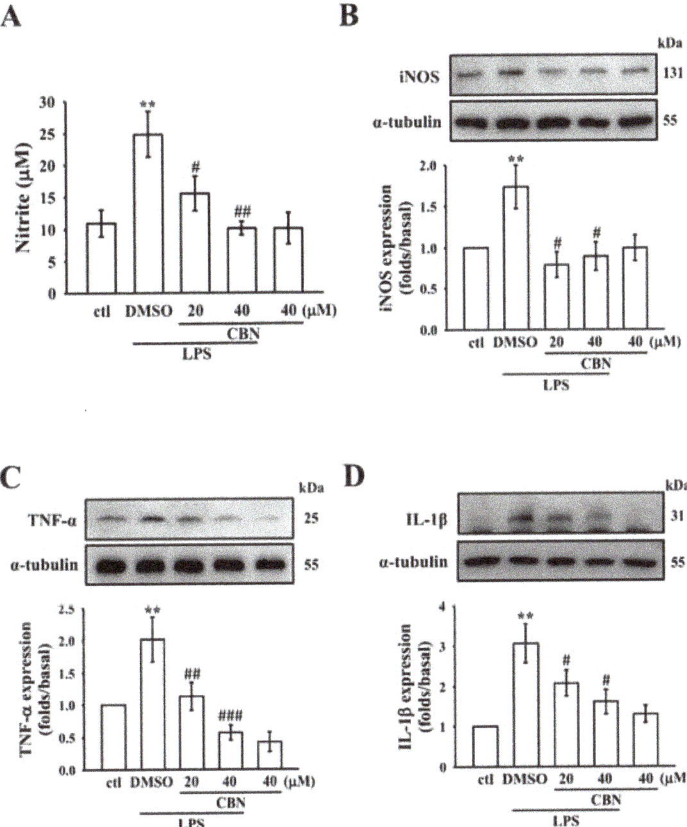

Figure 4. CBN inhibits NO production and expression of iNOS, TNF-α, and IL-1β in LPS-induced RAW cells. The content of NO (**A**) and the expression of iNOS (**B**), TNF-α (**C**), and IL-1β (**D**) proteins were assessed as defined in the Materials and Methods. Data presented are the means ± S.E.M. ($n = 4$); ** $p < 0.01$, LPS vs. control cells; # $p < 0.05$, ## $p < 0.01$, and ### $p < 0.001$, LPS vs. CBN-treated cells.

3.4.2. Effects on Inflammatory Cytokines

The anti-inflammatory activity of CBN was evaluated by its inhibitory effect against LPS-induced TNF-α and IL-1β expression in the RAW 264.7 macrophages. The expression of pro-inflammatory cytokines TNF-α and IL-1β was significantly increased in LPS-induced macrophage cells ($p < 0.001$). As shown in Figure 4C,D, CBN treatment significantly inhibited LPS-stimulated inflammatory cytokine expression, with greater inhibition occurring at higher CBN treatment concentrations.

3.4.3. Effects of CBN on Hepatic Histopathology

The production of inflammatory factors in liver tissues is complicated in liver injury. In the histological observation, collagen depositions were profusely distributed in the portal triad regions of liver tissues in LPS-induced mice (Figure 5(Ab)). Thick bundles of collagens were identified as blue and red in trichrome and Weigert's resorcin fuchsin staining, respectively (indicated by white arrows, Figure 5(Ab,e)). The existence of elastic fibers was also observed in the portal triad region of liver tissues in LPS-induced mice (indicated by black arrowheads, Figure 5(Ae)). As shown in Figure 5(Ab,e,h), the thickness of the portal vein was significantly increased in the LPS group compared with the control group. Remarkably, a high dosage of 20 mg/kg CBN treatment markedly reduced LPS-induced collagens and elastic fibers in liver tissues, as shown in Figure 5(Ac,f). In the reticulum staining, the intralobular stroma appeared as a network of reticular fibers between the sinusoids and the plates of hepatocytes (fine black fibers, Figure 5(Ah)). However, reticular fibers in the LPS-induced liver tissues in mice were more abundant than those in the control and CBN-treated mice (Figure 5(Ag,i)), as indicated by black arrows. These results indicated that administration of CBN could dampen the LPS-induced increment of elastic fibers following injury in liver cells.

3.5. CBN Reduces Liver Marker Enzymes ALT and AST

Serum ALT and AST transaminases were determined to examine LPS-induced liver function. A significant increase in activity of ALT and AST in LPS-induced mice compared with normal mice was observed ($p < 0.05$; Figure 5B,C). Treatment of CBN significantly reduced the activities of AST and ALT ($p <0.05$) at a high dose of 20 mg/kg, however, no significant effect was found at 10 mg/kg.

3.6. Effects of CBN on Liver MAPKs, NF-κB, Inflammatory Cytokines, and iNOS Protein Expression

The obtained results from the in vitro anti-inflammatory effects of CBN were further substantiated by examining the protein expression of MAPKs, NF-κB, inflammatory cytokines, and iNOS in liver tissues of LPS-induced mice. Similar to RAW cells, proteins were significantly elevated in LPS-exposed mouse liver tissues compared to normal mice (Figure 6). Pretreatment with CBN prominently reduced the expression of the phosphorylation of ERK, JNK, p65, TNF-α, IL-1β, and iNOS. This result indicates that CBN has a hepatoprotective effect via inhibiting these molecules.

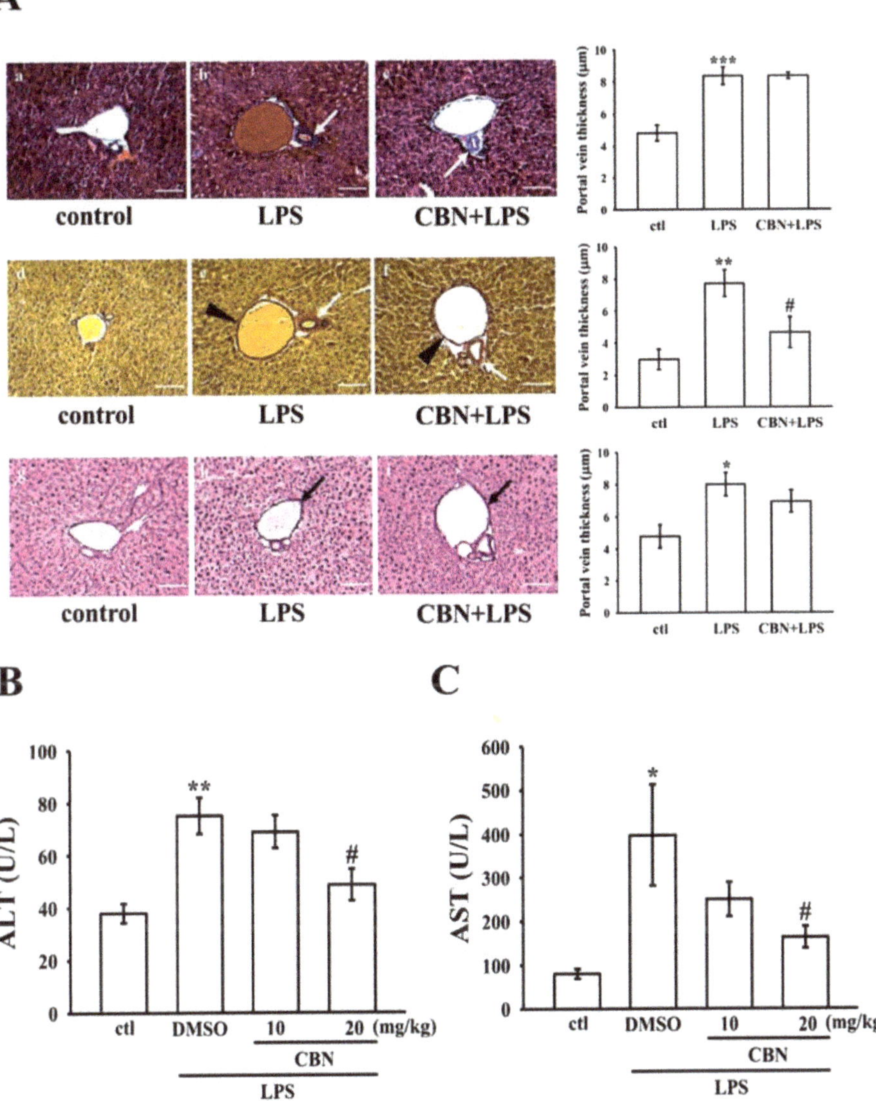

Figure 5. CBN dampens the LPS-induced hepatic injury in mice. (**A**) Effect of CBN on the histological alterations in liver tissue. Trichrome stain (**a–c**). The white arrow indicates the collagen deposition in the portal triad region of liver tissues. More abundant collagen fibers surrounded by arteriole and bile duct in LPS- and CBN + LPS-treated groups as compared to the control group. Weigert's resorcin fuchsin stain (**d–f**). Collagen fibers (red fibers indicated by white arrows) and elastic fibers (dark blue fibers indicated by black arrowheads) in control, LPS, and CBN + LPS groups. Reticulum stain (**g–i**). A network of fine black reticular fibers identified between the sinusoids and the plates of hepatocytes (indicated by black arrow) in the control, LPS, and CBN + LPS groups. The thickness of the portal vein was calculated using MShot Image Analysis System. Bar = 100 μm. (**B**,**C**) CBN reduces LPS-induced serum alanine aminotransferase (ALT) and aspartate aminotransferase (AST) activities. Illustrative views of control, LPS, CBN (20 mg/kg) + LPS groups are presented (magnification 20x). Data presented are the means ± S.E.M. (n = 6); * $p < 0.05$, ** $p < 0.01$, and *** $p < 0.001$, LPS vs. control group; # $p < 0.05$, LPS vs. CBN group. The statistical power range was from 0.8–0.9.

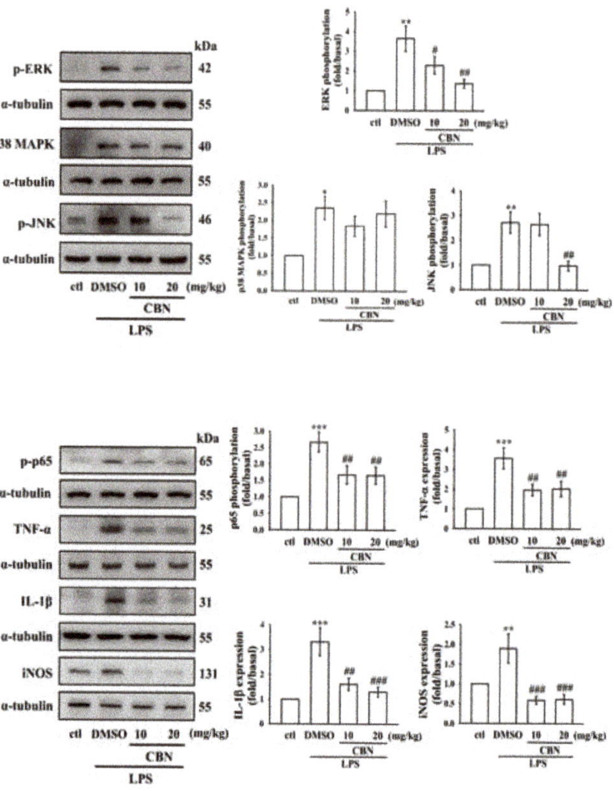

Figure 6. Effects of CBN on the expression of MAPK/NF-κB p65, TNF-α, and IL-1β and iNOS in LPS-induced liver tissue. Mice were treated with CBN (10 and 20 mg/kg) for 2 h and then induced by LPS (1 μg/mL) for 6 h. The expression of phosphorylated ERK, p38, JNK, and p65 and the protein expression of TNF-α, IL-1β, and iNOS were evaluated as described in the Materials and Methods. Data presented are the means ± S.E.M. ($n = 6$); * $p < 0.05$, ** $p < 0.01$, and *** $p < 0.001$, LPS vs. control group; # $p < 0.05$, ## $p < 0.01$, and ### $p < 0.001$, LPS vs. CBN group. The statistical power range was from 0.8–0.9.

4. Discussion

Studies have demonstrated that the production of inflammatory cytokines during the stimulation of macrophages plays a vital role in organ damage, including acute and chronic hepatic injury [24]. LPS induces liver injury by controlling oxidative stress and free radicals in hepatocytes [25]. Here, we showed that the ERK/JNK signaling molecules are important in the NF-κB-mediated induction of in vitro inflammatory cytokines, mediators, and in vivo liver injury, and that CBN administration is an effective modulator of inflammatory events. This study showed that free radicals are an important regulator of inflammation and that CBN treatment blocked hydroxyl radical formation and increased HO-1 expression in LPS-induced macrophage cells. These findings indicated that CBN offers its protective effect in LPS-induced macrophage inflammation at least in part through NF-κB- and MAPK-dependent mechanisms.

Reactive oxygen species activate superoxide anion, hydrogen peroxide, and hydrogen radicals. ROS are associated with different cellular and biological functions, such as controlling homeostasis, which is a significant factor for cell growth and survival [26]. However, the excess ROS formation will involve and play a significant role in the majority of pathophysiological events, including inflammation, via initiating intracellular pro-inflammatory mediators [27]. Moreover, ROS enhances MAPK and NF-κB activation in macrophages, resulting in the overexpression of genes, and can lead to the induction of inflammatory events [28]. A previous study demonstrated that ROS plays a dynamic role in septic shock and organ failure [26]. In addition, ROS is elevated in LPS-induced liver injury and antioxidants might be ideal agents to improve recovery from this injury [29]. In this study, we found that LPS significantly elevated the levels of hydroxyl radicals ($^{\bullet}OH$), which are inhibited by CBN in a concentration-dependent manner. Our result is substantiated with the findings of Jiang et al. [30] who found that LPS elevated the levels of free radicals, and that sophocarpine, an alkaloid, suppressed this elevation. Choi et al. reported that Bis (3-bromo-4,5-dihydroxybenzyl) ether, a bromophenol compound derived from a red alga, *Polysiphonia morrowii*, repressed LPS-induced inflammatory mediators by hindering the ROS-mediated ERK signaling pathway in RAW 264.7 cells [31]. Moreover, activation of HO-1 is a common response to oxidative stress as a cellular defense mechanism. Numerous lines of evidence show that expression of HO-1 induced by pharmacological treatment results in the reduction of cellular damage and inflammation. For instance, an anti-inflammatory role has been proposed for HO-1 [32]. This result is consistent with our finding that CBN enhances HO-1 expression in LPS-induced cells. Our results have shown that CBN ameliorates LPS-induced oxidative stress, as indicated by the suppressed $^{\bullet}OH$ production and increased HO-1 expression in RAW cells. These findings indicate that CBN could act as either a ROS inhibitor or antioxidant.

Mitogen-activated protein kinases play an important role in biological signal transduction into the nucleus from the cell membrane. MAPKs play a crucial role in various cellular processes, including gene expression, proliferation, cellular stress, and inflammatory response [33]. Studies have established that LPS could induce phosphorylation of ERK1/2, JNK, and p38 MAPKs in murine macrophages [34]. Further, the suppression of the MAPK pathway has been identified as an effective treatment to reduce inflammation [35]. NF-κB is involved in LPS-induced inflammatory signaling pathways, using inhibitory factor IκB in a resting condition. In LPS-activated cells, the NF-κB p65 subunit is detached due to IκB phosphorylation. Phosphorylated p65 translocates into the nucleus, where it regulates NF-κB-dependent target genes, including inflammatory mediators and cytokines [36]. Milani et al. [37] reported that β-carotene arrests nuclear translocation of NF-kB p65, which is correlated with its inhibitory effect on phosphorylation, and degradation of the NF-kB inhibitor. Similarly, β-carotene diminishes IκB phosphorylation and significantly hinders the nuclear translocation of NF-κB p65 [38]. Our earlier study revealed that CME-1, a novel polysaccharide, inhibited IκBα degradation and p65 Ser536 and MAPK phosphorylation in LPS-stimulated RAW 264.7 cells [39]. Our data support these results, showing that CBN inhibits NF-κB p65 nuclear translocation via hindering p65 phosphorylation on Ser536 and IκB degradation. Moreover, we used NF-κB (BAY 117082), ERK (PD98059), and JNK (SP600125) inhibitors to further explore the possible involvement of MAPK in NF-κB activation in CBN-mediated anti-inflammatory activity, and showed that CBN inhibits ERK and JNK in macrophage cells. The results revealed that that ERK1/2 and JNK signaling pathways mediate the LPS-induced NF-κB activation, inhibiting p65 phosphorylation. These findings also indicate the crucial role of ERK and JNK, but not p38MAPK, in the upstream regulator of NF-κB activation and they play vital role in the CBN-mediated anti-inflammatory effects.

LPS could induce inflammatory responses via elevating cytokines such as TNF-α, IL-6, and IL-1β, and hence inhibition of these cytokines could diminish the response of inflammation. The increased expression of iNOS can be attributed to the development of inflammation and liver injury, as this elevation can induce the production of NO, which is an additional stimulator involved in oxidative activity that in turn is involved in inflammation [35]. Moreover, NO is involved in a wide range of oxidative reactions, and LPS could induce a substantial increase in NO levels because of the elevated synthesis of iNOS. Hence, NO activation may be an initial marker of liver damage, and inhibition of this molecule could be a target for regulating inflammation [40]. In this study, CBN significantly reduced the expression of TNF-α, IL-1β, and iNOS in RAW cells and liver tissues, which confirmed CBN's anti-inflammatory and hepatoprotective effects. These results are consistent with an earlier study that reported anti-inflammatory effects of the polyphenol-enriched fraction (PEF) from *Acalypha wilkesiana* against LPS-induced inflammation [41]. It was found that PEF attenuated LPS-induced NO production and suppressed iNOS expression, and it also reduced the secretion of TNF-α, IL-1β, and IL-6 in LPS-stimulated macrophages. Our results also support those of Zhang et al. [42], who found that CBN suppressed TNF-α and IL-1β in the culture supernatant of THP-1 cells that had been stimulated by LPS.

Liver injury induced by LPS is linked with inflammatory mediators including superoxide, nitric oxide, TNF-α, IL-1β, IL-6, and other cytokines [43]. LPS stimulates NF-κB, leading to the initiation of many inflammatory genes, such as TNF-α and IL-1β [44]. Therefore, inactivation of NF-κB could attenuate LPS-induced liver injury. Studies have indicated that antioxidant and anti-inflammatory agents are valuable in LPS-induced hepatic injury [45]. This study indicated that administration of LPS caused increased ALT and AST activities in serum, measured as markers of liver injury. LPS-induced tissue injury results from an increase in the release of cytokines and oxidative stress [46], resulting the stimulation of apoptosis of hepatic cells and necrosis [47]. In this study, CBN pretreatment reduced the increases in ALT and AST enzymes and lowered the infiltration of inflammatory cells, fibrosis (elastic and reticular fibers), and collagen deposition induced by LPS in the liver tissues of mice, which were evidenced by trichrome, Weigert's resorcin fuchsin, and reticulum stains. Consistent with the in vitro findings, CBN also inhibits MAPKs and p65 phosphorylation, cytokines, and iNOS expression in the LPS-induced mouse liver. These results together specified the anti-inflammatory and hepatoprotective effect of CBN and its therapeutic potential for inflammatory diseases.

5. Conclusions

CBN exhibits compelling anti-inflammatory and hepatoprotective effects by preventing free radical formation and decreasing the expression of MAPK, especially ERK and JNK, followed by the suppression of NF-κB pathways. This in turn led to inhibition of NO, iNOS, TNF-α, and IL-1β in LPS-activated RAW cells and mouse liver. This study found that ERK and JNK act as upstream mediators of NF-κB activation in LPS-induced cells. CBN shows an antioxidant effect via increasing HO-1 expression in LPS-induced RAW cells. The detailed histological evaluation showed that CBN protects LPS-induced liver fibrosis as evidenced by decreased collagen, elastic, and reticular fibers. Though further study of the principal mechanisms is required, we concluded that CBN may offer a protective mechanism by controlling multiple signaling cascades, and this natural coumarin derivative could be used as a drug candidate for treating inflammation-mediated diseases.

Author Contributions: T.J., J.-R.S., and C.-H.H. designed and carried out the experiments, wrote original draft preparation. T.-H.F. performed histology work. S.-M.H., C.-C.C., C.-W.H., Y.-J.C., and W.-C.H. were involved in interpretation of data. P.S. and M.M. reviewed and edited the manuscript. All authors have read and agreed to the published version of the manuscript.

Funding: Ministry of Science and Technology of Taiwan (MOST107-2320-B-038-035-MY2 and MOST108-2320-B-038-031-MY3), Shin Kong Wu Ho-Su Memorial Hospital (2020SKHAND007 and 2021SK-HAND005), and Taipei Medical University (DP2-109-21121-01-N-08-03) are acknowledged for their grant support.

Institutional Review Board Statement: The Institutional Animal Care and Use Committee, Taipei Medical University, Taiwan (LAC-2016-0395) approved the animal experiments and care procedures.

Informed Consent Statement: Not applicable.

Data Availability Statement: The data presented in this study are available upon request on case-to-case basis.

Conflicts of Interest: The authors would like to confirm that there is no conflict of interest with this publication.

References

1. Medzhitov, R. Inflammation 2010: New adventures of an old flame. *Cell* **2010**, *140*, 771–776. [CrossRef]
2. Kankala, S.; Kankala, R.K.; Gundepaka, P.; Thota, N.; Nerella, S.; Gangula, M.R.; Guguloth, H.; Kagga, M.; Vadde, R.; Vasam, C.S. Regio selective synthesis of isoxazole–mercaptobenzimidazole hybrids and their in vivo analgesic and anti-inflammatory activity studies. *Bioorg. Med. Chem. Lett.* **2013**, *23*, 1306–1309. [CrossRef]
3. Shi, Q.; Cao, J.; Fang, L.; Zhao, H.; Liu, Z.; Ran, J.; Zheng, X.; Li, X.; Zhou, Y.; Ge, D.; et al. Geniposide suppresses LPS-induced nitric oxide, PGE2 and inflammatory cytokine by downregulating NF-kappaB, MAPK and AP-1 signaling pathways in macrophages. *Int. Immunopharmacol.* **2014**, *20*, 298–306. [CrossRef]
4. Karin, M.; Delhase, M. The I kappa B kinase (IKK) and NF-kappa B: Key elements of proinflammatory signaling. *Semin. Immunol.* **2000**, *12*, 85–98. [CrossRef]
5. Jurenka, J.S. Anti-inflammatory properties of curcumin, a major constituent of Curcuma longa: A review of preclinical and clinical research. *Altern. Med. Rev. J. Clin. Ther.* **2009**, *14*, 141–153.
6. Guha, M.; Mackman, N. LPS induction of gene expression in human monocytes. *Cell Signal.* **2001**, *13*, 85–94. [CrossRef]
7. Hommes, D.W.; Peppelenbosch, M.P.; van Deventer, S.J. Mitogen activated protein (MAP) kinase signal transduction pathways and novel anti-inflammatory targets. *Gut* **2003**, *52*, 144–151. [CrossRef] [PubMed]
8. Franceschi, C.; Campisi, J. Chronic inflammation (inflammaging) and its potential contribution to age-associated diseases. *J. Gerontol. A Biol. Sci. Med. Sci.* **2014**, *69*, S4–S9. [CrossRef] [PubMed]
9. Moret-Tatay, I.; Iborra, M.; Cerrillo, E.; Tortosa, L.; Nos, P.; Beltran, B. Possible biomarkers in blood for crohn's disease: Oxidative stress and micrornas-current evidences and further aspects to unravel. *Oxid. Med. Cell. Longev.* **2015**, *2016*, 2325162. [CrossRef]
10. Katakwar, P.; Metgud, R.; Naik, S.; Mittal, R. Oxidative stress marker in oral cancer: A review. *J. Can. Res. Ther.* **2016**, *12*, 438–446. [CrossRef]
11. So, B.E.; Bach, T.T.; Paik, J.H.; Jung, S.K. *Kmeria duperreana* (Pierre) Dandy extract suppresses LPS-induced iNOS and NO via regulation of NF-κB pathways and p38 in murin macrophage RAW 264.7 cells. *Prev. Nutr. Food Sci.* **2020**, *25*, 166–172. [CrossRef]
12. Manning, A.M.; Davis, R.J. Targeting JNK for therapeutic benefit: From junk to gold? *Nat. Rev. Drug. Discov.* **2003**, *2*, 554–565. [CrossRef]
13. Hao, G.; Wang, Z.G.; Fu, W.Y.; Yang, Y. Research progress on effect of coumarins compounds in anti-tumor. *Chin. J. Chin. Materia. Medica.* **2008**, *33*, 2016–2019.
14. Zhang, C.Y.; Zhang, B.G.; Yang, X.W. Studies on the chemical constituents of the root of Angelica pubescens f. biserrata. *Pharm. J. Chin. People's Lib. Arm.* **2007**, *23*, 241–245.
15. Li, R.Z.; He, Y.Q.; Qiao, M.; Xu, Y.; Zhang, Q.B.; Meng, J.R.; Gu, Y.; Ge, L.P. Studies of the active constituents of the Chinese drug 'Duhuo' Angelica pubescens. *Acta Pharm. Sin.* **1989**, *24*, 546–551.
16. Chen, Y.F.; Tsai, H.Y.; Wu, T.S. Anti-inflammatory and analgesic activities from roots of *Angelica pubescens*. *Planta Med.* **1995**, *61*, 2–8. [CrossRef] [PubMed]
17. Yang, X.W.; Guo, Q.M.; Wang, Y. Absorption and transport of 6 coumarins isolated from the root of *Angelica pubescens* f. biserrata in human Caco-2 cell monolayer model. *J. Chin. Integr. Med.* **2008**, *6*, 392–398. [CrossRef] [PubMed]
18. Hou, S.M.; Hsia, C.W.; Tsai, C.L.; Hsia, C.H.; Jayakumar, T.; Velusamy, M.; Sheu, J.R. Modulation of human platelet activation and in vivo vascular thrombosis by columbianadin: Regulation by integrin αIIbβ3 inside-out but not outside-in signals. *J. Biomed. Sci.* **2020**, *27*, 60. [CrossRef] [PubMed]
19. Chou, D.S.; Hsiao, G.; Shen, M.Y.; Tsai, Y.J.; Chen, T.F.; Sheu, J.R. ESR spin trapping of a carbon-centered free radical from agonist-stimulated human platelets. *Free Radic. Biol. Med.* **2005**, *39*, 237–248. [CrossRef] [PubMed]
20. Poss, K.D.; Tonegawa, S. Reduced stress defense in heme oxygenase 1-deficient cells. *Proc. Natl. Acad. Sci. USA* **1997**, *94*, 10925–10930. [CrossRef]
21. Jun, M.S.; Ha, Y.M.; Kim, H.S.; Jang, H.J.; Kim, Y.M.; Lee, Y.S.; Kim, H.J.; Seo, H.G.; Lee, J.H.; Lee, S.H.; et al. Anti-inflammatory action of methanol extract of *Carthamus tinctorius* involves in heme oxygenase-1 induction. *J. Ethnopharmacol.* **2011**, *133*, 524–530. [CrossRef] [PubMed]
22. Zhang, P.; Martin, M.; Michalek, S.M.; Katz, J. Role of mitogen-activated protein kinases and NF-κB in the regulation of proinflammatory and anti-Inflammatory cytokines by *Porphyromonas gingivalis* Hemagglutinin B. *Infect. Immun.* **2005**, *73*, 3990–3998. [CrossRef] [PubMed]
23. Salminen, A.; Huuskonen, J.; Ojala, J.; Kauppinen, A.; Kaarniranta, K.; Suuronen, T. Activation of innate immunity system during aging: NF-kB signaling is the molecular culprit of inflamm-aging. *Ageing Res. Rev.* **2008**, *7*, 83–105. [CrossRef] [PubMed]
24. Nolan, J.P. The role of intestinal endotoxin in liver injury: A long and evolving history. *Hepatology* **2010**, *52*, 1829–1835. [CrossRef] [PubMed]
25. Franco, R.; Cidlowski, J.A. Glutathione efflux and cell death. *Antioxid. Redox Signal.* **2012**, *17*, 1694–1713. [CrossRef]
26. Rhee, S.G. Cell signaling. H_2O_2, a necessary evil for cell signaling. *Science* **2006**, *312*, 1882–1883. [CrossRef]

27. Brune, B.; Dehne, N.; Grossmann, N.; Jung, M.; Namgaladze, D.; Schmid, T.; von Knethen, A.; Weigert, A. Redox control of inflammation in macrophages. *Antioxid. Redox Signal.* **2013**, *19*, 595–637. [CrossRef]
28. Son, Y.; Cheong, Y.K.; Kim, N.H.; Chung, H.T.; Kang, D.G.; Pae, H.O. Mitogen-activated protein kinases and reactive oxygen species: How can ROS activate MAPK pathways? *J. Signal Transduc.* **2011**, *2011*, 792639. [CrossRef]
29. Zapelini, P.H.; Rezin, G.T.; Cardoso, M.R.; Ritter, C.; Klamt, F.; Moreira, J.C.F.; Streck, E.M.; Pizzol, F.D. Antioxidant treatment reverses mitochondrial dysfunction in a sepsis animal model. *Mitochondrion* **2008**, *8*, 211–218. [CrossRef]
30. Jiang, Z.; Meng, Y.; Bo, L.; Wang, C.; Bian, J.; Deng, X. Sophocarpine attenuates LPS-induced liver injury and improves survival of mice through suppressing oxidative Stress, inflammation, and apoptosis. *Mediat. Inflamm.* **2018**, *2018*, 5871431. [CrossRef]
31. Choi, Y.K.; Ye, B.R.; Kim, E.A.; Kim, J.; Kim, M.S.; Lee, W.W.; Ahn, G.N.; Kang, N.; Jung, W.K.; Heo, S.J. Bis (3-bromo-4,5-dihydroxybenzyl) ether, a novel bromophenol from the marine red alga Polysiphonia morrowii that suppresses LPS-induced inflammatory response by inhibiting ROS-mediated ERK signaling pathway in RAW 264.7 macrophages. *Biomed. Pharmacother.* **2018**, *103*, 1170–1177. [CrossRef] [PubMed]
32. Inoue, S.; Suzuki, M.; Nagashima, Y.; Suzuki, S.; Hashiba, T.; Tsuburai, T.; Ikehara, K.; Matsuse, T.; Ishigatsubo, Y. Transfer of heme oxygenase 1 cDNA by a replication-deficient adenovirus enhances interleukin 10 production from alveolar macrophages that attenuates lipopolysaccharide-induced acute lung injury in mice. *Hum. Gene Ther.* **2001**, *12*, 967–979.
33. Thalhamer, T.; McGrath, M.A.; Harnett, M.M. MAPKs and their relevance to arthritis and inflammation. *Rheumatology* **2008**, *47*, 409–414. [CrossRef] [PubMed]
34. Bode, J.G.; Ehlting, C.; Häussinger, D. The macrophage response towards LPS and its control through the p38MAPK–STAT3 axis. *Cell. Signal.* **2012**, *24*, 1185–1194. [CrossRef] [PubMed]
35. Pelletier, J.P.; Fernandes, J.C.; Brunet, J.; Moldovan, F.; Schrier, D.; Flory, C.; Martel-Pelletier, J. In vivo selective inhibition of mitogen-activated protein kinase kinase $\frac{1}{2}$ in rabbit experimental osteoarthritis is associated with a reduction in the development of structural changes. *Arthritis Rheum.* **2003**, *48*, 1582–1593. [CrossRef] [PubMed]
36. Weng, L.; Zhang, H.; Li, X.; Zhan, H.; Chen, F.; Han, L.; Cao, X. Ampelopsin attenuates lipopolysaccharide-induced inflammatory response through the inhibition of the NF-κB and JAK2/STAT3 signaling pathways in microglia. *Intl. Immunopharmacol.* **2017**, *44*, 1–8. [CrossRef] [PubMed]
37. Milani, A.; Basirnejad, M.; Shahbazi, S.; Bolhassani, A. Carotenoids: Biochemistry, pharmacology and treatment. *Brit. J. Pharmacol.* **2017**, *174*, 1290–1324. [CrossRef]
38. Li, R.; Hong, P.; Zheng, X. β-carotene attenuates lipopolysaccharide-induced inflammation via inhibition of the NF-κB, JAK2/STAT3 and JNK/p38 MAPK signaling pathways in macrophages. *Anim. Sci. J.* **2019**, *90*, 140–148. [CrossRef]
39. Sheu, J.R.; Chen, Z.C.; Hsu, M.J.; Wang, S.H.; Jung, K.W.; Wu, W.F.; Pan, S.H.; Teng, R.D.; Yang, C.H.; Hsieh, C.Y. CME-1, a novel polysaccharide, suppresses iNOS expression in lipopolysaccharide-stimulated macrophages through ceramide-initiated protein phosphatase 2A activation. *J. Cell Mol. Med.* **2018**, *22*, 999–1013.
40. Tirapelli, L.F.; Batalhão, M.E.; Jacob-Ferreira, A.L.; Tirapelli, D.P.; Carnio, E.C.; Tanus-Santos, J.E.; Queiroz, R.H.; Uyemura, S.A.; Padovan, C.M.; Tirapelli, C.R. Chronic ethanol consumption induces histopathological changes and increases nitric oxide generation in the rat liver. *Tissue Cell.* **2011**, *43*, 384–391. [CrossRef]
41. Wu, H.; Pang, H.; Chen, Y.; Huang, L.; Liu, H.; Zheng, Y.; Sun, C.; Zhang, G.; Wang, G. Anti-Inflammatory effect of a polyphenol-enriched fraction from *Acalypha wilkesiana* on lipopolysaccharide-stimulated RAW 264.7 macrophages and acetaminophen-induced liver injury in mice. *Oxid. Med. Cell. Longevit.* **2018**, *2018*, 7858094. [CrossRef] [PubMed]
42. Zhang, C.; Hsu, A.C.Y.; Pan, H.; Gu, Y.; Zuo, X.; Dong, B.; Wang, Z.; Zheng, J.; Lu, J.; Zheng, R.; et al. Columbianadin suppresses lipopolysaccharide (LPS)-induced inflammation and apoptosis through the NOD1 pathway. *Molecules* **2019**, *24*, 549. [CrossRef] [PubMed]
43. Zhong, W.; Qian, K.; Xiong, J.; Ma, K.; Wang, A.; Zou, Y. Curcumin alleviates lipopolysaccharide induced sepsis and liver failure by suppression of oxidative stress-related inflammation via PI3K/AKT and NF-κB related signaling. *Biomed. Pharmacother.* **2016**, *83*, 302–313. [CrossRef] [PubMed]
44. Dan, C.; Jinjun, B.; Zi-Chun, H.; Lin, M.; Wei, C.; Xu, Z.; Ri, Z.; Shun, C.; Wen-Zhu, S.; Qing-Cai, J.; et al. Modulation of TNF-α mRNA stability by human antigen R and miR181s in sepsis-induced immunoparalysis. *EMBO Mol. Med.* **2015**, *7*, 140–157. [CrossRef] [PubMed]
45. Ajuwon, O.R.; Oguntibeju, O.O.; Marnewick, J.L. Amelioration of lipopolysaccharide-induced liver injury by aqueous rooibos (*Aspalathus linearis*) extract via inhibition of pro-inflammatory cytokines and oxidative stress. *BMC Complement. Altern. Med.* **2014**, *14*, 392. [CrossRef]
46. Lowes, D.A.; Webster, N.R.; Murphy, M.P.; Galley, H.F. Antioxidants that protect mitochondria reduce interleukin-6 and oxidative stress, improve mitochondrial function, and reduce biochemical markers of organ dysfunction in a rat model of acute sepsis. *Br. J. Anaesth.* **2013**, *110*, 472–480. [CrossRef]
47. Wang, Y.; Gao, L.N.; Cui, Y.L.; Jiang, H.L. Protective effect of danhong injection on acute hepatic failure induced by lipopolysaccharide and d-galactosamine in mice. *Evid. Based Complement. Alternat. Med.* **2014**, *2014*, 153902.

 *antioxidants* 

Article

Rebaudioside A Enhances Resistance to Oxidative Stress and Extends Lifespan and Healthspan in *Caenorhabditis elegans*

Pan Li [1,2], Zehua Wang [1,2], Sin Man Lam [1,3,*] and Guanghou Shui [1,2,*]

1 State Key Laboratory of Molecular Developmental Biology, Institute of Genetics and Developmental Biology, Chinese Academy of Sciences, Beijing 100101, China; lipan17@mails.ucas.ac.cn (P.L.); wangzh2014@genetics.ac.cn (Z.W.)
2 University of Chinese Academy of Sciences, Beijing 100049, China
3 LipidALL Technologies Company Limited, Changzhou 213022, Jiangsu, China
* Correspondence: smlam@genetics.ac.cn (S.M.L.); ghshui@genetics.ac.cn (G.S.); Tel.: +86-(10)-6480-6670 (S.M.L.); +86-(10)-6480-7781 (G.S.)

Abstract: Non-nutritive sweeteners are widely used in food and medicines to reduce energy content without compromising flavor. Herein, we report that Rebaudioside A (Reb A), a natural, non-nutritive sweetener, can extend both the lifespan and healthspan of *C. elegans*. The beneficial effects of Reb A were principally mediated via reducing the level of cellular reactive oxygen species (ROS) in response to oxidative stress and attenuating neutral lipid accumulation with aging. Transcriptomics analysis presented maximum differential expression of genes along the target of rapamycin (TOR) signaling pathway, which was further confirmed by quantitative real-time PCR (qPCR); while lipidomics uncovered concomitant reductions in the levels of phosphatidic acids (PAs), phosphatidylinositols (PIs) and lysophosphatidylcholines (LPCs) in worms treated with Reb A. Our results suggest that Reb A attenuates aging by acting as effective cellular antioxidants and also in lowering the ectopic accumulation of neutral lipids.

Keywords: *C. elegans*; non-nutritive sweetener; rebaudioside A; aging; oxidative stress resistance; TOR; lipidomics

Citation: Li, P.; Wang, Z.; Lam, S.M.; Shui, G. Rebaudioside A Enhances Resistance to Oxidative Stress and Extends Lifespan and Healthspan in *Caenorhabditis elegans*. *Antioxidants* **2021**, *10*, 262. https://doi.org/10.3390/antiox10020262

Academic Editors: Chiara Nediani and Monica Dinu
Received: 2 January 2021
Accepted: 2 February 2021
Published: 8 February 2021

1. Introduction

Obesity and its related spectrum of metabolic disorders have emerged as a global health epidemic in recent decades. It is widely believed that several risk factors, such as high-sugar diet or high-fat diet, partly contribute to the increasing occurrence of obesity and its related health problems, including type 2 diabetes, fatty liver disease, cardiovascular disease, and hypertension [1]. Non-nutritive sweeteners (NNSs) are increasingly used to replace traditional caloric sugars to reduce energy intake while maintaining tasting flavor in foods and beverages. NNSs are particularly useful alternative sweeteners for diabetics and individuals who need to control their blood sugar levels and minimize excessive weight gain [2]. Saccharin, aspartame, acesulfame potassium (Ace K), sucralose, neotame, rebaudioside A (Reb A), and cyclamate are food additives approved by the U.S. Food and Drug Administration (FDA) as Generally Recognized as Safe Status (GRAS) for consumption within their reasonable dose and daily intake limits [3–5]. The safety of NNSs, however, has remained somewhat controversial, as some of their unexpected impacts on metabolism are only beginning to unfold with recent investigation. For example, it was found that sucralose could release water and hydrogen chloride on heating to 125 °C [6], and promote the accumulation of reactive oxygen species (ROS) in mice. Sucralose also has adverse effects on glucose metabolism and gut hormones and induces intensely sweet taste preference that ultimately leads to irrational food intake [3,7,8]. Artificial NNSs application was also reported to elevate the risk of various cancers [9]. Unlike artificial NNSs, Reb A as a natural NNS has been found to possess a variety of biological functions beneficial to organisms.

Reb A is a natural product obtained from the extracts of *Stevia rebaudiana* (Bertoni), a perennial shrub in the sunflower family of Asteraceae (Compositae), which is native to the border area in Paraguay and Brazil. It was discovered that *Stevia* leaves are 30 times sweeter than sucrose, whereas stevia glycosides isolated from the leaves are 200–300 times sweeter than sucrose [10]. Stevia glycoside is also considered to be a safe natural NNS, with few side effects in animal experiments and human studies of hypertensive patients [11]. Reb A and stevioside isolated from *Stevia* share the basic structure of stevia glycosides [12,13]. Purified Reb A was approved by FDA in 2008 as GRAS. In contrast to most other NNSs, previous studies have demonstrated that Reb A is beneficial to metabolic health. For instance, Reb A was found to influence the cell cycle of human epidermal cells by increasing the distribution of cells in the S phase, while reducing that in G2/M phase [14]. Reb A also inhibits the inflammatory response in lipopolysaccharide-activated RAW264.7 mouse macrophage cells by impeding the secretion of interleukin-1α/-1β and other cytokines [15], and protects against tetrachloromethane-induced oxidative injury in human liver hepatocellular carcinoma (HepG2) cells [16]. Mice treated with long-term and low-dose Reb A were found to have altered gut microbiota composition with minimal effect on glucose metabolism and weight gain [17]. Reb A exerts anti-inflammatory, antioxidant, and antifibrotic effects on hepatic fibrosis induced by thioacetamide in rats [18]. Reb A was also found to significantly ameliorate murine non-alcoholic steatohepatitis and decrease hepatocyte triglyceride level [19], albeit the precise mechanism has remained obscured. The effect of Reb A on physiological aging, as well as on the healthspan and lifespan in model organisms, has not been previously investigated.

Caenorhabditis elegans (*C. elegans*) has been used as a model organism to study various biological processes, including cell polarity, cell signaling, cell cycle, gene regulation, senescence, autophagy, and metabolic processes [20]. *C. elegans* is also widely applied for rapidly screening metabolic responses to drugs or natural extracts, which provides the basis for mammalian biological experiments [21]. Moreover, worms represent an excellent model for aging studies. Their behavior and physical indices, such as pumping rate, lipofuscin accumulation, and locomotion can reflect aging rate and health status to a certain extent, facilitating investigation on the effects of different metabolites on age-related changes [22,23]. At present, several theories propose different underlying mechanisms of aging, which include the free radical damage theory, caloric restriction theory, and telomere senescence theory. The free radical aging hypothesis proposes that oxidative damages infringed upon cells and tissues may cause aging [24]. It is worth noting that reactive oxygen species (ROS) are considered to be harmful by-products from aerobic respiration, causing oxidative damage and accelerating the development of age-dependent diseases when accumulated in excess [25,26].

Lipids, especially polyunsaturated fatty acids (PUFAs), are susceptible to ROS attacks, and lipid peroxidation plays an important role in intracellular aging [27]. In addition to exerting structural functions, specific lipids, such as phosphatidylinositols (PIs) and phosphatidylinositol phosphates (PIPs), also partake in signal transduction, membrane trafficking, and other biological processes [28]. Mammalian target of rapamycin (mTOR) pathway integrates signals from growth factors and nutrients to facilitate cellular metabolic transition from catabolism to anabolism, promoting cell growth and cell cycle progression [29]. mTOR is activated in response to increases in cellular phosphatidic acids (PAs), which interact and stabilize the FK506-binding protein-12-rapamycin-binding (FRB) domains of mTOR complexes 1 (mTORC1) and mTORC2 [30].

Herein, we explored the effects of Reb A on the healthspan and lifespan of nematodes and investigated the potential mechanisms underlying Reb A-induced metabolic changes using a combination of transcriptomics and lipidomics approaches.

2. Materials and Methods

2.1. Strain and Culture Conditions

The wild-type strain N2 and *Escherichia coli* strain OP50-1 were obtained from the *Caenorhabditis* Genetics Center (CGC) at the University of Minnesota (Minneapolis, MN, USA), while mutant strains *xdEx1001* (P*daf-22*-PLIN1:GFP) were gifts from Dr Xun Huang's laboratory (IGDB, China). Streptomycin-resistant *Escherichia coli* strain OP50-1 as a unique food source can exclude interference from other bacteria. Nematodes were cultured on nematodes growth media (NGM) with active OP50-1 as the food source in biochemical incubators at 20 °C for all experiments.

All experiments were performed on age-synchronized worms and cultured to the L4 laval stage on normal NGM [31]. Synchronized cohorts of *C. elegans* were prepared using the bleaching method (10% NaClO/ddH$_2$O/10 M NaOH = 2.4:16.6:1). First, reproductive adult hermaphrodites were washed in 50 mL falcon tube using M9 buffer (5.8 g Na2HPO4, 3.0 g KH2PO4, 0.5 g NaCl, and 1.0 g NH4Cl dissolved in 1 L ddH$_2$O) containing 0.005% NP40, and further washed by M9 buffer three times to remove traces of NP40 and bacteria. The resultant worms and embryos were resuspended in the bleach solution, and embryos obtained after bleaching were rinsed three times with M9 buffer. The suspension was filtered through a 40 µm cell sieve to separate embryos from worm carcasses. Embryos were transferred onto NGM plates without food for 12 h to hatch and arrest at L1 stage, and synchronized L1 larva were washed off the plates and deposited onto NGM plates with active OP50-1 bacteria to grow to the L4 stage. L4 stage worms were used for subsequent experiments.

The Reb A NGM were prepared as follow: 1.67 mM Reb A (0.32 g Reb A powder (Aladdin Chemicals Co., Shanghai, China) in 200 mL ddH$_2$O) filtered with 0.2 µm cell fiter (Sartorius, Beijing, China) and stored in 4 °C prior to use. Stock solutions containing 7.5, 15, 22.5, and 30 mL of Reb A were added into 1.5 L NGM to get final concentrations of 0.0083, 0.017, 0.033, and 0.05 mM Reb A, respectively. Standard NGM was used as the control group. Both Reb A treatment group and control group used inactive OP50-1 (65 °C, 35 min) as the food source to reduce the effect of bacterial metabolism on nematode metabolism [32].

2.2. Pharyngeal Pumping Rate Assay

The pharyngeal pumping rate was monitored and quantified on the 5th and 10th day of adulthood, sequentially, under a stereozoom (Motic, Xiamen, China) as described previously [33]. Worms from Reb A (0.0083, 0.017, 0.033, and 0.05 mM), neotame (0.5, 1, 2, and 4 mM), and Raffiose (1, 2, 3, and 4 mM) treatment groups and a control group were evaluated. Fifteen worms from each group were randomly selected for the assay. At room temperature, the pharyngeal pumping frequency of each nematode was counted over a 20 s interval for three times, at room temperature.

2.3. Lipofuscin Assay

Assay of lipofuscin level in vivo was conducted, as previously described [34]. A total of 200 worms from the 0.0083 mM Reb A and the control group grown to the 8th day of adulthood were thoroughly washed with M9 buffer to remove bacteria. Worms were placed on 2% agarose pad and anesthetized with 10 mM levamisole. Images were captured by Observer Z1 laser scanning confocal microscope (Zeiss, Jena, Germany) under DAPI filter. Thirty worms from each group were captured and experiments were performed in triplicates.

2.4. Lifespan Assay

Hermaphrodites were crossed three times with males before lifespan assay. Approximately 200 L4 stage worms were used for the assay, as previously described [35]. Briefly, 200 worms from each group were transferred to 6 cm plates. During the reproduction period, worms were transferred to fresh plates each day to exclude hatching embryos.

Worms with abnormal behaviors such as crawling off the plate, died away from the agar and abnormal phenotypes such as "bagging", i.e., the hatching of live progeny inside the hermaphrodite, or "explosion", i.e., the eruption of intestines from the vulva, were censored from the analysis [36]. Percent survival on each plate was recorded daily until all the worms were dead. Worms that did not respond to a mechanical stimulus was scored as dead. A Kaplan–Meier lifespan analysis was conducted, and *p* value was calculated using a log-rank test.

2.5. Swimming Assay

Swimming assays were performed as previously described [37]. At least 30 worms on the 5th and 10th day of adulthood were randomly selected and transferred to a glass slide containing 10 μL of M9 buffer. They were allowed to acclimate to liquid medium for 10 s, and then body swing for 30 s was scored. The experiment was repeated three times.

2.6. Osmotic Avoidance Assay

Osmotic avoidance assay was conducted at 20 °C, as previously described [38]. Synchronized L4 stage worms were transferred onto 9 cm NGM plates with or without 0.0083 mM Reb A treatment. On reaching Day 1 of adulthood, hermaphrodites were placed on a blank NGM with a small drop of 0.3 M glycerin solution dissolved in M9 buffer in the worm's forward path. If the worm stopped moving forward, positive osmotic avoidance was scored. Each nematode was tested 5 times and the percent osmotic avoidance of each nematode was calculated. A total of 20 worms were tested for each group.

2.7. Brood Size Assay

Swimming assays were performed, as previously described [39]. Two worms at L4 stage were transferred to each 3 cm NGM plates with or without 0.0083 mM Reb A. Worms were transferred to fresh plates daily over the next 7 days, and the total number of embryos laid was counted. The experiment was repeated three times.

2.8. Body Length and Body Width

This assay was carried, out as previously described [40]. L4 stage worms were cultured on NGM plates with or without Reb A (0.0083, 0.017, 0.033, and 0.05 mM), and photos were taken with an Olympus MVX10 fluorescence microscope (Olympus, Beijing, China) on the first day of adulthood. Photos of 50 worms were evaluated from each group, and the experiment was carried out in triplicate.

2.9. Heat Stress Assay

Heat stress assay was performed, as previously described [41]. Briefly, 100 L4 stage worms were put onto 6 cm NGM plates with or without 0.0083 mM Reb A, and allowed to grow to the 5th day of adulthood at 20 °C. The temperature was shifted to 35 °C for 5 h heat shock. After heat shock, worms were allowed to recover at 20 °C for 96 h and the number of alive and dead worms were scored. Experiments were conducted in triplicate and the results were presented as the percentage of surviving nematodes.

2.10. Acute Oxidative Stress Assay

Acute oxidative stress assay was performed, as previously described with modification [42]. A total of 1000 worms at L4 stage were cultured on 9 cm NGM plates with or 0.0083 mM Reb A to the 6th day of adulthood at 20 °C. Worms were washed off plates with M9 buffer and about 100 worms were transferred to transparent 96-well plates containing 200 mM paraquat dissolved in double deionized water (ddH_2O) in each well. Worms were treated for 6 h, and the lid cover was opened at 1 h intervals to supply oxygen. After 6 h, the percent survival from each group was scored.

2.11. Accumulation of Cellular Reactive Oxygen Species (ROS)

This assay was carried out, as previously described [43]. ROS accumulation was evaluated in worms on the 6th day of adulthood in the presence of 7 mM paraquat treatment for 2 h. At the end of treatment, 190 μL of worm suspension was added to individual wells in a 96-well black plate each containing 10 μL of 10 mM H2DCF-DA (MCE, Shanghai, China). The fluorescence intensity of each well was measured with 485 nm excitation and 535 nm emission filters at 30 min intervals for a total duration of 6 h via multifunction reader (BioTek, Winooski, VT, USA). Worms were scored under stereozoom in each well and data were normalized to worm number.

2.12. Food Intake Assay

This assay was carried out in liquid culture medium, as previously described [44]. Worms were cultured in 190 μL of S-complete buffer (5.8 g NaCl, 50 mL 1 M phosphate buffer solution dissolved in 924 mL ddH_2O, and after sterilization added 3 mL 1 M $MgSO_4$, 6 mL 0.5 M $CaCl_2$, 6 mL 100× trace metal solution, 1 mL 5 mg/mL cholesterol, and 10 mL 1 M potassium citrate) with 50 μg/mL carbenicillin and 0.1 μg/mL fungizone in black, flat-bottom, optically clear 96-well plates. Each well contained 15 nematodes and 10 μL of 6 mg/mL OP50-1. Age-synchronized nematodes were seeded at L1 larva and grown at 20 °C. Plates were covered with sealers (Biorad, Rockville, MD, USA) to prevent evaporation. A final concentration of 12 mM 5-fluoro-2′-deoxyuridine (FUDR) was added to prevent egg laying. Reb A was added at L4 stage in the treatment group. Absorbance at 600 nm (OD600) in each well was measured using microplate spectrophotometer (EON, Winooski, VT, USA). Measurements were taken every 24 h starting from 1st of adulthood. Before measuring OD600, each plate was placed onto a plate shaker for 25 min. Living animals in each well were scored microscopically on the basis of movement on the 4th day of adulthood. For comparisons between different treatments, food intake was expressed as a percentage relative to worms in the control group.

2.13. RNA Extraction and Transcriptome Analysis

RNA was extracted using RNA simple Total RNA Kit (TIANGEN, Beijing, China). Briefly, 3000 worms were collected on 10th day of adulthood, snap-frozen using liquid nitrogen and stored at −80 °C. Two spoons of sterilized ceramic beads and 500 μL Trizol buffer were added to each sample and worms were homogenized on a bead ruptor (OMNI, Seattle, WA, USA), followed by adding 500 μL Trizol buffer and incubated at room temperature for 5 min. Chloroform was added for phase separation, and the colorless aqueous phase was transferred to new 1.5 mL tubes following centrifugation at 4 °C, 12000 rpm for 10 min. Anhydrous ethanol was slowly added to aqueous phase and sample suspensions were transferred to CR3 columns for RNA purification. The library construction and sequencing were performed at Shanghai Biotechnology Corporation with an Illumina HiSeq 2500 instrument (Illumina, San Diego, CA, USA). RNA integrity was assessed using the RNA Nano 6000 Assay Kit of the Agilent Bioanalyzer 2100 system (Agilent Technologies, Santa Clara, CA, USA). Gene ontology (GO) enrichment analysis of the differentially expressed genes (DEGs) was implemented by the GOseq R packages based Wallenius non-central hyper-geometric distribution [45], which can adjust for gene length bias in DEGs. KEGG [46] is a database resource for understanding high-level functions and utilities of the biological system, such as the cell, the organism, and the ecosystem, from molecular-level information, especially large-scale molecular datasets generated by genome sequencing and other high-throughput experimental technologies (http://www.genome.jp/kegg/, accessed on 25 January 2021). We used KOBAS software to test the statistical enrichment of DEGs in KEGG pathway database.

2.14. Postfix Nile Red (NR) Staining

Postfix Nile red (NR) staining on the 5th and 10th day of adulthood were performed, as described [47]. Worms were washed using 1 mL PBST (PBS with 0.01% Triton X-100) and centrifuged at 560× g for 1 min. Supernatant was discarded. Worms were washed repeatedly for three times until the supernatant turned clear. The worms were incubated with 100 µL of 40% isopropanol, and then centrifuged at 560× g for 1 min. Worms were incubated in darkness for 2 h in 600 µL of working solution (1 mL 100% isopropanol containing 6 µL 5 mg/mL Nile red powder). After centrifugation at 560× g for 1 min, supernatant was discarded and 600 µL PBST was added. The worms were incubated for 30 min to remove excess NR staining. After removing most of the staining solution, the fixed worms were mounted onto 2% agarose pads for microscopic observation and photography with Observer Z1 Laser-scanning confocal microscope (Zeiss, Jena, Germany). Image J software was used to split picture to three color channels, and relative fluorescence intensity under the red channel was calculated.

2.15. Oxygen Consumption Rate (OCR) Measurement

The oxygen consumption rate (OCR) measurement was carried out, as previously described [48]. We used Seahorse Xfe96 (Agilent, Santa Clara, CA, USA) to detect 10th day adult OCR under normal culture condition and under oxidative stress. Briefly, after hydrating the probe and system equilibration, we put 3–20 worms into each sample wells for OCR measurement. The results obtained were normalized to worm number in each well to reflect the worm mean OCR.

2.16. Malondialdehyde (MDA) Content Assay

The malondialdehyde (MDA) content was quantitated by MDA kit (Solarbio, Beijing, China) using worms on the 6th day of adulthood after exposure to oxidative stress induced by 7 mM paraquat. Sample absorbance at 450, 532, and 600 nm were measured using multifunctional reader (BioTek, VT, USA). Piece bicinchoninic acid (BCA) (Thermo Fisher, Waltham, MA, USA) kit was used to quantify MDA content.

2.17. Lipid Extraction and MS Analysis

Lipids were extracted from 10,000 worms on the 10th day of adulthood, as previously described using a modified version of the Bligh and Dyer's protocol [49]. Lipidomics analyses were conducted on an 1260 HPLC (Agilent, CA, USA) coupled with a QTRAP 5500 (Sciex, Framingham, MA, USA), as previously described [50,51]. Lipids were quantitated by referencing to spiked internal standards, which included PA-34:0(17:0/17:0), PC-d_{31}-34:1(16:0/18:1), PE-d_{31}-34:1(16:0/18:1), PG-d_{31}-34:1(16:0/18:1), d_7-PI33:1(15:0/18:1), PS-d_{31}-34:1(16:0/18:1), LPC-d_4-26:0, LPE-C17:1, LPG C17:1, LPI C17:1, LPS-C17:1, CL 22:1(3)/14:1, DAG 16:0/16:0-d_5, DAG 18:1/18:1-d_5 from Polar Lipids (Avanti, AL, USA) and TAG $(16:0)_3$-d_5, TAG $(14:0)_3$-d_5, TAG $(15:0)_3$-d_{29} from isotopes (CDN, Montreal, QC, Canada).

2.18. Statistical Analysis

All results are presented as mean ± SEM, and statistical analyses were performed using SPSS software 24.0 (SPSS Inc., Chicago, IL, USA) and presented by GraphPad Prism 7.0 software (GraphPad software, San Diego, CA, USA). For two group comparison, unpaired two tailed *t*-test was applied. Multiple group comparisons were conducted using one-way analysis of variance (ANOVA), and pairwise statistical significance was evaluated by Dunnet's test.

3. Results

3.1. Reb A Extends Lifespan in C. elegans

C. elegans exhibits several changes in physiological phenotypes during senescence. For instance, pharyngeal pumping rate and lipofuscin accumulation are well-known hallmarks of aging for *C. elegans*. Pharyngeal pumping rate increases gradually throughout the developmental larval stages and peaks at the L4 stage, and then decreases gradually during aging [52]. The accumulation of lipofuscin, which comprises highly oxidized proteins and lipids, is also observed with aging [22]. Preliminary screening of a few sweeteners demonstrated the potential of Reb A with regard to anti-aging, while treatment with neotame or low caloric sweetener raffinose exerted adverse effects on nematodes (Figure S1). On the fifth day of adulthood, Reb A (0.0083, 0.017, and 0.033 mM) treated worms showed significantly elevated pharyngeal pumping rate, whereas in 10-day-old adults, only worms treated with 0.0083 mM Reb A exhibited increased pharyngeal pumping rates relative to the control (Figure 1A). In addition, a preliminary investigation using various concentrations of Reb A indicated that 0.0083 mM Reb A treatment prolonged the lifespan of nematodes to the greatest extent (Figure S2). Therefore, we had subsequently chosen to investigate the metabolic effects of treating worms with 0.0083 mM Reb A. We further examined in vivo lipofuscin accumulation with aging by detecting blue autofluorescence (via a DAPI filter) but observed no significant changes in lipofuscin autofluorescence in 8-day-old adults between treatment and control groups (Figure 1B). The results from the lifespan assay indicated that supplementation with 0.0083 mM Reb A significantly extended longevity ($p < 0.001$), maximal lifespan (15.58% increase), and medium lifespan (17.6% extension) in *C. elegans* as compared with the control medium (Figure 1C). These results revealed that Reb A can delay aging and extend longevity in *C. elegans*.

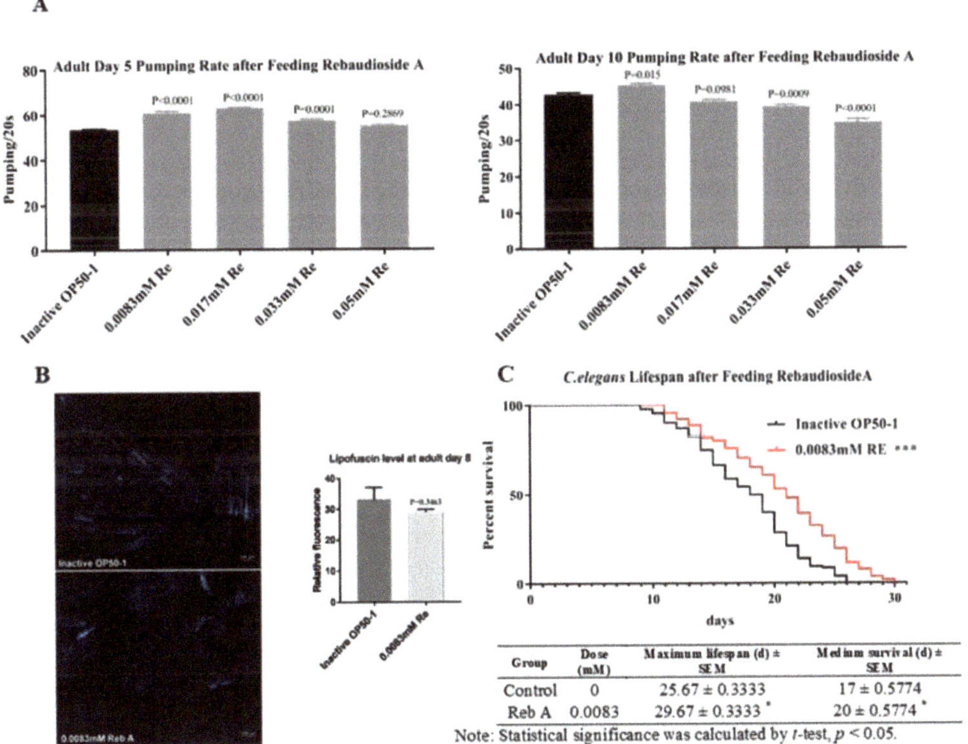

Group	Dose (mM)	Maximum lifespan (d) ± SEM	Medium survival (d) ± SEM
Control	0	25.67 ± 0.3333	17 ± 0.5774
Reb A	0.0083	29.67 ± 0.3333 *	20 ± 0.5774 *

Note: Statistical significance was calculated by *t*-test, $p < 0.05$.

Figure 1. Reb A extends lifespan in *C. elegans*. (**A**) Pumping rate in worms on the 5th and 10th day of adulthood in wild-type N2 strain on both control nematodes growth media (NGM) and NGM supplemented with rebaudioside A (Reb A) at designated concentrations (0.0083, 0.017, 0.033, and 0.05 mM). One-way ANOVA was used for multiple group comparisons and pairwise significance was evaluated by Dunnet's test; (**B**) Lipofuscin fluorescence in worms on the 8th day of adulthood grown under control NGM and 0.0083 mM Reb A NGM. Statistical significance was calculated by unpaired *t*-test; (**C**) Survival plot and lifespan assay comparing worms grown on control NGM and that supplemented with 0.0083 mM Reb A. Statistical significance was calculated by log-rank (Mantel–Cox) test, * $p < 0.05$, *** $p < 0.001$. Maximum and medium survival were shown in the table appended below.

3.2. *Reb A Extends Healthspan in C. elegans*

In addition to lifespan, healthspan has emerged as an increasingly important parameter for evaluation of anti-aging capacity [53]. Thus, we also detected additional physical parameters, including swimming behavior and osmotic avoidance behavior, which are effective indicators of healthspan in *C. elegans*. Swimming behavior was assessed via measuring the number of body swings in nematode in liquid medium. *C. elegans* treated with 0.0083 mM Reb A exhibited enhanced swimming behavior (Figure 2A); 0.0083 mM of Reb A supplementation elevated swimming ability in young *C. elegans* (5-day-old adults), and also in aged worms (10-day-old adults). Thus, Reb A significantly delayed deterioration in swimming movement caused by aging. Osmotic avoidance behavior is regulated by multi-directional induction ash sensory neurons (ASH), which mediate evasion from hypertonic solutions as well as other noxious chemicals and mechanical stimuli [37,54]. Our result showed that supplementation with 0.0083 mM Reb A significantly improved osmotic avoidance behavior (Figure 2B), which implied that Reb A enhances ASH sensitivity. However, Reb A supplementation had no significant effects on brood size, body length, and body width (Figure 2C,D). Therefore, the above results showed that Reb A supplementation increased pharyngeal pumping, swimming movement, and osmotic avoidance behavior, which indicated the maintenance of healthspan in *C. elegans*.

3.3. *Reb A Enhances Resistance to Oxidative Stress in C. elegans*

As lifespan extension is closely associated with enhanced survival under exposure to different stressors [55], next, we examined whether Reb A supplementation altered the capacity of worms to withstand heat stress and oxidative stress. After exposing 5-day-old adult worms to heat stress (i.e., 35 °C for 5 h) and following a recovery period at 20 °C for 96 h, there was no obvious difference in survival rates between Reb A-treated group and control group (Figure 3A). Next, we examined the effect of Reb A treatment on oxidative stress resistance. Worms were exposed for 6 h to 200 mM paraquat, an intracellular free radical-generating compound that induces acute oxidative stress [36]. Significant increase in survival rate was observed in the 0.0083 mM Reb A treatment group (Figure 3B). Therefore, while Reb A treatment had no effect on the resistance to heat shock, Reb A alleviates acute oxidative stress induced by paraquat in *C. elegans*.

ROS are highly reactive oxygen molecules that induce genotoxicity and physiological damages that destroy the innate mechanisms of stress resistance, which may lead to DNA damages, altered gene expression, perturbed cellular signaling, lipid peroxidation, and imbalances in protein homeostasis, ultimately resulting in cell senescence and death [41]. Excessive accumulation of ROS is also associated with aging and a myriad of aging-related diseases, including cardiovascular diseases and cancers. Therefore, we postulated that Reb A-enhanced resistance to oxidative stress may be attributed to a higher capacity to cope with intracellular ROS. A common fluorescent probe H_2DCFDA was used to quantitate intracellular ROS. It was noted that 0.0083 mM Reb A supplementation significantly reduced cellular ROS following exposure to oxidative stress (i.e., 7 mM paraquat for 2 h) in 6-day-old adult worms as compared with the control group (Figure 3C). Malonaldehyde (MDA), one of the end products of lipid peroxidation triggered by ROS, is an important biomarker of cellular oxidative stress [56]. We further examined the influence of Reb A supplementation on MDA content in worms treated with 7 mM paraquat (Figure 3D). MDA content was slightly reduced in Reb A-treated worms but failed to reach statistical significance ($p = 0.1858$). Thus, these results showed that Reb A supplementation at 0.0083 mM resulted in increased resistance to oxidative stress induced by paraquat treatment, mainly attributed to reductions in cellular ROS production upon oxidative stress exposure.

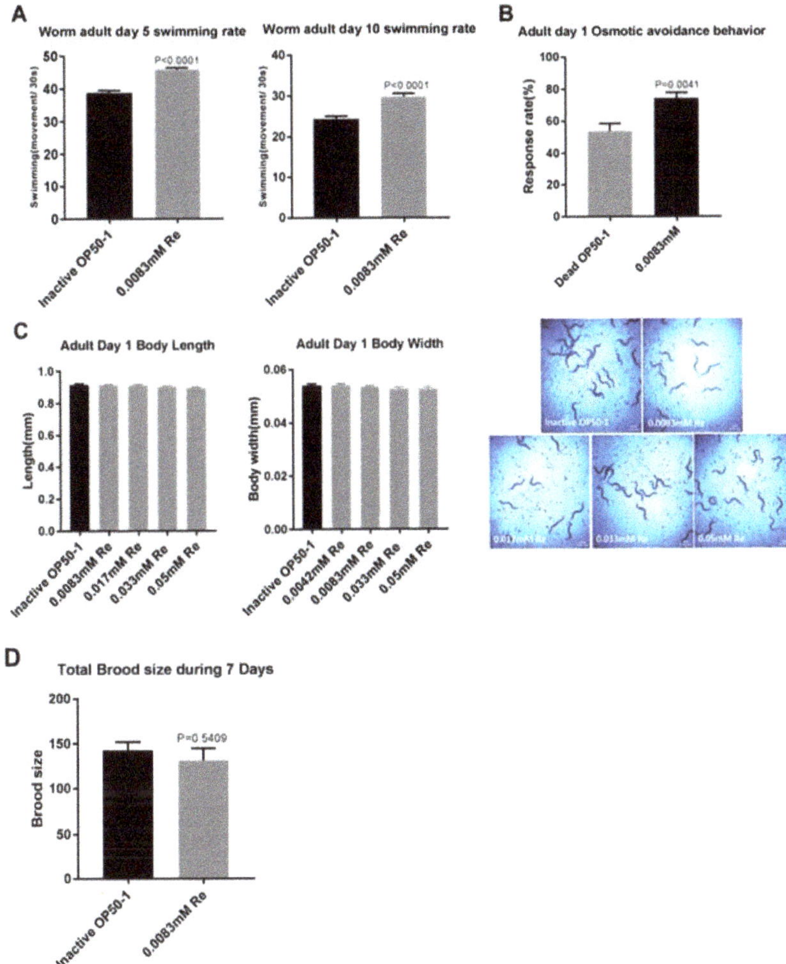

Figure 2. Reb A extends healthspan in *C. elegans*. (**A**) Swimming movement in *C. elegans* on the 5th and the 10th day of adulthood on control NGM and NGM supplemented with 0.0083 mM Reb A; (**B**) Osmotic avoidance behavior in *C. elegans* on the 1st day of adulthood on control NGM and NGM supplemented with 0.0083 mM Reb A. Statistical significance was calculated by unpaired *t*-test; (**C**) Body length and width of *C. elegans* on the 1st day of adulthood grown on control NGM and NGM supplemented with different concentrations of Reb A. Statistical significance was calculated by Dunnet's test; (**D**) A comparison on total brood size over a seven-day reproductive period.

The mitochondrial free radical theory of aging suggests that the efficiency of the respiratory chain decreases with aging, accompanied by increasing electron leakage from the respiratory chain and greater production ROS production. Excess accumulation of ROS also causes mitochondrial dysfunction [57]. In addition, mitochondrial respiration is decreased gradually during aging in *C. elegans*. We observed no significant difference in OCR values between the 0.0083 mM Reb A group and the control group under non-stressed conditions (Figure 3E,F). As Reb A was shown to reduce ROS accumulation under oxidative stress, we utilized Seahorse XFe96 to measure OCR under oxidative stress in Reb A-treated and control groups. The results showed that 0.0083 mM Reb A treatment significantly increase OCR under oxidative stress, suggesting the protective effects of Reb A on mitochondrial respiration (Figure 3G,H).

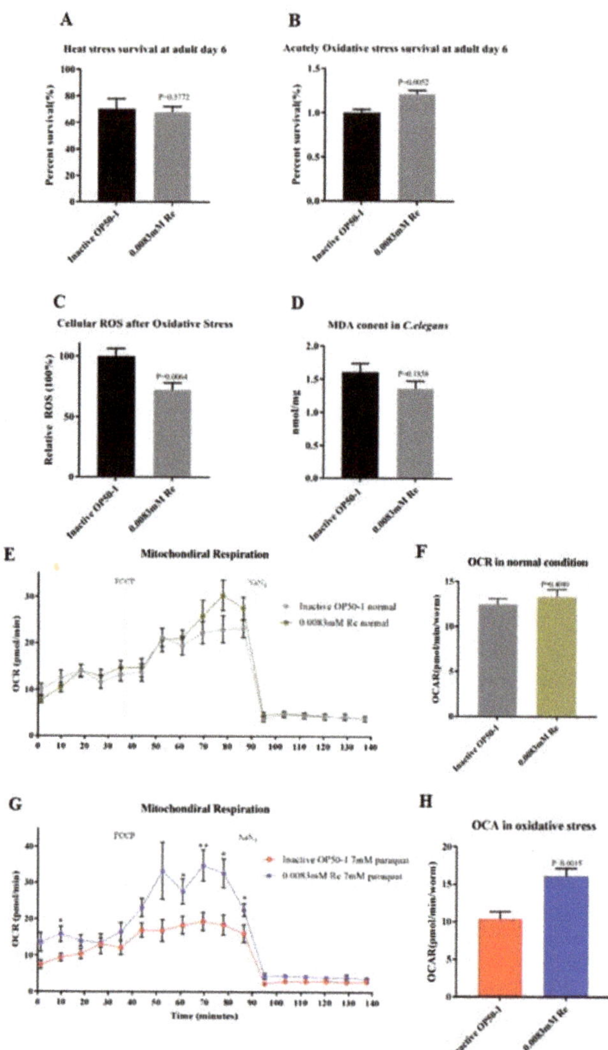

Figure 3. Reb A enhances resistance to oxidative stress in *C. elegans*. (**A**) Heat stress survival in the 6-day-old adult worms from Reb A and control groups; (**B**) Acute oxidative stress (treatment with 200 mM paraquat for 2 h) survival in 6-day-old adult worms from Reb A and control groups; (**C**) Relative cellular ROS level after treatment with 7 mM paraquat in the 6-day-old adult worms from Reb A and control groups; (**D**) MDA content in the 6-day-old adult worms from Reb A and control groups following oxidative stress exposure; (**E**) Under normal condition, basal respiration, maximal respiration, and spare capacity of 0.0083 mM Reb A-treated 6-day-old adults were not significantly different from those of the control group ($n = 16$ wells); (**F**) OCR was not statistically different between the 6-day-old adult worms treated with 0.0083 mM Reb A and the control group; (**G**) The 6-day-old adult worms from the 0.0083 mM Reb A group showed increased basal respiration and maximal respiration after 2 h of 7 mM paraquat treatment as compared with the control group ($n = 12$ wells), * $p < 0.0021$, ** $p < 0.0332$; (**H**) OCR under oxidative stress was significantly increased in 6-day-old adult worms treated with 0.0083 mM Reb A as compared with the control group. Statistical significance was calculated by unpaired *t*-test.

3.4. Reb A Inhibited the CeTOR Signaling Pathway

We showed that Reb A treatment is anti-aging and brings forth longevity extension and higher resistance against oxidative stress in *C. elegans*. However, the mechanism by which Reb A extends lifespan in nematodes remains unclear. Dietary restriction (DR) was previously found to significantly increase longevity and healthspan in yeast, nematodes, and mammals [37]. To confirm whether Reb A supplementation extends lifespan mediated by DR, we analyzed food intake of nematodes in liquid culture. Reb A treatment started from L4 larval stage to the fourth day of adulthood, and OD600 value of the culture medium was measured by spectrophotometer. There was no significant difference in food intake among 0.0083 mM Reb A group ($p = 0.9478$), 0.0042 mM Reb A group ($p = 0.7740$), and control groups (Figure 4A), which indicated that Reb A supplementation prolongs lifespan via DR-independent mechanism.

Next, we investigated changes in transcriptomes between the 10-day-old adult worms grown under control NGM and NGM supplemented with 0.0083 mM Reb A. Using fold change $\geq$1.5 and *t*-test *p*-values <0.05 as thresholds to select differentially expressed genes (DEGs) among 20,490 genes profiled between the treatment and control groups, we found 152 DEGs, which include 68 upregulated DEGs, and 84 downregulated DEGs in the Reb A supplemented group relative to the control group (Figure 4B,C). Bidirectional clustering analysis of DEGs based on the fragments per kilobase million (FPKM) values showed maximum horizontal distance between the control group and the Reb A treatment group (Figure 4B).

The DEG annotation analysis is helpful for the interpretation of gene functions. Therefore, DEGs with corrected $p < 0.05$ were used for GO function annotation and KEGG pathway analysis. In this study, among the top 20 enriched GO terms, 12 biological processes were represented, including dTDP biosynthetic process, nucleotide phosphorylation, and dTTP biosynthetic process, and so on (Figure 4D).

Nematodes lifespan is known to be regulated by various classical signaling pathways, including insulin/insulin-like growth factor (IIS) pathway, sirtuin 1 signaling pathway, target of rapamycin (TOR) signaling pathway, mitochondrial related genes, and dietary restriction related genes [58]. The KEGG pathway annotation revealed that genes implicated in mTOR pathway and pyruvate metabolism were most significantly altered following Reb A supplementation (Figure 4E). TOR is a serine/threonine kinase that regulates growth, development, and behavior by modifying protein synthesis, autophagy, and various cellular processes in response to nutrient changes [59].

Therefore, we further examined the expression of genes implicated in CeTOR signaling pathway using qPCR, such as *daf-15*, *rict-1*, and *let-363*. The expressions of *daf-15*, which encodes Raptor in TORC1 expression, *clk-2* and *ife-1* were significantly reduced in Reb A treatment groups (Figure 4F). The main upstream signaling pathways of CeTOR includes AMPK signaling pathway, the PI3K/Akt signaling pathway, and MAPK signaling pathway. mRNA levels of *Age-1*, *pdk-1*, and *akt-2* along the PI3K/Akt signaling pathway, which promotes the activation of CeTOR signaling pathway, were significantly reduced in worms treated with Reb A (Figure 4G). Inhibition of the TOR signaling pathway was shown to prolong lifespan in worms, which also mediated larval development, lipid storage, mRNA translation, and autophagy [60]. Indeed, we found that the expression of autophagy-related genes *unc-51*, *pha-4*, and *lgg-1* were upregulated, and the expression of *daf-16*, which encodes DAF-16/FOXO transcription factor, was also significantly increased in Reb A-treated worms. The expression of genes related to stress resistance regulation, such as *mtl-1*, *sod-2*, *sod-3*, and *gcs-1*, were significantly also upregulated in Reb A-treated worms (Figure 4H). Thus, the qPCR results corroborated transcriptomics results and suggest that Reb A treatment prolongs the lifespan of nematodes by inhibiting CeTOR. However, whether Reb A supplementation can bring about phenotype rescue in CeTOR pathway gene mutants awaits further verification.

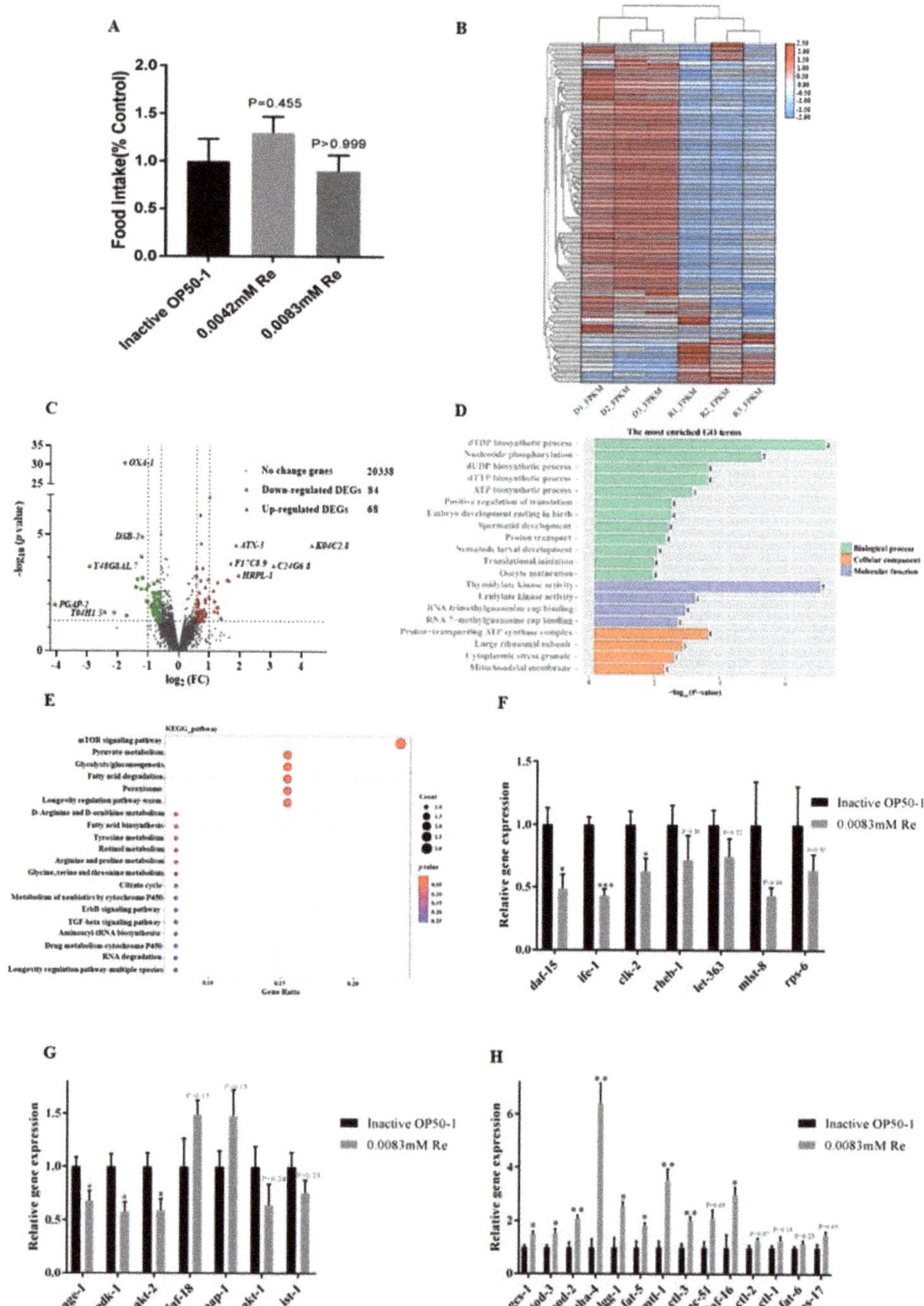

Figure 4. Reb A inhibited CeTOR signaling pathway. (**A**) Food intake showed no significant difference in worms grown under control NGM and NGM supplemented with both 0.0042 mM and 0.0083 mM Reb A; (**B**) Bidirectional clustering heat map of differentially expressed genes (DEGs) expression based on FPKM data of RNA-seq transcriptome; (**C**) Volcano plot

illustrates DEGs in Reb A treatment groups relative to the control group. Red dots are upregulated genes while green dots were downregulated genes; (**D**) Top 20 GO terms based on transcriptome of Reb A-treated worms relative to controls. The vertical axis corresponds to the GO terms divided into different categories, and the horizontal axis displays the value of $-\log_{10}$ (p-value). The number on the right of each GO term indicates the number of DEGs. Green represents biological processes, red represents cellular components, and blue represents molecular functions; (**E**) Top altered pathways with Reb A treatment in *C. elegans*. The vertical axis corresponds to the KEGG pathways, and the horizontal axis displays the enriched value expressed as the ratio of DEG to the total gene number in each pathway. The size and color of dots indicates the DEG gene number and the p-value, respectively; (**F**) Expression level of genes in CeTOR signaling pathway, which include *daf-15*, *ife-1*, *clk-2*, *rheb-1*, *let-363*, *mlst-8*, and *rps-6*; (**G**) Gene expression levels in PI3K/Akt signaling pathway, and selected genes include *age-1*, *pdk-1*, *akt-2*, *daf-18*, *aap-1*, *akt-1*, and *ist-1*; (**H**) The expression levels of genes related to longevity signaling pathways of nematode include *gcs-1*, *sod-3*, *sod-2*, *pha-4*, *lgg-1*, *fat-5*, *mtl-1*, *ctl-3*, *unc-51*, *daf-16*, *ctl-2*, *fat-6*, and *lips-17*. Unpaired two-tailed test was used to examination of statistical significance, p value represented by GP type, *** $p < 0.0002$, ** $p < 0.0021$, * $p < 0.0332$, and others showed precise p values.

3.5. Reb A Supplementation Lowers Lipid Storage in C. elegans

Inhibition of the TOR signaling pathway also leads to upregulation of *lipl-4* expression and enhanced lipolysis [61]. In the current study, the qPCR results showed that both the expressions of lipid desaturase-related genes *fat-5* and lipolysis-related gene *lips-17* were significantly increased in the Reb A groups (Figure 4H). Thus, we examined whether Reb A treatment modified lipid accumulation in *C. elegans*. We further investigated overall lipid storage via NR staining and lipid droplet morphology using the PLIN1:GFP (P*daf-22*-PLIN1:GFP) strain. Consistent with a previous report [20], Reb A reduced lipid storage, in both 5-day-old and 10-day-old adult worms (Figure 5A,B). The expression of *Drosophila* PLIN1:GFP in *C. elegans* labels intestinal lipid droplets under DAF-22 promoter P*daf-22*-PLIN1:GFP. After hybridization of PLIN1:GFP (P*daf-22*-PLIN1:GFP) transgenic strain with wild type N2 nematodes for two generations, fluorescence-labeled worms were selected for expansion. We found that nematodes supplemented with Reb A at 0.0083 mM possess smaller lipid droplets than both the control group and the 0.033 mM Reb A supplementation group (Figure 5C,E). Quantitative P*daf-22*-PLIN1:GFP ring size revealed lipid droplets in worms treated with 0.0083 mM Reb A were significantly reduced in size as compared with those treated with higher concentration of Reb A at 0.033 mM (Figure 5F). Thus, 0.0083 mM Reb A can lower lipid storage in *C. elegans*. Our result is consistent with previous findings that Reb A treatment reduces hepatic TG in mice with non-alcoholic fatty liver disease [20].

3.6. Reb A Alters Lipid Metabolism in C. elegans

Effects of Reb A treatment on lipid storage prompted us to investigate fine changes in the lipidome of Reb A-treated worms. We quantitated phospholipid profiles of the 10-day-old adults after supplementation with Reb A. Changes in the lipid classes of phosphatidylcholines (PCs), phosphatidylethanolamines (PEs), phosphatidylserines (PSs), PIs, PAs, and phosphatidylglycerols (PGs), cardiolipins (CLs), lysophosphatidylethanolamines (LPEs), lysophosphatidylserines (LPSs), lysophospholipinositols (LPIs), and LPCs were illustrated as barplots (Figure 6A). Lipidomics revealed that PAs and PIs were significantly reduced in worms treated with 0.0083 mM Reb A relative to the control (Figure 6A), i.e., LPC-16:1, LPC-20:4, and LPC-20:5 were significantly reduced (Figure 6B), and PA-36:1, PA-36:2, PA-36:3, PA-38:7, PA-40:7, and PA-40:8 were significantly decreased (Figure 6C). As for PIs, mostly polyunsaturated PIs including PI-37:4, PI-37:5, PI-37:6, PI-37:7, PI-38:4, PI-38:6, PI-39:5, and PI-39:6 were significantly reduced in Reb A-treated worms (Figure 6D). Therefore, Reb A alters the phospholipid profiles of *C. elegans*. How Reb A affects membrane phospholipid composition and its possible connection with perturbed neutral lipid storage (i.e., triacylglycerols TAGs that constitute the bulk of lipid droplets) in *C. elegans*, however, warrants further investigation in future studies. In this light, a recent study demonstrated that enhanced de novo biosynthesis of PCs along the TAG-DAG-PC axis underlies the drastic reduction in circulating neutral lipids such as TAGs in GCK-MODY patients [62]. Reductions in PAs upon Reb A treatment may, thus, indicate a reduced supply of lipid intermediates for downstream biosynthesis of neutral lipids, while lowered PIs may indicate perturbations in the PI3K signaling pathway, respectively. High-coverage lipidomics that encompass various forms of fatty acyl derivatives may help to reveal the relevant mechanisms underlying perturbed lipid metabolism and lipid droplet morphology in Reb A-treated worms [63,64], which denotes a meaningful future direction for in-depth interrogation on the working mechanism of Reb A.

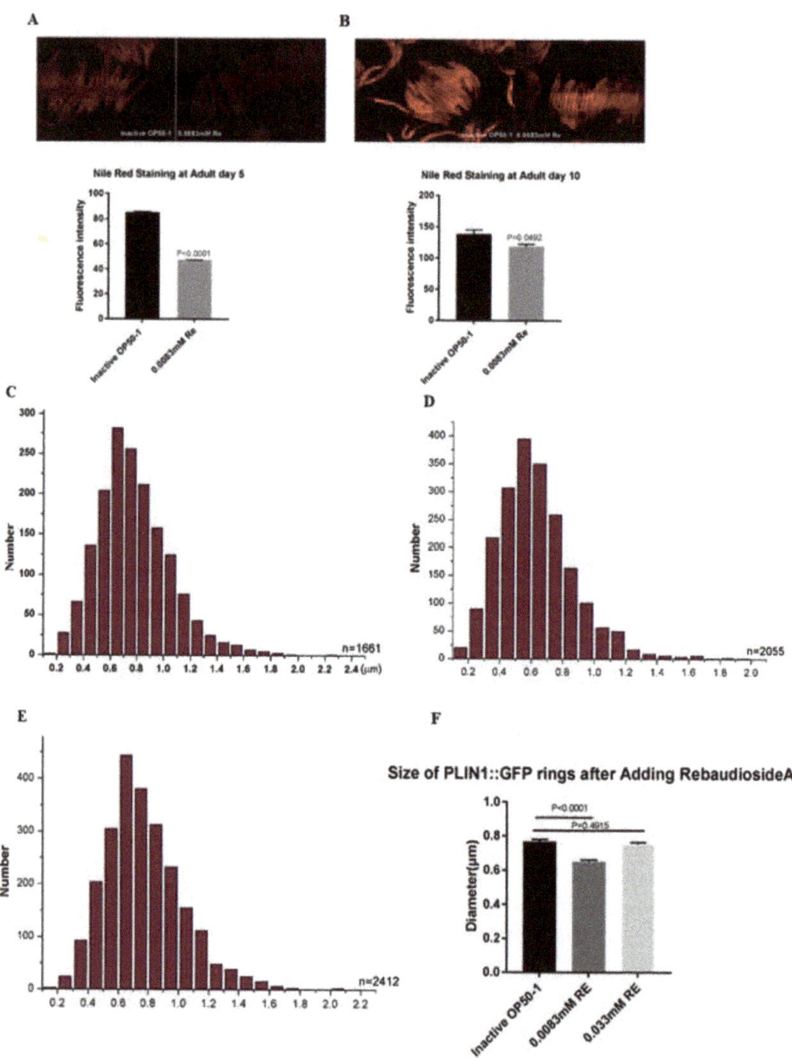

Figure 5. Reb A supplementation lowers lipid storage in *C. elegans*. (**A**) Nile red staining in 5-day-old adults from the 0.0083 mM Reb A and control groups; (**B**) Nile red staining in 10-day-old adults from the 0.0083 mM Reb A treatment and control groups. Statistical significance was calculated by unpaired two-tailed *t* test; (**C**) Distribution of lipid droplet diameters in 1st *xdEx1001* strain in control group; (**D**) Distribution of lipid droplet diameter of 1st *xdEx1001* strain in 0.0083 mM Reb A treatment group; (**E**) The adult lipid droplet diameter distribution of 1st *xdEx1001* strain in 0.033 mM Reb A treatment group; (**F**) Quantification of the size of PLIN1:GFP rings in Reb A treatment groups relative to the control group. Statistical significance was calculated by ANOVA with post hoc Dunnet's test.

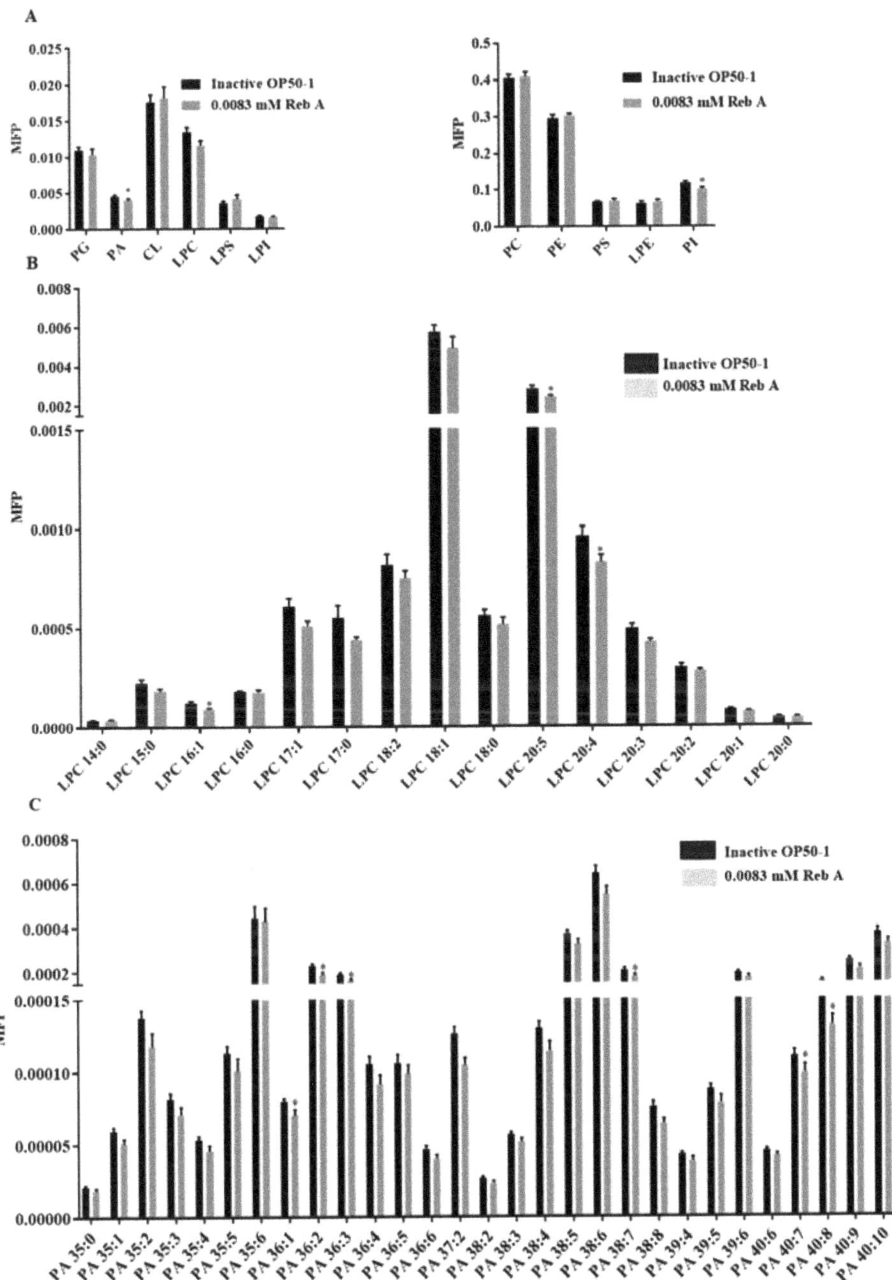

Figure 6. *Cont.*

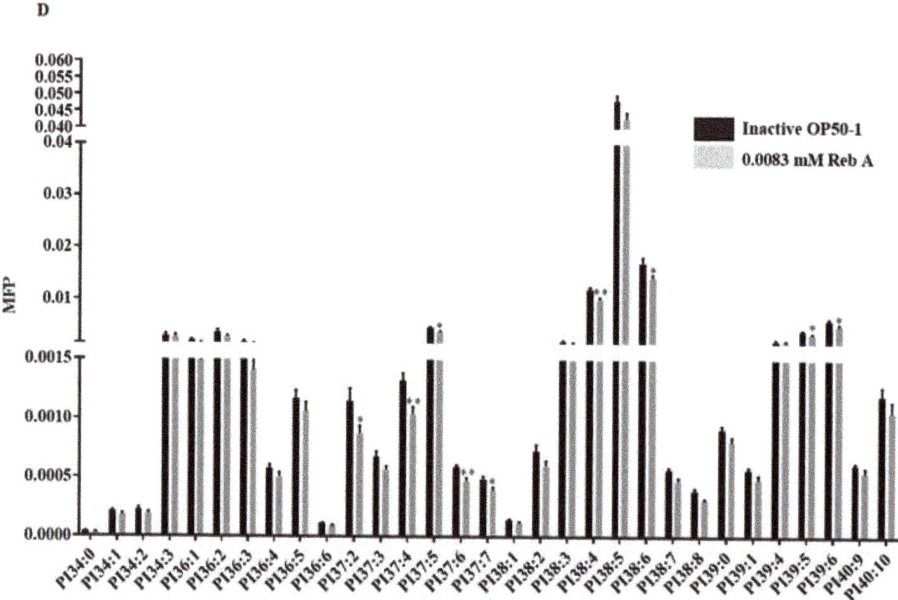

Figure 6. Reb A affects alters lipid metabolism of *C. elegans*. (**A**) The content of phosphatidylcholines (PGs), phosphatidic acids (PAs), cardiolipins (CLs), lysophosphatidylcholines (LPCs), LPS, lysophospholipinositols (LPIs), phosphatidylcholines (PCs), phosphatidylethanolamines (PEs), phosphatidylserines (PSs), lysophosphatidylethanolamines (LPEs), and phosphatidylinositiols (PIs) in the 10-day-old adults were analyzed by lipidomics, presented as molar fractions normalized to total polar lipids (MFP); (**B**) Profiles of LPCs in 10-day-old adult worms; (**C**) Profiles of PAs in 10-day-old adult worms; (**D**) Profiles of PIs in 10-day-old adult worms. Statistical significance was calculated by unpaired two-tailed *t* test. ** $p < 0.001$ and * $p < 0.05$.

4. Conclusions

In this paper, we systematically investigated the potential physiological effects of Reb A supplementation on *C. elegans*. We found that Reb A treatment is beneficial towards the lifespan and healthspan of worms, and also improves their associated lipid profiles i.e., lowering neutral lipid accumulation. We showed that the average lifespan and maximum lifespan of nematodes treated with 0.0083 mM Reb A were significantly prolonged, with the maintenance of a younger physiological state. Treatment with Reb A also significantly reduces ROS level and enhances resistance to acute oxidative stress in nematodes.

Transcriptome analysis uncovered DEGs enriched in mTOR signaling pathway in Reb A-treated worms. The qPCR analysis further confirmed that the expression of genes related to the TOR and PI3K/Akt signaling pathways were inhibited, while autophagy-related genes were elevated upon Reb A treatment. Therefore, Reb A may activate autophagy by inhibiting TOR and PI3K/Akt signaling pathways at the molecular level. Finally, we found that lipid storage and levels of PAs and polyunsaturated PIs were significantly reduced in nematodes treated with Reb A. In summary, our study has systematically shown that Reb A, a natural non-nutritive sweetener, prolongs lifespan, enhances oxidative stress resistance, and improves lipid metabolism in vivo in the model organism *C. elegans*, which serves as a ground for future studies exploring the potential medicinal and beneficial effects of Reb A as a replacement for caloric sugars in human foods and beverages.

Supplementary Materials: The following are available online at https://www.mdpi.com/2076-3921/10/2/262/s1, Figure S1: NNSs pumping rate in *C. elegans*, Figure S2: Different concentration of Reb A effect on lifespan in *C. elegans*.

Author Contributions: Conceptualization, G.S. and P.L.; methodology, P.L., Z.W. and S.M.L.; validation, P.L. and Z.W.; formal analysis, P.L., Z.W., and S.M.L.; investigation, P.L. and Z.W.; data curation, P.L. and Z.W.; writing—original draft preparation, P.L.; writing—review and editing, S.M.L. and G.S.; visualization, P.L.; supervision, G.S. and S.M.L.; project administration, G.S.; funding acquisition, G.S. All authors have read and agreed to the published version of the manuscript.

Funding: This research was funded by the National Key R&D Program of China (2018YFA0800901, 2018YFA0506902) and the National Natural Science Foundation of China (92057202, 31871194).

Institutional Review Board Statement: Not applicable.

Informed Consent Statement: Not applicable.

Data Availability Statement: Data are available upon reasonable request to the corresponding author.

Conflicts of Interest: Sin Man Lam is an employee of LipidALL Technologies.

References

1. Pradhan, S.; Shah, U.H.; Mathur, A.; Sharma, S. Experimental evaluation of antipyretic and analgesic activities of aspartame reply. *Indian J. Pharmacol.* **2016**, *43*, 486–487.
2. Walbolt, J.; Koh, Y. Non-nutritive sweeteners and their associations with obesity and Type 2 diabetes. *J. Obes. Metab. Syndr.* **2020**, *29*, 114–123. [CrossRef] [PubMed]
3. Ahmad, S.Y.; Friel, J.K.; Mackay, D.S. Effect of sucralose and aspartame on glucose metabolism and gut hormones. *Nutr. Rev.* **2020**, *78*, 725–746. [CrossRef] [PubMed]
4. Pereira, M.A. Sugar-sweetened and artificially-sweetened beverages in relation to obesity risk. *Adv. Nutr.* **2014**, *5*, 797–808. [CrossRef]
5. Uebanso, T.; Ohnishi, A.; Kitayama, R.; Yoshimoto, A.; Nakahashi, M.; Shimohata, T.; Takahashi, A. Effects of low-dose non-caloric sweetener consumption on gut microbiota in mice. *Nutrients* **2017**, *9*, 560. [CrossRef] [PubMed]
6. Eisenreich, A.; Gürtler, R.; Schäfer, B. Heating of food containing sucralose might result in the generation of potentially toxic chlorinated compounds. *Food Chem.* **2020**, *321*, 126700. [CrossRef]
7. Alcántar-Fernández, J.; Navarro, R.E.; Salazar-Martínez, A.M.; Pérez-Andrade, M.E.; Miranda-Ríos, J. *Caenorhabditis elegans* respond to high-glucose diets through a network of stress-responsive transcription factors. *PLoS ONE* **2018**, *13*, e0199888. [CrossRef]
8. Kundu, N.; Domingues, C.C.; Patel, J.; Aljishi, M.; Ahmadi, N.; Fakhri, M.; Sylvetsky, A.C.; Sen, S. Sucralose promotes accumulation of reactive oxygen species (ROS) and adipogenesis in mesenchymal stromal cells. *Stem Cell Res.* **2020**, *11*, 1–7. [CrossRef]
9. Praveena, S.M.; Cheema, M.S.; Guo, H.-R. Non-nutritive artificial sweeteners as an emerging contaminant in environment: A global review and risks perspectives. *Ecotoxicol. Environ. Saf.* **2019**, *170*, 699–707. [CrossRef] [PubMed]
10. Prakash, I.; Chaturvedula, V.S.P. *Steviol Glycosides: Natural Non-Caloric Sweeteners (Phytochemistry)*; Springer: Cham, Switzerland, 2016.
11. Fitch, C.; Keim, K.S. Position of the academy of nutrition and dietetics: Use of nutritive and nonnutritive sweeteners. *J. Acad. Nutr. Diet.* **2012**, *112*, 739–758. [CrossRef]
12. Ohta, M.; Sasa, S.; Inoue, A.; Tamai, T.; Fujita, I.; Morita, K.; Matsuura, F. Characterization of novel steviol glycosides from leaves of *Stevia rebaudiana* morita. *J. Appl. Glycosci.* **2010**, *57*, 199–209. [CrossRef]
13. Chatsudthipong, V.; Muanprasat, C. Stevioside and related compounds: Therapeutic benefits beyond sweetness. *Pharmacol. Ther.* **2009**, *121*, 41–54. [CrossRef]
14. Schiano, C.; Grimaldi, V.; Franzese, M.; Fiorito, C.; De Nigris, F.; Donatelli, F.; Soricelli, A.; Salvatore, M.; Napoli, C. Non-nutritional sweeteners effects on endothelial vascular function. *Toxicol. In Vitro* **2020**, *62*, 104694. [CrossRef] [PubMed]
15. Choi, D.H.; Cho, U.M.; Hwang, H.S. Anti-inflammation effect of Rebaudioside A by inhibition of the MAPK and NF-κB signal pathway in RAW264.7 macrophage. *J. Appl. Biol. Chem.* **2018**, *61*, 205–211. [CrossRef]
16. Wang, Y.; Li, L.; Wang, Y.; Zhu, X.; Jiang, M.; Song, E.; Song, Y. New application of the commercial sweetener rebaudioside a as a hepatoprotective candidate: Induction of the Nrf2 signaling pathway. *Eur. J. Pharmacol.* **2018**, *822*, 128–137. [CrossRef]
17. Nettleton, J.E.; Klancic, T.; Schick, A.; Choo, A.C.; Shearer, J.; Borgland, S.L.; Chleilat, F.; Mayengbam, S.; Reimer, R.A. Low-dose *Stevia* (*Rebaudioside* A) consumption perturbs gut microbiota and the mesolimbic dopamine reward system. *Nutrients* **2019**, *11*, 1248. [CrossRef] [PubMed]
18. Casas-Grajales, S.; Ramos-Tovar, E.; Chávez-Estrada, E.; Alvarez-Suarez, D.; Hernández-Aquino, E.; Reyes-Gordillo, K.; Cerda-García-Rojas, C.M.; Camacho, J.; Tsutsumi, V.; Lakshman, M.R.; et al. Antioxidant and immunomodulatory activity induced by stevioside in liver damage: In Vivo, In Vitro and in silico assays. *Life Sci.* **2019**, *224*, 187–196. [CrossRef]
19. Xi, D.; Bhattacharjee, J.; Salazar-Gonzalez, R.-M.; Park, S.; Jang, A.; Warren, M.; Merritt, R.; Michail, S.; Bouret, S.; Kohli, R. Rebaudioside affords hepatoprotection ameliorating sugar sweetened beverage- induced nonalcoholic steatohepatitis. *Sci. Rep.* **2020**, *10*, 1–11. [CrossRef] [PubMed]

20. Kaletta, T.; Hengartner, M.O. Finding function in novel targets: *C. elegans* as a model organism. *Nat. Rev. Drug Discov.* **2006**, *5*, 387–399. [CrossRef]
21. O'Reilly, L.P.; Luke, C.J.; Perlmutter, D.H.; Silverman, G.A.; Pak, S.C. *C. elegans* in high-throughput drug discovery. *Adv. Drug Deliv. Rev.* **2014**, *70*, 247–253.
22. Pincus, Z.; Mazer, T.C.; Slack, F.J. Autofluorescence as a measure of senescence in *C. elegans*: Look to red, not blue or green. *Aging* **2016**, *8*, 889–898. [CrossRef]
23. Golegaonkar, S.; Tabrez, S.S.; Pandit, A.; Sethurathinam, S.; Jagadeeshaprasad, M.G.; Bansode, S.; Sampathkumar, S.G.; Mukhopadhyay, A. Rifampicin reduces advanced glycation end products and activates DAF-16 to increase lifespan in *Caenorhabditis elegans*. *Aging Cell* **2015**, *14*, 463–473. [CrossRef]
24. Harman, D. Aging: A theory based on free radical and radiation chemistry. *J. Gerontol.* **1956**, *11*, 298–300. [CrossRef]
25. Moribe, H.; Konakawa, R.; Koga, D.; Ushiki, T.; Nakamura, K.; Mekada, E. Tetraspanin is required for generation of reactive oxygen species by the dual oxidase system in *Caenorhabditis elegans*. *PLoS Genet.* **2012**, *8*, e1002957. [CrossRef]
26. Ewald, C.Y. Redox Signaling of NADPH oxidases regulates oxidative stress responses, immunity and aging. *Antioxidants* **2018**, *7*, 130. [CrossRef]
27. De Diego, I.; Peleg, S.; Fuchs, B.; Tamayo, I.D.D. The role of lipids in aging-related metabolic changes. *Chem. Phys. Lipids* **2019**, *222*, 59–69. [CrossRef] [PubMed]
28. Wang, R.; Li, B.; Lam, S.-M.; Shui, G. Integration of lipidomics and metabolomics for in-depth understanding of cellular mechanism and disease progression. *J. Genet. Genom.* **2020**, *47*, 69–83. [CrossRef]
29. Menon, D.; Salloum, D.; Bernfeld, E.; Gorodetsky, E.; Akselrod, A.; Frias, M.A.; Sudderth, J.; Chen, P.-H.; De Berardinis, R.; Foster, D.A.; et al. Lipid sensing by mTOR complexes via de novo synthesis of phosphatidic acid. *J. Biol. Chem.* **2017**, *292*, 6303–6311. [CrossRef] [PubMed]
30. Shamalnasab, M.; Gravel, S.-P.; St-Pierre, J.; Breton, L.; Jäger, S.; Aguilaniu, H. A salicylic acid derivative extends the lifespan of *Caenorhabditis elegans* by activating autophagy and the mitochondrial unfolded protein response. *Aging Cell* **2018**, *17*, e12830. [CrossRef] [PubMed]
31. Zheng, S.-Q.; Ding, A.-J.; Li, G.-P.; Wu, G.-S.; Luo, H.-R. Drug absorption efficiency in *Caenorhbditis elegans* delivered by different methods. *PLoS ONE* **2013**, *8*, e56877. [CrossRef]
32. Huang, C.; Xiong, C.; Kornfeld, K. Measurements of age-related changes of physiological processes that predict lifespan of *Caenorhabditis elegans*. *Proc. Natl. Acad. Sci. USA* **2004**, *101*, 8084–8089. [CrossRef] [PubMed]
33. Liao, V.H.-C.; Yu, C.-W.; Chu, Y.-J.; Li, W.-H.; Hsieh, Y.-C.; Wang, T.-T. Curcumin-mediated lifespan extension in *Caenorhabditis elegans*. *Mech. Ageing Dev.* **2011**, *132*, 480–487. [CrossRef]
34. Lin, C.; Zhang, X.; Su, Z.; Xiao, J.; Lv, M.; Cao, Y.; Chen, Y. Carnosol improved lifespan and healthspan by promoting antioxidant capacity in *Caenorhabditis elegans*. *Oxidative Med. Cell. Longev.* **2019**, *2019*, 5958043. [CrossRef] [PubMed]
35. Wilkinson, D.S.; Taylor, R.C.; Dillin, A. Chapter 12-analysis of aging in *Caenorhabditis elegans*. *Methods Cell Biol.* **2012**, *107*, 353–381. [PubMed]
36. Fang, E.F.; Waltz, T.B.; Kassahun, H.; Lu, Q.; Kerr, J.S.; Morevati, M.; Fivenson, E.M.; Wollman, B.N.; Marosi, K.; Wilson, M.A.; et al. Tomatidine enhances lifespan and healthspan in *C. elegans* through mitophagy induction via the SKN-1/Nrf2 pathway. *Sci. Rep.* **2017**, *7*, 46208. [CrossRef]
37. Li, G.; Gong, J.; Lei, H.; Liu, J.; Xu, X.Z.S. Promotion of behavior and neuronal function by reactive oxygen species in *C. elegans*. *Nat. Commun.* **2016**, *7*, 13234. [CrossRef] [PubMed]
38. Rangsinth, P.; Prasansuklab, A.; Duangjan, C.; Gu, X.; Meemon, K.; Wink, M.; Tencomnao, T. Leaf extract of *Caesalpinia mimosoides* enhances oxidative stress resistance and prolongs lifespan in *Caenorhabditis elegans*. *BMC Complement. Altern. Med.* **2019**, *19*, 164. [CrossRef] [PubMed]
39. Farias-Pereira, R.; Park, Y. Cafestol increases fat oxidation and energy expenditure in *Caenorhabditis elegans* via DAF-12-dependent pathway. *Food Chem.* **2020**, *3*. [CrossRef]
40. Rathor, L.; Pant, A.; Awasthi, H.; Mani, D.; Pandey, R. An antidiabetic polyherbal phytomedicine confers stress resistance and extends lifespan in *Caenorhabditis elegans*. *Biogerontology* **2017**, *18*, 131–147. [CrossRef]
41. Sun, H.; Li, L.J.; Zhang, A.H.; Zhang, N.; Sun, W.J. Liuwei Dihuang Wan, a traditional Chinese medicinal formula, protects against osteoporosis. *Anal. Chim. Acta* **2013**, *4*, 1–6.
42. Sarasija, S.; Norman, K.R. Measurement of ROS in *Caenorhabditis elegans* using a reduced form of fluorescein. *Bio-Protocol* **2018**, *8*, 1–11. [CrossRef] [PubMed]
43. Gomez-Amaro, R.L.; Valentine, E.R.; Carretero, M.; Leboeuf, S.E.; Rangaraju, S.; Broaddus, C.D.; Solis, G.M.; Williamson, J.R.; Petrascheck, M. Measuring food intake and nutrient absorption in *Caenorhabditis elegans*. *Genetics* **2015**, *200*, 443–454. [CrossRef] [PubMed]
44. Young, M.D.; Wakefield, M.J.; Smyth, G.K.; Oshlack, A. Gene ontology analysis for RNA-seq: Accounting for selection bias. *Genome Biol.* **2010**, *11*. [CrossRef]
45. Kanehisa, M.; Araki, M.; Goto, S.; Hattori, M.; Hirakawa, M.; Itoh, M.; Katayama, T.; Kawashima, S.; Okuda, S.; Tokimatsu, T.; et al. KEGG for linking genomes to life and the environment. *Nucleic Acids Res.* **2007**, *36*, D480–D484. [CrossRef] [PubMed]
46. Mao, X.; Cai, T.; Olyarchuk, J.G.; Wei, L. Automated genome annotation and pathway identification using the KEGG Orthology (KO) as a controlled vocabulary. *Bioinformatics* **2005**, *21*, 3787–3793. [CrossRef]

47. O'Rourke, E.J.; Soukas, A.A.; Carr, C.E.; Ruvkun, G. *C. elegans* major fats are stored in vesicles distinct from lysosome-related organelles. *Cell Metab.* **2009**, *10*, 430–435.
48. Koopman, M.; Michels, H.; Dancy, B.M.; Kamble, R.; Mouchiroud, L.; Auwerx, J.; Houtkooper, R.H. A screening-based platform for the assessment of cellular respiration in *Caenorhabditis elegans*. *Nat. Protoc.* **2016**, *11*, 1798–1816. [CrossRef]
49. Lam, S.M.; Wang, Z.; Li, J.; Huang, X.; Shui, G. Sequestration of polyunsaturated fatty acids in membrane phospholipids of *Caenorhabditis elegans* dauer larva attenuates eicosanoid biosynthesis for prolonged survival. *Redox Biol.* **2017**, *12*, 967–977. [CrossRef]
50. Lam, S.M.; Tong, L.; Reux, B.; Duan, X.; Petznick, A.; Yong, S.S.; Khee, C.B.S.; Lear, M.J.; Wenk, M.R.; Shui, G. Lipidomic analysis of human tear fluid reveals structure-specific lipid alterations in dry eye syndrome. *J. Lipid Res.* **2014**, *55*, 299–306. [CrossRef]
51. Song, J.W.; Lam, S.M.; Fan, X.; Cao, W.J.; Wang, S.Y.; Tian, H.; Li, B.; Jiang, T.J.; Wang, R.; Shui, G.; et al. Omics-driven systems interrogation of metabolic dysregulation in COVID-19 pathogenesis. *Cell Metab.* **2020**, *32*, 188–202. [CrossRef]
52. Diomede, L.; Rognoni, P.; Lavatelli, F.; Romeo, M.; Di Fonzo, A.; Foray, C.; Fiordaliso, F.; Palladini, G.; Valentini, V.; Perfetti, V.; et al. Investigating heart-specific toxicity of amyloidogenic immunoglobulin light chains: A lesson from *C. elegans*. *Worm* **2014**, *3*, e965590. [CrossRef]
53. Bansal, A.; Zhu, L.J.; Yen, K.; Tissenbaum, H.A. Uncoupling lifespan and healthspan in *Caenorhabditis elegans* longevity mutants. *Proc. Natl. Acad. Sci. USA* **2015**, *112*, 277–286. [CrossRef] [PubMed]
54. Lee, E.C.; Kim, H.; Di Tano, J.; Manion, D.; King, B.L.; Strange, K. Abnormal osmotic avoidance behavior in *C. elegans* is associated with increased hypertonic stress resistance and improved proteostasis. *PLoS ONE* **2016**, *11*, e0154156. [CrossRef]
55. Lu, M.; Tan, L.; Zhou, X.G.; Yang, Z.L.; Zhu, Q.; Chen, J.N.; Wu, G.S. Secoisolariciresinol diglucoside delays the progression of aging-related diseases and extends the lifespan of *Caenorhabditis elegans* via DAF-16 and HSF-Oxid. *Med. Cell Longev.* **2020**, *2020*, 1293935. [CrossRef]
56. Zhang, J.; Shi, R.; Li, H.; Xiang, Y.; Xiao, L.; Hu, M.; Ma, F.; Ma, C.W.; Huang, Z. Antioxidant and neuroprotective effects of Dictyophora indusiata polysaccharide in *Caenorhabditis elegans*. *J. Ethnopharmacol.* **2016**, *192*, 413–422. [CrossRef] [PubMed]
57. López-Otín, C.; Blasco, M.A.; Partridge, L.; Serrano, M.; Kroemer, G. The hallmarks of aging. *Cell* **2013**, *153*, 1194–1217. [CrossRef] [PubMed]
58. Shen, P.; Yue, Y.; Zheng, J.; Park, Y. *Caenorhabditis elegans*: A convenient In Vivo model for assessing the impact of food bioactive compounds on obesity, aging, and Alzheimer's disease. *Annu. Rev. Food Sci. Technol.* **2018**, *9*, 1–22. [CrossRef]
59. Blackwell, T.K.; Sewell, A.K.; Wu, Z.; Han, M. TOR signaling in *Caenorhabditis elegans* development, metabolism, and aging. *Genetics* **2019**, *213*, 329–360. [CrossRef] [PubMed]
60. Johnson, S.C.; Rabinovitch, P.S.; Kaeberlein, M. mTOR is a key modulator of ageing and age-related disease. *Nature* **2013**, *493*, 338–345. [CrossRef]
61. Lapierre, L.R.; Hansen, M. Lessons from *C. elegans*: Signaling pathways for longevity. *Trends Endocrinol. Metab.* **2012**, *23*, 637–644. [CrossRef]
62. Wang, X.; Lam, S.M.; Cao, M.; Wang, T.; Wang, Z.; Yu, M.; Li, B.; Yhang, H.; Ping, F.; Shui, G.; et al. Localized increases in CEPT1 and ATGL elevate plasmalogen phosphatidylcholines in HDLs contributing to atheroprotective lipid profiles in hyperglycemic GCK-MODY. *Redox Biol.* **2021**, *40*, 101855. [CrossRef] [PubMed]
63. Lam, S.M.; Wang, Z.; Li, B.; Shui, G. High-coverage lipidomics for functional lipid and pathway analyses. *Anal. Chim. Acta* **2021**, *1147*, 199–2101. [CrossRef] [PubMed]
64. Lam, S.M.; Zhou, T.; Li, J.; Zhang, S.; Chua, G.H.; Li, B.; Shui, G. A robust, integrated platform for comprehensive analyses of acyl-coenzyme As and acyl-carnitines revealed chain length-dependent disparity in fatty acyl metabolic fates across Drosophila development. *Sci. Bull.* **2020**, *65*, 1840–1848. [CrossRef]

MDPI
St. Alban-Anlage 66
4052 Basel
Switzerland
Tel. +41 61 683 77 34
Fax +41 61 302 89 18
www.mdpi.com

Antioxidants Editorial Office
E-mail: antioxidants@mdpi.com
www.mdpi.com/journal/antioxidants

www.ingramcontent.com/pod-product-compliance
Lightning Source LLC
LaVergne TN
LVHW071251170726
843515LV00006BA/1377

* 9 7 8 3 0 3 6 5 4 3 6 3 5 *